Islamic Science and the Making of the European Renaissance

Transformations: Studies in the History of Science and Technology
Jed Z. Buchwald, general editor

Jesuit Science and the Republic of Letters
Mordechai Feingold, editor

Ships and Science: The Birth of Naval Architecture in the Scientific Revolution, 1600–1800
Larrie D. Ferreiro

Wireless: From Marconi's Black-Box to the Audion
Sungook Hong

The Path Not Taken: French Industrialization in the Age of Revolution, 1750–1830
Jeff Horn

Harmonious Triads: Physicists, Musicians, and Instrument Makers in Nineteenth-Century Germany
Myles W. Jackson

Spectrum of Belief: Joseph von Fraunhofer and the Craft of Precision Optics
Myles W. Jackson

Affinity, That Elusive Dream: A Genealogy of the Chemical Revolution
Mi Gyung Kim

Shifting Ontologies: Materials in Eighteenth-Century Science
Ursula Klein and Wolfgang Lefèvre

American Hegemony and the Postwar Reconstruction of Science in Europe
John Krige

Conserving the Enlightenment: French Military Engineering from Vauban to the Revolution
Janis Langins

Picturing Machines 1400–1700
Wolfgang Lefèvre, editor

Heredity Produced: At the Crossroads of Biology, Politics, and Culture, 1500–1870
Staffan Müller-Wille and Hans-Jörg Rheinberger, editors

Secrets of Nature: Astrology and Alchemy in Early Modern Europe
William R. Newman and Anthony Grafton, editors

Historia: Empiricism and Erudition in Early Modern Europe
Gianna Pomata and Nancy Siraisi, editors

Nationalizing Science: Adolphe Wurtz and the Battle for French Chemistry
Alan J. Rocke

Islamic Science and the Making of the European Renaissance
George Saliba

Islamic Science and the Making of the European Renaissance

George Saliba

The MIT Press
Cambridge, Massachusetts
London, England

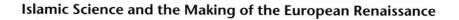

For information on quantity discounts, please email special_sales@mitpress.mit.edu.

Set in Stone Serif and Stone Sans by Graphic Composition, Inc. Printed and bound in the United States of America.

Library of Congress Cataloging-in-Publication Data

Saliba, George.
Islamic science and the making of the European Renaissance / George Saliba.
p. cm.
Includes bibliographical references and index.
ISBN-13: 978-0-262-19557-7 (hardcover : alk. paper)
1. Science—Islamic countries—History. 2. Islam and science. 3. Science, Medieval.
4. Civilization—Western—Islamic influences. I. Title.
Q127.I742.S35 2007
509.17'67—dc22

2006023618

10 9 8 7 6 5 4 3 2

Contents

Preface

This is essentially an essay in historiography. It critiques the current trends of the historiography of Islamic and Arabic science and attempts to make use of the latest historical findings in order to propose a new historiography that could better explain the scientific developments and, in a more general sense, the major trends in the intellectual history of Islamic civilization. It touches on periodization, on the relation of science to the general intellectual environment, on the social and political dimensions of scientific production, and on the relationship between the technical scientific details, in a particular discipline and the social support and recognition of those disciplines.

The main ideas discussed herein have already been articulated, in a preliminary manner, in my book *al-Fikr al-ʿIlmī al-ʿArabī* (Balamand University Press, Lebanon, 1998). Now they are available—more extensively developed in some major respects—to those who do not read Arabic. The main features of the thesis expressed earlier in *al-Fikr* are supported here by fuller evidence. Furthermore, the new literature that has appeared since the publication of *al-Fikr*, especially that which bears on that book's main thesis, is critiqued in this volume. This can, then, be seen as a critique of the contents of that literature, and of the conclusions reached therein. The scrutiny to which those conclusions are now subjected is necessitated by new evidence that has raised doubts about their validity.

The terms "Islamic science" and "Arabic astronomy," used extensively in this book, call for an explanatory comment. "Islamic science," is intended to designate those sciences that were developed in the Islamic civilization and which did not fall within the sphere of disciplines usually designated with the Arabic expression *al-ʿulūm al-islāmīya* (Islamic sciences). The latter group usually dealt with religious Islamic thought proper and thus is not of central concern in this volume. In contrast, the "Islamic sciences" studied here were considered as part of the "foreign" or "rational" sciences (*ʿulūm al-awāʾil* or *al-ʿulūm al-ʿaqlīya*), or even the "philosophical" sciences (*al-ʿulūm al-falsafīya* or *al-ḥikmīya*), in classical Islamic times, and did not in any way designate the religious, juridical, exegetical, linguistic, or Qurʾanic sciences that were usually separately classified as *al-ʿUlūm al-naqlīya* (the transmitted sciences). "Islamic" is therefore used in this more complex civilizational sense and not in the religious sense.

The term "Arabic" finds its justification in two major ways: First, Arabic was for a long time the scientific language of the Islamic civilization, from the eighth and ninth centuries to our own times, in much the same way as it was the language of the religious sciences as well, irrespective of the geographic area where those sciences were written or studied. These conditions, which prevailed throughout most of Islamic history, opened various avenues for people of various races and religious backgrounds to participate in the production of this civilization. Those same people may have spoken Persian, Syriac, or even later Turkish and Urdu at home. And yet they mostly expressed their intellectual production, and especially the scientific part of it, in Arabic, much as Ibn Maymūn (Maimonides) wrote most of his philosophical and medical works in Arabic while reserving Hebrew for his religious and juridical production. Second, the history of the discipline of astronomy is used in this book as a template to illustrate the periodization and the ups and downs of Islamic scientific thought in general. And the kind of astronomy that was most prevalent in the Islamic civilization, and that was also most vibrant, was the new astronomy that was called *ʿilm al-hayʾa* (science of the configuration [of the world] = Astronomy), a coined Arabic phrase that had no Greek equivalent. It was this astronomy that continued to be written almost exclusively in Arabic from the ninth century on. This is also the astronomy that forms the main focus of this book. Furthermore, there were no times, throughout Islamic intellectual history, when the term "Arabic Astronomy" could have been possibly taken to mean that this astronomy was in any way restricted to the geographical domain of the Arabic-speaking regions, or that Arabic was the exclusive language of that discipline. The manner in which this term is used here simply means that Arabic was clearly the language in which most of the works in this discipline were written, as is evidenced by the vast majority of the surviving texts.

Although this book is written in English, and may later appear in other European languages, its ultimate message may resonate differently with readers who feel a sense of kinship with the Islamic civilization, whatever their racial, national, linguistic, or religious affiliation. It is to these readers that the issues discussed here would make the most sense, irrespective of whether they would want to refer to this production as Islamic or Arabic. And to the same readers I extend the invitation to participate in the discussion that I hope this book will generate.

But I must quickly caution those readers not to read this book as an expression of the greatness of the Islamic scientific tradition, although it was indeed one of the greatest of such traditions, but to read it as an invitation to reflect on the sense of their own history, especially in these "postcolonial" and yet deeply "colonial" times for the Muslim and Arab worlds. I sincerely wish to invite such readers to consider ultimately the kind of history that could be written when one de-emphasizes the usual political and religious histories that are often narrated *ad nauseam*, and privileges instead the scientific production and the complex social, economic, and intellectual conditions that allowed that production to come into existence.

If there is a lesson to be learned here from the history of science for our modern times, and if there is any hope to learn something about the social, political, and economic mechanisms that allow scientific production to prosper, for purposes of modern development in almost all developing countries, irrespective of their religious or cultural legacies, it should be grounded in this kind of history of science that keeps an eye on the technical intricacies of scientific thought itself, and at the same time investigates the social, political, and economic mechanisms that allowed, and may still allow, this thought to flourish. This book is intended to shed light on such issues.

I now turn to the most pleasant task of acknowledging all the help I have had along the way that made this book possible. In that regard, my deepest thanks should go first and foremost to M. François Zabbal, of the Institut du Monde Arabe (Paris), for making the first expression of this book possible when he invited me to give its early contents as a series of lectures under the auspices of La Chaire de l'Institut du Monde Arabe during the spring of

2004. But of course there are many others whose names have to be withheld for fear of missing some in the midst of the multitude.

However, in the actual process of turning the contents of those early lectures into a book, I cannot but pay tribute to particular people without whose advice and encouragement this book would have never seen the light of day. Among those who made this book possible are Jed Buchwald, the editor of the Transformations series, and my colleague and friend Noel Swerdlow. The encouragement I received from those two people and the extremely valuable criticisms were incalculable, and have certainly gone a long way in saving me from many slips and errors. Whatever errors and misstatements remain are completely my own and no one else should be implicated in my folly.

A special expression of gratitude should also go to M. Alain Segonds, of the Belles Lettres, of Paris, who gave the manuscript a very thorough reading when it was still in its earliest stages, and who suggested a number of corrections that helped me greatly in sharpening the arguments. I am also indebted to my students and to the various groups of people who attended my public lectures over the years, when I first began to explore the germs of those ideas, which were not yet fully formulated, and are now further developed in this book. Those people patiently listened to what must have sounded to them like half-baked thoughts, and always pushed me to develop those ideas further in order to reach the form they have now reached.

In the same breath I also thank all those who abandoned the most beautiful Parisian spring evenings and flocked faithfully to the Institut du Monde Arabe in Paris, every Tuesday night, for six weeks in May and June of 2004, to participate in the formal presentation of the lectures upon which this book is based. Of those people, I specially thank those who raised the various challenging questions that forced me to reconsider a great number of issues and to rearticulate them much more precisely. But those questions could not have been raised had it not been for the most diligent team of simultaneous translators who rendered my unwritten English lectures into coherent French, a feat that continues to amaze me.

All those people are in no way responsible for the inevitable ambiguities that may still persist in the proposed formulation of this new Arabic scientific historiography. For the very nature of this proposal leaves it vulnerable to the experimental hazards incurred by its novelty.

I am also indebted to my friends and colleagues, both in the United States and in France, whose areas of expertise bordered very closely on the history of Arabic science, and who paid me the utmost complement by attending the lectures at IMA and by pointing out, to my benefit, the strengths and weaknesses of the arguments I made there. It is those subtle correctives that can no longer be separated from my main train of thought, nor can they be footnoted separately, that have now become part and parcel of my own convictions and, of course, inform my latest thinking on the subject. In this global sense, I thank them for those correctives. But I must single out my dear friends and colleagues: Professor Muhsin Mahdi of Harvard, who graced me with his presence at some of the lectures even when he was not feeling well, and M. Maroun Aouad, of the CNRS in Paris, for having given me the pleasure of good arguments over the years, and who has never failed to point out my follies with utmost politeness. He can obviously notice now that he has not been all that successful in curing me completely. The follies that still persist in this book can easily attest to that. But if any of the arguments I make here can make a small dent in changing peoples' thoughts about the nature of Arabic and Islamic science then all those arguments would not have been in vain, and I gladly accept the responsibility of their failure when they did.

The manuscript editor at The MIT Press was kind to listen to me and follow my instructions to "go lightly."

Last but not least, I should also thank all those wonderful people at the Kluge Center of the Library of Congress who made the production of this book possible by offering me, during my sabbatical year, a working space that I have described as the closest portal to heaven as I will ever see.

1 The Islamic Scientific Tradition: Question of Beginnings I

This chapter and the next address one of the most interesting aspects of Islamic civilization: the rise of a scientific tradition that was crucial to the development of universal science in pre-modern times. These chapters are connected by a common title, to indicate their interdependence. This first chapter surveys the various theories that have confronted the question of why and when this scientific tradition came into existence. It begins with a detailed account of the theories. The critique that follows addresses their failure to account for the facts as we know them from the primary scientific and historical sources of early Islamic times; it also lays the foundation for an alternative explanation of those facts in the next chapter. Because of this structure, the reader may encounter many unanswered questions in the first chapter, and will be repeatedly asked to await the answers that will come in the second.

There is hardly a book on Islamic civilization, or on the general history of science, that does not at least pretend to recognize the importance of the Islamic scientific tradition and the role this tradition played in the development of human civilization in general. Authors differ in how much space they allocate to this role, but they all seem to agree on a basic narrative, to which I will refer as *the classical narrative*. The main outline of this narrative goes back to medieval and Renaissance times and has been repeated over and over again.

The narrative seems to start with the assumption that Islamic civilization was a desert civilization, far removed from urban life, that had little chance to develop on its own any science that could be of interest to other cultures. This civilization began to develop scientific thought only when it came into contact with other more ancient civilizations, which are assumed to have been more advanced, but with a particular nuance to "advanced." The ancient civilizations in question are the Greco-Hellenistic civilization on the western edge of, and overlapping with, the geographical domain of the Islamic civilization, and the Sasanian (and by extension the Indian) civilization to the east and the southeast. These surrounding civilizations are usually endowed with considerable antiquity, with high degrees of scientific production (at least at some time in their history), and with a degree of intellectual vitality that could not have existed in the Islamic desert civilization.

This same narrative never fails to recount an enterprise that was indeed carried out during Islamic times: the active appropriation of the sciences of those ancient civilizations through the willful process of translation. And this translation movement is said to have encompassed nearly all the scientific and philosophical texts that those ancient civilizations had ever produced.

The classical narrative then goes on to recount how those translations took place during the early period of the Abbasid times (circa 750–900 A.D.) and how they quickly generated a veritable golden age of Islamic science and philosophy.

In this context, very few authors would go beyond the characterization of this Islamic golden age as anything more than a re-enactment of the glories of ancient Greece, and less so the glories of ancient India or Sasanian Iran. Some would at times venture to say that Islamic scientific production did indeed add to the accumulated body of Greek science a few features, but this addition is usually not depicted as anything the Greeks could not have done on their own had they been given enough time. Nobody would, for example, dare to suggest that the scientists who worked in Islamic times could have produced a new kind of science (in contrast with the science that was practiced in classical Greek times), or to imply that those scientists may have come to realize, from their later Islamic vantage point, that the very same Greek science, which became available to them through the long process of translation, was in itself deficient and fraught with contradictions.

The classical narrative, however, persists in imagining that the Islamic science that was spurred by these extensive translations was short-lived as an enterprise because it soon came into conflict with the more traditional forces within Islamic society, usually designated as religious orthodoxies of one type or another. The anti-scientific attacks that those very orthodoxies generated are supposed to have culminated in the famous work of the eleventh-twelfth-century theologian Abū Ḥāmid al-Ghazālī (d. 1111). The major work of Ghazālī that is widely cited in this regard is his *Tahāfut al-Falāsifa* (Incoherence of the Philosophers), which is sometimes also mistakenly referred to as *tahāfut al-falsafa* (incoherence of philosophy).

By sheer luck and proverbial serendipity, the Latin West was beginning to awaken around the same time. And this awakening set in motion a translation movement that identified and translated major Arabic philosophical and scientific texts into Latin during a period that has come to be known at times as the Renaissance of the twelfth century. Some of the texts that were translated into Latin during this period had already been translated from much earlier Greek and Sanskrit texts into Arabic. I am thinking in particular of such major Greek works as the *Almagest* of Ptolemy (d. ca. 150 A.D.) and the *Elements* of Euclid (d. ca. 265 B.C.), which had been translated into Arabic more than once during the ninth century, and of the passage of the Indian numerals via Arabic to Europe, where they came to be known as "Arabic" numerals.

The classical narrative goes on to postulate that from then on Europe had no need for Arabic scientific material, and that the Islamic scientific tradition was beginning to decline under the onslaught of the works of Ghazālī and thus was no longer deemed important by other cultures. In the grand scheme of things, the European Renaissance was then characterized as a deliberate attempt to bypass the Islamic scientific material, in another act of "appropriation" so to speak, and to reconnect directly with the Greco-Roman legacy, where almost all science and philosophy began, and where the European Renaissance could find its wellsprings.

Critique of the Classical Narrative

In what follows, I would like to subject this classical narrative to some criticism and to point to some of the problems that it fails to solve, before I propose, in the next chapter, an alternative narrative that, I believe, accounts for the historical facts in a much more comprehensive fashion. I do so because the classical narrative leaves us with some unresolved problems that we cannot afford to leave unsettled if we ever wish to understand the actual process by which Islamic science came into being when it did, and in a more general fashion the process by which science, in general, is born and nourished in any society. But in order to do that, first I have to deconstruct some of the basic tenets of this classical narrative.

That Islamic civilization was isolated in a desert environment is an oversimplification. As is well known, Islamic civilization came into being around the cities of Mecca and Medina and around northern Arabian tribal areas and cities that were not exactly desert steppes. Within that environment the pre-Islamic Arabian civilization had already developed some basic astronomical and medical sciences that survived well into Islamic times. In a chapter that was written about 15 years ago but not published until 2001, I tried to summarize the scientific knowledge of pre-Islamic Arabia, and came to the conclusion that the sciences that could be documented there were not much different in quality from the sciences of the surrounding regions of Byzantium, Sasanian Iran, or even India.[1]

But most importantly, the classical narrative leaves us with yet more serious and inexplicable problems, both with regard to the beginnings of Islamic science and with regard to its decline and eventual demise. In the case of beginnings, the classical narrative creates the impression that the birth of Islamic science took place during the early period of the Abbasid times, mainly during the latter part of the eighth century and the early part of the ninth, as a result of one or more of the following processes of transformation:

(1) Contact between the nascent Islamic civilization and the more ancient civilizations of Byzantium and Sasanian Iran is supposed to have taken place when the domain of Islamic civilization expanded outside the Arabian Peninsula and came to inherit the domains of those earlier civilizations or to share great geographic spans with them.[2] This "contact theory" had the distinct advantage of explaining the birth of Islamic science as a result of outside forces, a disposition already signaled by a particular reading of the classical Arabic sources. Those sources speak, for example, of the "ancient sciences" when they wished to describe the sciences that were brought into Islamic civilization from outside, or when they wished to contrast those sciences with the "Islamic sciences" (usually understood as the religious sciences that grew within the civilization). At times the two sciences are posited as being in direct opposition.

The downside of this theory is that it cannot furnish an explanation for the high quality of Greek scientific and philosophical texts that were translated into Arabic during this contact period of early Abbasid times when the contemporary surrounding cultures of the time had not been participating in the production of such texts for centuries before the advent of Islam.[3] In other words, the scientific and philosophical texts, usually designated by the term "ancient sciences" in the classical Arabic sources, contained material that was already written in the classical period of Greek civilization, and most of them were indeed produced before the third or the fourth century A.D. As far as we can tell, and as far as the sources demonstrate, no similar activities continued to take place in Byzantine[4] or Sasanian civilization that could have put those texts in circulation and thus made them readily available to the translators who worked in the extensive translation movement of early Abbasid times. When we examine that translation movement, we find translators such as Ḥunain b. Isḥāq (d. 873) searching for classical Greek scientific texts all over the old Byzantine domain, and sometimes failing to find what was needed.[5] Under such conditions, when books were not taught or used in wide circulation, how could contact have produced any positive and effective transfer of knowledge? The classical narrative has no convincing answer to such a straightforward question.

Besides, for scientific contacts to be successful it is only natural to assume that both cultures had to have been at similar levels of development so that ideas from one culture could easily find a home in the other.

(2) Those who were conscious of the downside of the contact theory, and of its failing to document contemporary scientists of Byzantium or Sasanian Iran who could have produced texts similar to the ones that were being sought by the translators of Abbasid times (that is, texts of the quality of ancient more classical Greek scientific and philosophical texts), thought they could avoid that pitfall by proposing another form of transfer that I shall call the *pocket* transmission theory.[6]

In this new theory, assumptions were made about the survival of ancient scientific and philosophical texts in a few cities in Byzantium or in the then-defunct Sasanian Empire. In those cities, classical Greek scientific texts were supposed to have been preserved. Antioch (the cradle of early Christianity), Ḥarrān (the site of many legends recorded in later Islamic sources), and Jundīshāpūr (where academies, hospitals, and observatories were supposed to have flourished) were all mentioned at one time or another as major repositories of ancient classical Greek texts.

But preserving second-century texts for hundreds of years, or even making new copies of them when there was need for them in Baghdad (as was done during the ninth century[7]), could not guarantee that there would be people who could understand these texts when they were being sought for translation,[8] during Abbasid times, about 700 years after they were written. Moreover, scientific and philosophical ideas usually flourish through open discussions. And it would be highly unlikely that enough such discussions were taking place between the fourth and the eighth century to affect another incoming culture. After all, there were some major reversals in scientific knowledge during that intervening period—for example, Cosmas Indicopleustes (c. 550 A.D.) proposed a flat Earth about 800 years after Eratosthenes measured Earth's circumference.[9] Knowing of the treatment of the mathematician Hypatia, who fell in between two competing powers of her time (the church and the state), and of her violent death at the hands of a mob of church followers who used her learning against her in the form of rumors in their political struggle, makes the kind of folk and popular science that was propagated by Cosmas more characteristic of Byzantine science than of the more sophisticated science of earlier classical Greek times. In that light it becomes unimaginable that any Byzantine scholar of that period could have produced anything of the sophistication of Ptolemy, Euclid, or Galen, or even fully understood what those giants had written.

Furthermore, neither in Antioch nor in Ḥarrān nor in Jundīshāpūr could one find a single scientist or philosopher of any importance who could have produced any work that could demonstrate his or her sophisticated understanding of the classical Greek scientific and philosophical texts, let alone match them in brilliance. Sure, one may find some references to such folk scientific ideas as names of stars, calendar approximations, or some astrological prognostications, of the type we see in the works of the Syriac scientists mentioned below, or even the works of Paulus Alexandrinus.[10] One may even find some elementary medical texts, or texts dealing with weather prognostication and star configurations, or even texts containing pharmacological material (mostly in the form of home remedies). But nothing of the caliber of the classical Greek scientific texts could be found.

Besides, how could it be possible for one or two cities in any empire to acquire and maintain a viable scientific tradition when there was no concrete evidence of such a flourishing tradition in any of those cities, nor was there any evidence that the rest of the cities of the empire could have produced anything of the sort? If the capital city could not have those sciences available, and if those who went through great hardships to study them (as had happened with Leon the Mathematician) were so poorly viewed by their own students, then how could those sciences be available at less important centers, so that they would exert any perceivable influence on a foreign culture that came in contact with them? If those pockets could exist for hundreds of years in such isolation, and still maintain a sophisticated degree of philosophical and scientific production similar to what was reached in the classical times before the third century A.D., that would be an unparalleled phenomenon that would require much documentation by the proponents of such a theory before it could be really accepted.

And yet there is some debatable evidence of sorts. In his account of the transmission of philosophy to the lands of Islam,[11] the philosopher al-Fārābī (d. 950) recounts the story of how philosophy was transmitted from Greece to Alexandria, and from there to Antioch, Ḥarrān, and Marw, and finally to Baghdad. But a close examination of that story (which became the basis of a famous article by Max Meyerhof, "Von Alexandrien nach Baghdad"[12]) makes one appreciate Paul Lemerle's remark about it: "Je ne suis pourtant pas certain qu'on puisse accepter sans retouche la séduisante construction de M. Meyerhof."[13] The story certainly seems to reveal more about Fārābī's desire to connect himself to the long philosophical line stretching back to Aristotle than about his desire to produce an accurate historical account of the actual transmission of philosophy from Greece to the Islamic civilization. This is corroborated by the fact that it is the same Fārābī, and in the same story, who recounts the persecution of the philosophers (and we should understand that as including scientists, since science, at the time, was really natural philosophy) at the hands of the Byzantine emperors as well as the Christian church. In it he only mentions the very brief respite from persecution that occurred during the very short rule of Flavius Claudius Julius (361–363). More pointedly, it was Fārābī too, who recounts, in the same story, the persecution of philosophy at the hands of Christianity (consistent with what we just mentioned of the fate of Hypatia and others). And in that regard, Fārābī asserts, in no ambiguous terms, that philosophy was finally freed only when it reached the lands of Islam.

If this were the case, and there is much evidence to corroborate the account of persecution as we have already seen, then how could classical

Greek philosophy maintain a rigorous tradition, in cities far apart, and at such times when the official policy of the state was to suppress that very same tradition, and when the only support that was ever given to philosophy was during a three-year reign of an emperor who was fought on every ground and was indeed called "the apostate"? With all those questions, and with this kind of evidence that is used for its support, one need not say anything more about the inability of this theory to explain the transmission of Greek science into Arabic.

(3) Then there are those who propose a more nuanced theory of transmission of the Greek philosophical sciences to Islamic civilization by postulating a transmission that went through the Syriac medium first. And this theory too has some evidence to support it. In this context people cite the works of the Syriac writers Paul the Persian (c. 550) and Sergius of Rasʿaina (d. 536), and the slightly later writers Severus Sebokht (c. 660) and George, Bishop of the Arabs (c. 724). The theory asserts that those people brought the Greek tradition into Syriac first, only to make it available for Arabic translations later on.

And all those Syriac authors produced works that could be described as scientific, with some degree of seriousness. But when those works are examined carefully, they turn out to be of the same quality as the ones that were produced in the larger Byzantine Empire; that is, they were elementary relative to the classical Greek texts. Paul's work did not seem to extend beyond the elementary treatises on logic,[14] and Sergius did not apparently venture with his astronomical explorations much beyond the *Apotelesmatica* of Paulus Alexandrinus (c. 378), from which he adopted a very elementary approximative method for calculating the positions of the sun and the planets.[15] The method was so crude that it could nowhere be compared with the more exacting methods of Ptolemy's *Almagest* and *Handy Tables*. The fact that Sergius knew of such august works of the classical Greek tradition is duly attested by his references to them, but only to say that they were to be sought only by those who needed higher precision. He seemed to have satisfied himself with the work of Paulus Alexandrinus.

The slightly more sophisticated works of Severus Sebokht (for example, his treatise on the use of the astrolabe[16]), and those of George, Bishop of the Arabs,[17] are not much closer to the classical Greek scientific texts, and in general they exhibit the more historically understandable standard of being of about the same quality as the contemporary Byzantine sources from which they seem to have derived their inspiration. And why should it be otherwise? Why should the poorer Byzantine subjects, as the Syriac-speaking subjects were, know more than the more sophisticated and much richer Byzantine overlords?

In fact we get echoes of this social class distinction, and the enmities that went with it, from the works of Severus Sebokht himself, who does not shy away from bragging against the Byzantine Greeks by asserting that his own ancestry extended all the way back to Babylonia, and that there were other nations, like the Indians, who could outsmart the Greeks in science.[18] He cites as evidence of the Indians' superiority their knowledge of the decimal system, with which, he says, "they calculate with nine figures only."[19]

All this evidence illustrates that the Syriac route of transmission, at least during pre-Islamic and early Islamic times, could not have been much more reliable than the contact or the pocket theory of transmission. And yet the rise of the more sophisticated Islamic scientific tradition in early Islamic times owes a great deal to the acquisition of the Greek scientific legacy and the direct translations of major classical Greek scientific and philosophical texts. How did this happen? The following chapter will, I hope, shed some light on this.

Having resorted to the three methods of transmission that are often mentioned by the proponents of the classical narrative, we find ourselves at a loss to explain how this transmission took place. This, to say nothing of the motivation of the early Abbasid caliphs for the acquisition of these ancient sciences, which had been already abandoned for about 700 years before those early Abbasids began to translate them. Why the sudden awakening? And why were the Abbasids so motivated toward the beginning of the ninth century to finance, patronize, and undertake such a major operation, or even make it "a regular state activity,"[20] as is often stressed by the classical narrative but rarely explained? It is hoped that the following chapter will shed some light on this subject too.

The early Abbasids' involvement in the activity of transmission remains to be explained, even if all those problems regarding the manner in which the "ancient sciences" were transmitted to the Islamic civilization were all resolved once and for all, and even if the classical narrative that generated them was abandoned. For there would still remain a second and more important problem: that of the timing of this transmission, which the classical narrative locates toward the beginning of the Abbasid times. Why at that time in particular and not during the earlier 100-year rule of the the Umayyads? What was so special about the Abbasids? Here the classical narrative offers three plausible explanations for that starting point, two of them corollaries of one another:

(1) It is very well known, as is repeatedly emphasized by the classical narrative, that the general character of the Abbasid dynasty allowed the ascendancy of the "Persian elements" of the Islamic empire. For, after all, the argument goes, the Abbasids rose in rebellion first in Transoxania, and they did so against the Umayyads, who were in turn characterized by the classical narrative that bases itself on many other classical Arabic sources as champions of the "Arab elements" of the empire. In fact one finds some echoes of such contentions in the classical Arabic sources themselves.

It is true that the Abbasids, who came to power with the swords of the central Asian troops, brought along with them clients who ruled on their behalf in the Transoxanian provinces, and thus depended greatly on the loyalty of those central Asian troops, many of whom were of Turkic and Persian origins. It is also true that the men who occupied the high positions of government, at least in the early Abbasid times, and at the ranks of viziers and the like, such as the members of the Barmakid family, were themselves of Persian descent. And despite the devastating demise of the Barmakids toward the beginning of the ninth century (when the whole family was simply wiped out from positions of power[21]) other Persian families such as the Nawbakhts simply replaced them in the high positions of government.

That the sources speak of Persians, Turks, and Arabs (among others) during the early Abbasid period indicates that these sources, from which the classical narrative derived its inspiration, began to reflect, at that particular time, the racial makeup of the people in power. That phenomenon itself must be explained rather than be stipulated in such essentialist terms, as the classical narrative seems to do with that particular historical setting.

In other words, and even if we privilege the classical narrative with some analytical power, then we still have to explain why the "Persian elements" of the Islamic empire would resort to translating Greek scientific and philosophical sources and not restrict themselves to translating Persian sources, for example. Dimitri Gutas, in his recent book *Greek Thought, Arabic Cul-*

ture,[22] offers a plausible explanation. Gutas refers to what he claims was the prevailing ideology of the time, reflected in a source that was quoted in the *Fihrist* of al-Nadīm (c. 987), and which asserted that all sciences began in Persia and that those sciences were translated into Greek at the time of Alexander's invasion of Persia, thus leaving the Persians deprived of their legacy after the cataclysmic devastation that befell them at the hands of Alexander. So when those Persians came to power, inexplicably only during Abbasid times and not before during Sasanian times when they were the full masters of the lands east of the Euphrates and sometimes even west of it, they awakened to that ancient legacy and decided to reclaim it. Thus, starting with al-Manṣūr, the second Abbasid caliph who enjoyed a relatively long reign, to al-Mahdī, and Hārūn al-Rashīd, and then of course to Al-Maʾmūn, who epitomized this trend, one caliph after the other doggedly persisted in reclaiming this Greek scientific heritage. They also patronized the more literary Persian translations, simply because there were no more sciences left in Persian after their abandonment from the time of Alexander's plunder.

This explanation fits well with the then-prevailing trend in the classical sources just mentioned, in which the "Persian elements" were made responsible for this large-scale Abbasid enterprise. It does not explain, however, the lack of real interest in such reclamation of original Persian sciences from the Greeks during the times of the Sasanians, when they were the masters of the domain, and in constant warfare with the Greeks. In fact, the same reports that speak of the reclamation of the Persian sciences from Greek during Abbasid times also speak of earlier Sasanian attempts to reclaim Persian sciences, but mainly from India and China, and from the Greeks only as an afterthought. These reclamation efforts remain unsubstantiated.[23]

Searching for evidence of the actual scientific texts that were produced or translated during Sasanian times, one could certainly find at least one astronomical work, the so-called *Zīj-i Shahriyār,* which was later translated from Persian into Arabic. And since the *Zīj* itself was composed during Sasanian times, this does indeed indicate an interest in scientific works in the Sasanian Empire. Unfortunately the *Zīj* is no longer extant. But from the few citations of it in later Arabic sources, it seems to have been more indebted to Indian astronomical sources than to Greek ones,[24] and thus this particular, almost unique, source does not attest to the interest in Sasanian Iran in reclaiming "their" Greek heritage. Rather it points in the other direction.

Other astrological texts, such as the *Anthologia* of Vettius Valens[25] and the

Carmen Astrologicum of Dorotheus Sidonius,[26] were indeed reclaimed from Greek into ancient Persian, and were later translated into Arabic during Abbasid times. But even those astrological texts can hardly be called a reclamation of the Greek sciences on the scale or sophistication in which they were reclaimed during Abbasid times. A look at the second of those texts and the fragments that have been quoted of the first reveals that they were mainly books of descriptive astrology and not the more sophisticated and demanding horoscopic astrology, which could be attained only after the translation of the more sophisticated texts such as Ptolemy's *Handy Tables.* Such tables would indeed enable one to cast a horoscope.

Furthermore, when one surveys the texts that were translated during the Abbasid times, one finds a major qualitative difference between the texts that were translated then and the texts that were translated before, either into Syriac or into Pahlevi. In the earlier times, such elementary, mainly descriptive texts were translated into the various languages. In the later Abbasid times, most of the books that were sought for translation were on the whole theoretical in nature and were much more sophisticated in content. In contrast, one finds in the later period such translations as the *Almagest* of Ptolemy, Euclid's *Elements,* the *Arithmetica* of Diophantus, the *Conics* of Apollonius, and the *Arithmetic* of Nicomachus, and also more descriptive yet analytically theoretical texts, such as the *Tetrabiblos* of Ptolemy. There is no record that even the *Tetrabiblos* was been translated into Syriac or Pahlevi in pre-Abbāsid times. The Syriac text that is designated as the *Tetrabiblos* at the Bibliothèque Nationale de France [Syr. 346, fols. 1–35] is in fact a paraphrase, and a poor one at that, and not a translation of the type that was done in the case of the more theoretical texts that were translated into Arabic during the Abbasid times. And yet we do not even know when this paraphrase was produced.

Therefore, when the classical narrative seeks the motivation for the translation activity in the dominance of the "Persian elements" in the Abbasid empire, and in their desire to reclaim what they thought was theirs of the Greek sciences, that explanation creates more difficulties than it resolves, for it remains completely silent about the lack of concrete evidence for such motivation at the time of the supreme Persian ascendancy during Sasanian times. Furthermore, the legend of the translation of Persian sciences into Greek at the time of Alexander is not such a reliable story that it could be used as an explanatory basis for the translation movement that took place during the early Abbasid times. In fact, the story itself is a part of the phenomenon of the translation movement itself and a feature of the intellectual life of early Abbasid times and not the explanatory cause of it. In all likelihood, the story was created after the facts and thus itself needs to be explained.

(2) Another motivation for the translation activity during early Abbasid times, which is often cited by proponents of the classical narrative, is the ascension of al-Maʾmūn to power in 813, and his reliance on the Muʿtazilite school of *Kalam* as a state theology. This particular caliph is often endowed with an interest in the philosophical sciences and a preoccupation with introducing the Muʿtazilite doctrines in the realm, so much so that he began to see dreams that justified his disposition. In one of those dreams he is supposed to have seen Aristotle himself,[27] and to have had the chance to interrogate the great master about the great ethical and philosophical issues of the day. He asked Aristotle, for example, "What is good?" Aristotle is supposed to have replied "That which is good in the mind." And when asked "What next?" Aristotle is supposed to have answered "That which is good in the law." When al-Maʾmūn persisted in asking "What next?" Aristotle is supposed to have added "That which is considered good by the people." But when he again asked "What next?" Aristotle stopped and said "There is no next." In another account, Aristotle is supposed to have continued to advise al-Maʾmūn to treat those "who advised him about gold like gold" (an apparent reference to alchemists), and then he is supposed to have said "and you should adhere to the oneness of God (*ʿalaika bi-l-tawḥīd*)." The last phrase is an obvious reference to the Muʿtazilite doctrine, as those people were called "the people of oneness" (*ahl al-tawḥīd*) on account of their insistence on God's oneness, which did not even allow the Qurʾan, God's speech, to have been co-existent with Him at the beginning of time.

(3) The third motivation is also associated with the Muʿtazilites and their connection with al-Maʾmūn, who made their doctrine official state doctrine. This policy was also followed by two of his successors and eventually led to a type of inquisition often referred to in the sources as the *miḥna* (testing/interrogation, inquisition),[28] hardly an enlightened open environment for scientific inquiry. In this *miḥna* people were supposed to declare that the Qurʾan was created in time, specifically in agreement with the Muʿtazilite

doctrine that insisted on God's oneness in the beginning. People who refused to adopt such a doctrine, including the great jurist Aḥmad b. Ḥanbal (d. 855), were put in jail.[29] This climate is supposed to have energized philosophical thinking during that period of Abbasid rule, or at least so the classical narrative goes, and thus it must have motivated the acquisition of the major Greek philosophical texts, and thus opened the doors for the vast translations that followed. In other words, the classical narrative asserts that once the doctrinal debates within Islamic society reached their peak to become part of state policy, the state must have encouraged the translations of all those philosophical and scientific texts in order to buttress its intellectual position.

This explanation could have been plausible had it been supported by the facts. In this regard, the historical sources tell us that the Muʿtazilite connection with the state was indeed very short-lived, and when the caliph al-Mutawakkil came to power (847 A.D.) he not only reversed the policies of al-Maʾmūn but went on to support the Muʿtazilite opponents, at this time called *ahl al-ḥadīth* (people of tradition—meaning people who sought legal justifications in the traditions of the prophet, and less so in human reasoning as the Muʿtazilites had done). And yet it was during the reign of this last caliph that the greatest amount of translations from Greek sources were ever accomplished and mostly by the prolific translator of the time, the famous Ḥunain ibn Isḥāq (d. 873), who worked as a physician at al-Mutwakkil's court. The books that were translated from Greek, mostly during the time of al-Mutawakkil, far outweigh those that were patronized by al-Maʾmūn. In fact I know of only one surviving book that is expressly designated as having been translated at the order of al-Maʾmūn, but I am not sure whether that designation was there on the book when it was first translated in 829 or whether it was added later by an owner or some other librarian trying to give its history.[30]

The classical sources do in fact speak of all sorts of scientific activities that were patronized by al-Maʾmūn, some apparently verifiably real such as the mission he sent to the desert of Sinjār to measure the length of one degree along the Earth's meridian,[31] and to conduct some astronomical observations. Other, perhaps more fanciful, stories such as the missions he sent to Constantinople to acquire Greek scientific manuscripts or Greek scientists speak to some interest this caliph may have had in such matters.[32] But it is never clear whether those activities were indeed ordered by al-Maʾmūn himself or by bureaucrats working in his administration. The role of the bureaucrats will become clearer in the next chapter. For now, the same historical sources, report that the later bureaucrats, who worked in al-Mutawwakil's administration, were themselves the ones who sponsored and paid for a great number of books to be translated. They also executed a great number of scientific and technological projects.[33] In fact I do not know of a single book that was translated for al-Mutawakkil himself, despite the great intellectual activities that took place during his reign, but I know of a great number of books that were translated for three brothers, known collectively as Banū Mūsā, who worked at his court, and sometimes at great risk. I shall have reason to return to this aspect of the translation movement in the following chapter when I explain the alternative narrative regarding the rise of science in early Islamic times. For now, I continue with the critique of the classical narrative.

Other Problems with the Classical Narrative

When it comes to details, the classical narrative cannot account for the very scientific facts that have been preserved either in the classical historical sources of the period or in the scientific texts themselves. For example, more than one historian tells us[34] that when the caliph al-Manṣūr wished to build the city of Baghdad, in 762 A.D., he assembled three astrologers and charged them with casting the horoscope for the future city. They were supposed to choose the time for the foundation so that no potentate would be killed in the city. The horoscope itself is preserved in the *Chronology* of Bīrūnī, and in several other sources. Most sources agree that the astrologers who were assigned that task included Nawbakht (a Persian astrologer who became the progenitor of the Nawbakht family of astrologers, which served caliphs for a whole century), Ibrāhīm al-Fazārī, and Māshāʾallāh al-Fārisī. Bīrūnī states explicitly that it was Nawbakht who determined the day for the foundation of the city to coincide with the propitious 23rd of July of that year.

If the ancient Greek sciences were supposed to have been brought into Arabic by the Persian-leaning elements of the Abbasid dynasty, even if we grant that this interest started with al-Manṣūr himself, and if we grant that they could recruit for the purpose of the horoscope the Persian astronomers Nawbakht and Māshāʾallāh, then who was this Ibrāhīm al-Fazārī, obviously an Arab from the tribe of Fazāra, who was also invited to join them, and where did he acquire the kind of advanced astronomical knowledge that he would have needed for casting such a horoscope at that early time in the Abbāsid reign? Where did his usual collaborator Yaʿqūb b. Ṭāriq learn his own astronomy so that he could produce, together with Fazārī, a translation of the Sanskrit *Sidhanta* (*al-Sindhind*), which was completed during the caliphate of al-Manṣūr (754–775 A.D.)?[35] Later sources always joined those two names together,[36] so it is sometimes difficult to determine who did what. For the purposes of the Baghdad horoscope, we may stipulate that Fazārī may have learned his craft in Persia. But the sources are silent on that, and we do not know much about the Persian astronomy of the time beyond the existence of the *Shariyār zīj* (which was quoted in later sources). Furthermore, the historical sources that connect the two assert that this very same Fazārī and/or Ibn Ṭāriq also wrote a theoretical astronomical work called *Tarkīb al-aflāk,* which seems to have been lost. The same Fazārī is also credited with the authorship of his own *zīj,* in which he used the "arab years" (*ʿalā sinīy al-ʿArab*).[37] Writing a theoretical astronomical text, transferring a *zīj* to a different calendar with a completely different intercalating scheme, and producing astronomical instruments such as astrolabes—as we are also told about these men—could not have been done by amateur astronomers. Who educated Fazārī and Ibn Ṭāriq in all these fields of astronomy? And even if we believe that the three astrologers also used the Persian *Zīj-i Shahriyār* for the purposes of the horoscope, we should also ask about another Arab, ʿAlī b. Ziyād al-Tamīmī, from the tribe of Tamīm, who was supposed to have translated this *zīj* into Arabic.[38] Who taught al-Tamīmī how to translate a *zīj,* and when he did so did he also transfer it into Arab years (as we are told Fazārī had done)?

All this evidence indicates that there was a class of people, who were already in place by the time the Abbasids took over from the Umayyad dynasty, who were competent enough to use sophisticated astronomical instruments, to cast horoscopes, to translate difficult astronomical texts, and to transfer their basic calenderical parameters, as well as to compose theoretical astronomical texts such as *Tarkīb al-aflāk*. Such activities could not have been accomplished by people who were just learning how to translate under the earliest Abbasids, as the classical narrative claims.

The situation gets more complicated, again on the level of details, when we look at the works that were produced about 75 years later by people like al-Ḥajjāj b. Maṭar (fl. ca. 830), who translated the two most sophisticated Greek scientific texts: Euclid's *Elements* and Ptolemy's *Almagest*. We know, for example, that al-Ḥajjāj finished his translation of the *Almagest* in the year 829, as is attested in the surviving copy now kept at the Library of Leiden University (Or. 680). And when we look at this translation we are immediately struck by two most startling phenomena: the language of the text is impeccably good Arabic, technical terms and all; and the Arabic translation even corrects the "mistakes" of the original Greek *Almagest*. Who taught al-Ḥajjāj the technical terms, and who taught him how to correct the mistakes of the *Almagest*? Neither of these questions is resolvable if we continue to believe the classical narrative that dates the beginning of the serious translations to the time of al-Maʾmūn (813–833). Early translations usually struggle with technical terminology, and usually do not go beyond the letter of the text and would never dare correct its mistakes, if they could understand the text in the first place.

Furthermore, we know that al-Ḥajjāj's translation of those scientific works was not the first. In fact, we are explicitly told by some sources that those two books were already translated under the patronage of Khālid al-Barmakī, the vizier of Harūn al-Rashīd (d. 809), and maybe by al-Ḥajjāj himself, and by others that they were translated during the time of al-Manṣūr (754–775).[39] But the farther back these translations are pushed, the more complicated the story becomes, for the question of the development of technical terminology would still persist and actually becomes even more difficult to answer. In any event, the text as it is now preserved in the 829 A.D. translation reveals a maturity that could not have come from one generation of translators. And thus we must allow for a longer period of translation so that more than one generation of translators would create enough output to produce technical terminology and teach the sophisticated mathematics and linguistic skills that were required to render the *Almagest*, the *Elements*, and similar books into the kind of coherent Arabic in which they are preserved.

During the same early period—that is, during the reign of al-Maʾmūn—we also witness the creation of the new discipline of algebra by Muḥammad b. Mūsā al-Khwārizmī (fl. ca. 830),[40] already in a mature format—treating, for example, the field of second-degree equations in its most general form. This happened before the translation of the work of Diophantus and other Greek sources. This does not mean that classical Greek sources, or for that matter ancient Babylonian sources, did not include algebraic problems, but the coinage of the new term for algebra (*al-jabr*), and the statement of the discipline in general as different from arithmetic,[41] required a kind of maturity that could not have come with the first generation of translators if we assume that translations began with the early Abbasid times as the classical narrative stipulates. Under such circumstances we are entitled to ask " Who taught al-Khwārizmī to do what he did?"

Similarly, a few years later, or even contemporaneously with Khwārizmī, we witness the creation of the discipline of *Hay'a,* as in *'ilm al-hay'a,* which also did not have a Greek parallel. And that too could not have come about, as it did in the work of Qusṭā b. Lūqā (fl. ca. 850), which is still preserved in an Oxford Manuscript,[42] during the first generation of translations. Moreover, it is remarkable to note that Qusṭā himself, like other accomplished translators of his time, was already composing his own new scientific books, like his book of *Hay'a* just mentioned, while he was still translating older, more common Greek scientific texts. Ḥunain did the same, and so did many others in this period. All that could not have come about at the hands of people who were translating for the first time, and needing to create the new technical terminology for their translations as well as their original compositions. In Quṣṭā b. Lūqā's Arabic translation of the *Arithmetica* of Diophantus there is a clear adoption of the algebraic language that was developed by the Arabic-writing algebraists of Quṣṭā's time, as is evident from Qusṭā's reference to the title of Diophantus's work as *ṣinā'at al-jabr* (Art of Algebra), a term that does not exist in Greek, and as was discussed by Rashed.[43] This kind of liberty with the translation clearly demonstrates the dynamic nature of the translation process of the early ninth century. Classical Greek scientific texts could easily be acclimatized within the current Arabic sciences of the time, thus transforming the translation process into a simultaneous creative process as well.

Furthermore, the remarkable advances that were made by Ḥabash al-Ḥāsib (fl. ca. 850) in the field of trigonometry and mathematical projection go far beyond what was known from the Indian and the Greek sources, and they could not have been accomplished by someone who was only a beneficiary of an early stage of translation. Ḥabash devised new ways of projecting planespheric astrolabes that preserved such fundamental features as directions to a specific point on the globe (in this case Mecca) and the distances to that point.[44] Such projections were not known from any earlier civilization, and their existence must give rise to questions regarding the possibility of the production of such results by people who would have been still struggling with the creation of new technical terms if they were contemporaries with the early generation of translators.

This generation of early mathematicians and astronomers must have also developed the Indian numeral system to such an extent that by the next century we note the first appearance of decimal fractions together with the decimal point in a manuscript completed in Damascus in 952 by Uqlīdisī. [45]

In sum, such results as the new algebra and trigonometry, the new *hay'a* as well as the new methods of projection and the introduction of the Indian numerals and the development of decimal fractions, could not have all been produced at the same time with no previous works in those domains or in domains directly related to them. As a result, if the classical narrative insists on the beginning of the translation movement with the coming of the Abbasid Empire, and for reasons that were only motivated by the desire of the Abbasid caliphs, these questions will have to be answered before such claims can be accepted.[46]

Scientific Instruments and Observational Astronomy

In the field of scientific instrumentation, like the production of new types of mathematical projections that were created by Ḥabash as was already stated, those instruments could not have been created *ex nihilo,* as the classical narrative would want us to accept. In the case of Ḥabash's astrolabe, the new projections seemed to be related to the new Islamic requirements of facing Mecca while praying five times a day and performing a pilgrimage at least once in a lifetime. Yet such developments still required a remarkable sophistication in the application of geometric and trigonometric methods. Under normal circumstances, all these features would not usually come at once, but would rather progress slowly over time.

Similarly, the scientists of the same generation of Ḥajjāj, Khwārizmī, and Ḥabash and their colleagues seem to have also taken it upon themselves to double-check the observational results that were reported in the Greek and Indian sources from which they were trying to get their own inspiration. And there too, we find remarkable results already achieved in this very early period that indicate a much longer acquaintance with those fields. The observation that determined that the inclination of the ecliptic was not

23;51,20° (as was reported in Ptolemy's *Almagest*[47]) or 24°[48] (as was reported in the Indian sources), but that it was about 23;30° (as was determined during the first half of the ninth century[49]). That could not have come about as a result of the efforts of inexperienced astronomers who were conducting those observations for the first time. Such precision could only be achieved by mature astronomers who knew exactly what they were doing. That their value for the inclination is still in circulation today is a testament to the ingenuity of those ninth-century observers.

In the same vein, the determination of the new value for the precession parameter as 1°/66 years[50] or for the value of the solar equation, or the motion of the solar apogee—supposed to be fixed by Ptolemy—also could not have come about at the hands of inexperienced astronomers who were trying their hands on the discipline for the first time just as the major texts of that discipline were being translated. All these results must presuppose a longer acquaintance with such methods of observations, such new notions of precision, and such reflection on the function of instruments in determining new parameters. In sum, they must presuppose a much longer period of instruction and acquaintance with such concepts before the efforts would begin to yield such fruits.

Add to that the critique of the Greek observational as well as theoretical approaches to astronomy that were leveled by Muḥammad b. Mūsā b. Shākir[51] and his brothers Aḥmad and Ḥasan. Muḥammad, the first of the three brothers, would critique Ptolemy for his incoherent description of the physical operations among the celestial spheres, and would deem such motions physically impossible. And the three brothers together, or someone in their circle, would critique the method by which Ptolemy determined the position of the solar apogee.[52] These are not efforts that could happen all at once without previous experience with observational techniques, acquaintance with instruments, critical judgment of the sources of error, a developed concept of precision, and a well-thought-out connection between the observations and the theoretical results that were being achieved. People who were still struggling to translate texts for the first time could not normally achieve such maturity.

Problems with the End

Not only does the classical narrative fail to solve the problems I have been discussing so far, which are connected with the beginnings of scientific activities in Islamic civilization; it also fails to account for the questions raised during the later centuries. In particular, the decline of Islamic science, which was supposed to have been caused by the religious environment that was generated by Ghazālī's attack on the philosophers or by his introduction of the "instrumentalist" vision, does not seem to have taken place in reality. On the contrary, if we only look at the surviving scientific documents, we can clearly delineate a very flourishing activity in almost every scientific discipline in the centuries following Ghazālī. Whether it was in mechanics,with the works of Jazarī (1205)[53]; or in logic, mathematics, and astronomy, with the works of Athīr al-Dīn al-Abharī (c. 1240),[54] Mu'ayyad al-Dīn al-ʿUrḍī (d. 1266),[55] Naṣīr al-Dīn al-Ṭūsī (d. 1274),[56] Quṭb al-Dīn al-Shīrāzī (d. 1311),[57] Ibn al-Shāṭir (d. 1375),[58] al-Qushjī (d. 1474),[59] and Shams al-Dīn al-Khafrī (d. 1550)[60]; or in optics, with the works of Kamāl al-Dīn al-Fārisī (d. 1320)[61]; or in Pharmacology, with the works of Ibn al-Baiṭār (d. 1248)[62]; or in medicine, with the works of Ibn al-Nafīs (d. 1288),[63] every one of those fields witnessed a genuine original and revolutionary production that took place well after the death of Ghazālī and his attack on the philosophers, and at times well inside the religious institutions.

It is not only that the classical narrative could not actually account for this prolific scientific production, at a time when the whole Islamic world was supposed to have been gripped by religious fervor, as the classical narrative dictates. Its failure went even further. It warped the production of those scientists when it deemed their results insignificant, and when it noted that those results were not translated into Latin during the medieval period, and thus concluded that the European Renaissance was achieved independently of what was taking place in these later centuries of the Islamic world. The works of this world that fell in between European medieval times and the time of the Renaissance could not be included in the general kind of history of science that the classical narrative could assimilate. As a result, the schism between what was happening in the Islamic world and what happened in the Latin West between the Middle Ages and the Renaissance grew deeper and deeper with the application of the classical narrative to the history of science. At the end the chasm was so deep that the relationship between those two worlds could no longer be understood, if its study was ever attempted.

With the European renaissance perceived as an independent European enterprise, and with the trajectory of scientific developments focusing on what took place in renaissance Europe, we also lost sight of the very exciting activities that took place at the borders between the Islamic and Byzantine civilizations. With the classical narrative emphasizing the importance of Arabic sources, only in as much as those sources could lead to the recovery of classical Greek antiquity—itself the object of the Renaissance as is commonly held—the outflow of scientific ideas from the lands of Islam to the Byzantine territories through the translations that went back from Arabic into Greek (Byzantine Greek at this time), starting at least as early as the tenth century and continuing till the fall of the Byzantine empire in the fifteenth century, still have not been accounted for. As a result, a whole chapter of scientific activities migrating across cultures remains almost completely lost to this day. Had it not been for the few maverick efforts of Neugebauer,[64] Pingree,[65] Tihon,[66] and their colleagues, and most recently Mavroudi,[67] no one would have known that there was such a rich chapter of scientific exchanges between Islam and Byzantium in a completely unexpected direction. This exchange, as it is becoming more and more apparent may have played a very important role in transmitting scientific ideas from Islamic civilization to the European renaissance, and thus must change the very image of the renaissance itself when it is fully accounted for.

Of the problems associated with the classical narrative, we must note that the insistence on the independence of the European renaissance from outside influences also keeps us from appreciating the role of such distinguished Renaissance scientists as Guillaume Postel (1510–1581), whose handwritten annotations on Arabic astronomical texts, still preserved in European libraries, must raise the question about the very nature of the astronomical activities of the European renaissance. When we look at some of the Arabic astronomical manuscripts that were owned by Postel and were annotated in his own hand, and remember that Postel may have very well used those same manuscripts to deliver his lectures in Latin at the institution that later became the Collège de France, we are then forced to ask "Whose science was Arabic science in Renaissance Europe?"[68] All these problems must be resolved, not only in order to understand the extent to which Islamic science was integral to the science of the Renaissance, but also in order to understand the very nature of the Renaissance science itself.

In the same vein, if we ignore, as the classical narrative urges us to do, the theoretical contacts between the land of Islam and Renaissance Europe, such as the transmission of mathematical theorems used in astronomical theories, then the sudden appearance of those theorems in Latin Renaissance texts will also remain unaccounted for and incomprehensible. We already know that astronomers of the Islamic world had used those very theorems for a few centuries. We shall have occasion to return to this very fertile area of research when we consider the relationship between Copernicus's mathematical astronomy and his Islamic predecessors.

The case of the discipline of astronomy in particular is very relevant here for yet another reason. For it was this discipline in specific that seems to have suffered the most as a result of the popularity and the hegemony of the classical narrative. On the one hand, we note a remarkable activity, of the highest order of mathematical and technical rigor, that kept on flourishing in the Islamic world after the death of Ghazālī, so much so that I have dubbed this post-Ghazālī period as the golden age of Islamic astronomy, and yet none of those results that were reached during that period had a chance of being considered by the proponents of the classical narrative as being worthy of attention, let alone consider their influence on Renaissance Europe. In fact, as we shall see later, some of the results achieved in this period were so badly understood by the very few orientalists who ventured to study them, that their significance was not understood properly, both to the disadvantage of the historian of Islamic science as well as the historian of Renaissance science.

For example, when the great orientalist Baron Carra De Vaux attempted to understand the most important chapter in the astronomical work of Naṣīr al-Dīn al-Ṭūsī, *al-tadhkira* (book II, chapter 11), in order to make the results of this chapter available to Paul Tannery for his classic *Recherches sur l'histoire de l'astronomie ancienne,*[69] De Vaux had this to say: "Le chapitre dont nous allons donner la traduction suffira peut-être à faire sentir ce que la science musulmane avait de faiblesse, de mesquinerie, quand elle voulait être originale."[70] He continued: "La portée de ce chapitre n'est donc pas très grande; il mérite neanmoins d'etre lu à titre de curiosité."[71] This was said of the chapter that was most relevant to the astronomy of Copernicus, who himself used the results that were already established in it by Ṭūsī to construct a very essential component of his own astronomy of the *De Revolutionibus*. As a result of the frame of mind that was generated by the classical narrative, the real significance of this chapter to the revolution against Ptolemaic astronomy, and to the work of Copernicus that was yet to come, is completely lost to the historian who insisted that no new results could have been produced after Ghazālī's attack on the philosophers.

Still in the field of astronomy, and to detail further the amount of damage done by the hegemony of the classical narrative in intellectual history, take the remarkable work of another orientalist, Francois Nau, who edited and translated the work of Bar Hebraeus (1286), *Livre de l'ascension de l'esprit sur la forme du ciel et de la terre*.[72] Without doubt, this is the most innovative work in Syriac. Composed around 1279, it was heavily influenced by the Arabic astronomical revolution that was taking place during the thirteenth century. While editing and translating that work, Nau could not understand the "strange things" (*sharbe noukroyoye,* choses étrangères) that were relevant to the "nature of the spheres of the moon"[73] when these things were in fact lists of objections to Ptolemaic astronomy of which even Bar Hebraeus was aware, although he was not a practicing astronomer. Similar terminology was used by Bar Hebraeus to describe the problem of the equant, which was more associated with the "upper" planets (Saturn, Jupiter, Mars, and Venus) in Ptolemy's astronomy.[74] These "strange" things that Bar Hebraeus was pointing to were in fact in the same tradition of objections against Ptolemaic astronomy and had already been listed and codified in Arabic sources from the ninth century on. They were most elaborately codified in the famous extant work of Ibn al-Haitham (d. 1049) called *al-Shukūk ʿalā Baṭlamyūs* (*Dubitationes in Ptolemaeum*).[75]

Furthermore, Nau could not have been aware of the interdependence between the text of Bar Hebraeus and the texts of his contemporaries Muʾayyad al-Dīn al-ʿUrḍī and Naṣīr al-Dīn al-Ṭūsī as well as others. The works of those Arabic-writing astronomers had not yet been studied by the time when Nau was writing, except for the one chapter of Ṭūsī's work which was translated by De Vaux and which had no parallel in the work of Bar Hebraeus. But most probably, those post-Ghazālī works were not studied because the proponents of the classical narrative did not deem them important enough since they came from the period during which no important works were supposed to have been written. This is a typical example of a self-fulfilling prophecy.

Similar things happened in the field of medicine. To name only one more instance of the damage the classical narrative has inflicted upon the post-Ghazālī texts, I draw attention to the work of the famous Ibn al-Nafīs of Damascus and Cairo, who dared check the work of the great Greek physician Galen and dared say that there was a medical problem in that work. Galen had stipulated that the blood was purified in the heart by being passed from the right ventricle to the left one through a passage between the two ventricles. Ibn al-Nafīs protested loudly, around the year 1241, that there was no such a passage between the two ventricles of the heart. He went on to say that the body of the heart at that point was solid and does not allow a visible passage as "most people had said," nor an invisible one, as was stated by Galen. After rejecting the authority of Galen, by only using the evidence that he must have seen with his own eyes, he went on to articulate the need for the blood to pass through the lungs before it could be cleaned and passed on to the left ventricle so that it could be pumped through the body again. Of course this finding appears later in the works of Michael Servetus (ca. 1553) and Realdo Colombo (ca. 1559),[76] to be further refined and re-articulated by Harvey in 1627 and become the famous pulmonary or lesser circulation of the blood. The important point I wish to make here is that Ibn al-Nafīs's objections went unnoticed by proponents of the classical narrative, because those proponents did not expect to find such original thought at such a late date in the post Ghazālī period. As a result those objections were deprived of being contextualized in their normal Islamic habitat where such similar medical and philosophical objections against Galen had already been raised before by such people as Abū Bakr al-Rāzī (d. 925) in his famous book *al-Shukūk ʿalā Jālīnūs* (Doubts contra Galen),[77] or against the astronomical works of Ptolemy as was done in the just-cited work of Ibn al-Haitham.

Arguments are still raging about the importance of Ibn al-Nafīs's findings and their relevance to the European scientists of the sixteenth and seventeenth centuries, all because the classical narrative had simply exercised such a hold on people's minds, and for so long, that it now seems to make it almost impossible to think outside its boundaries. This is the kind of damage that this classical narrative has already caused to our understanding of the post-Ghazālī texts, as well as the texts of the European Renaissance itself.

2 The Islamic Scientific Tradition: Question of Beginnings II

The Alternative Narrative

The detailed critique of the classical narrative, in the preceding chapter, was undertaken for the sole purpose of liberating the historical and scientific sources from the stronghold of presupposed ideas. And now that we have seen the inadequacy of this classical narrative, I think it is time to abandon it altogether in favor of an alternative narrative that can explain the texts and the facts of history slightly better. In this, as throughout this book, I will rely more heavily on the discipline of astronomy to illustrate its progress in light of another narrative that could explain its various phases more appropriately. I choose astronomy, not only because this discipline was invariably held as the queen of the sciences in almost every culture, but because this field continued to witness a steady progress from its very beginnings in early Islamic times till the sixteenth century and thereafter. I presume that the narrative that could account for the history of astronomy could ultimately have its effectiveness tested when it is also used to account for the history of other disciplines. One may continue to re-examine the alternative narrative in light of the evidence that the other disciplines may bring forth, and repeat the process until we reach the day when we can hopefully construct a narrative that can truly help us understand the fundamental role of science in Islamic civilization. Only then could we securely and more confidently relate the role of Islamic science to the role played by other sciences in other cultures.

I am aware that what we now know of individual Islamic scientific disciplines still represents the very tip of the iceberg, and thus this tip may yield a defective picture when taken to represent the whole iceberg. But I do believe that we know enough, at least in the discipline of astronomy, so that we can use it as a template through which we can build a more accurate narrative regarding the place of science in Islamic culture. I invite colleagues who work in other disciplines of the same culture, especially those disciplines that experienced a sustained growth over the centuries, to test this new narrative against the facts that they can gather in their own disciplines and to commence a dialogue on how best to explain the role of the various aspects of Islamic science. I firmly believe that it will be very difficult to speak of one Islamic science that had this or that characteristic, but much more feasibly speak of various disciplines experiencing different trajectories throughout the long history of Islamic civilization. And it is the latter story that we should attempt to detail.

The roots of the alternative narrative that I propose here should be sought in the historical sources themselves. This, despite the fact that one cannot find many such sources that theorize about the beginnings of scientific activities per se. Yet we still find some who did something close to that or others from which we can cull such elementary attempts at theorizing. It is those sources in particular that I would like to interrogate, and to emphasize at this point, in order to keep the historical context as close as possible to the events we are trying to disentangle.

The foremost theoretician of the early Islamic period is a man whose personal biography is slightly obscure, but whose work, which has survived almost in its entirety, is filled with nuggets that seem to have escaped the attention of modern students of Islamic intellectual history as well as the modern historians of Islamic science. The person in question is Abū al-Faraj Muḥammad b. Abī Yaʿqūb Isḥāq al-Nadīm, also known as al-Warrāq, the paper and/or bookseller. From the evidence of his name alone we cannot tell whether Abū al-Faraj himself was the one who acquired the title *al-Nadīm* (boon companion), or whether the title had already belonged to his father Abū Yaʿqūb. I opt for the first, since we know nothing about the father. Moreover, the kind of work Abū al-Faraj himself had produced, in which he combined anecdotal and serious narrative history, amply qualified him to the companionship of any caliph. Yet, one can still find many people referring to him as Ibn al-Nadīm (the son of the boon companion). We do not know much about his birth or death dates, but what interests us here is his remarkable work *al-Fihrist,* a book that he completed, according to his own statement, in the year 377 A.H. = 987–988 A.D.[1] In it, al-Nadīm tries to explain the intellectual history of Islamic civilization, up to his time, by surveying the intellectual production, in all conceivable disciplines that were known in early Islam, that he himself had come across or about which he had already heard. The book is arranged in ten treatises (*maqālas*), each devoted to one of the distinct intellectual fields that were recognized in his time. The seventh of those treatises, which concerns us directly here, deals with the subject of the "ancient sciences," or in his own words "contains the accounts of the philosophers and the ancient sciences and all the books that were composed in those domains." And it is in this treatise that we find the following accounts about the origins of the scientific activity in early Islam.

I give these accounts as a prelude to the introduction of the alternative narrative because I wish to claim that this alternative narrative was already proposed in a round-about elementary way by al-Nadīm himself. Only that up till now no one has gone through the pain of developing it further. This exercise does not only promise to yield a better understanding of al-Nadīm's work itself, but can give us the tools with which to understand the scientific developments that were only narrated by al-Nadīm and by bio-bibliographers who followed him later in the eleventh and thirteenth centuries.

The Historical Account of the Rise of Science in Early Islamic Times According to al-Nadīm

I wish to preface al-Nadīm's account by the remark that the problems relating to the beginnings of scientific activity in early Islamic times which were discussed in the previous chapter, as well as the phenomenal rise of science during that period, did not go unnoticed by early generations of intellectuals who lived within the earliest centuries of Islam. In fact, such topics must have become topics of debate to be picked up by all those who were interested in explaining the appearance of scientific production in Islamic civilization. And as it is well known, the very concept of "ancient sciences" itself—as distinguished from the "Islamic sciences"—was itself coined at this time, and must have quickly become a major topic of discussion to intellectual historians of Islam as far back as the sources seem to record. But the ninth and tenth centuries in particular are especially significant for this discussion, for as we shall soon see, the very terms "ancient" and "Islamic" sciences, or "rational" versus "traditional" meant something quite particular in this period.

Ninth-century sources, and then more elaborately tenth-century ones, spoke about such phenomena and offered their own explanations of them. But the most sophisticated account of the rise of science in early Islamic civilization and the motivation for it, that I know of, is this very account, given in the *Fihrist* of al-Nadīm.

In the introductory section of the Seventh Treatise of al-Nadīm's *Fihrist,*[2] a treatise devoted to the "ancient sciences" and their importation into Islamic civilization as we just said, al-Nadīm tries to survey the various opinions that were commonly held on the subject during his own time. Here he acts more as an intellectual historian who tries to explain historical events rather than a historian who simply records them. In his own style of holding the reader's attention, he arranges those accounts in the form of anecdotes, more like short stories (he actually calls each of them a *ḥikāya* (story), but in each case he reports the transmission of science from one culture to another as if he was trying to lay down the theoretical foundation for the phenomenon of the transmission of science in general. Without explicitly saying so, he certainly hoped to use these various stories in order to explain the introduction of the "ancient sciences" into Arabic.

The first two stories are attributed to the scientists themselves, that is, those who made a living from their knowledge of the "ancient sciences." By scientists, al-Nadīm seems to indicate those professionals who made a living from the "ancient sciences," and who were most likely to know the history of their profession better than anyone else. In itself, that was a very reasonable assumption, one would think. But unfortunately al-Nadīm gives no indication that he knew that the assumption itself was open to the internal party biases of the professionals. In al-Nadīm's opinion, those scientists deserved the lion's share of the interpretive narrative that governs the history of their discipline.

Because this part of the discussion touches upon the interpretive aspect of the rise of science in Islamic civilization, at least as far as al-Nadīm was concerned, and because it is both theoretically, as well as historically, very important for our discussion, I will therefore quote al-Nadīm's account in some detail at this point.

Al-Nadīm takes his first story from a book called *Kitāb al-Nahmaṭān*. The book itself is no longer extant in full, and seems to have survived only in fragments, such as the one quoted here by al-Nadīm.[3] Its author was Abū Sahl al-Faḍl b. Nawbakht, simply called Abū Sahl by al-Nadīm. He was apparently the same Abū Sahl who was the astrologer of Hārūn al-Rashīd, and the son of Nawbakht who participated in the casting of the horoscope of Baghdad during the time of al-Manṣūr, as we saw before. While we know that Nawbakht, the father, may have died at the end of al-Manṣūr's reign in 775, we do not know how much longer the son Abū Sahl outlived, if he did, Harūn al-Rashīd, who died in 809. In any event, in al-Nadīm's account according to the text of *al-Nahmaṭān* is quoted as such (with some paraphrasing to avoid the flowery language):

The Story of Abū Sahl b. Nawbakht[4]

The kinds of sciences from which the science of the stars takes its indications of future things were already known and described in the books of the Babylonians. It was from those books that the Egyptians had learned their craft, and the Indians have also employed it in their own country. That was at the time when those ancient people had not yet committed sins and evil deeds, and had not yet sunk as deeply into the ignorance that caused their minds to become confused and their dreams to abandon them. Their confusion led to the loss of their religion, and thus they became totally lost and completely ignorant. They remained in that condition for a while until some of their descendants experienced an awakening that allowed them to remember the past sciences and the conditions of bygone days and how things used to be governed and consequences used to be drawn about the state of the inhabitants and the positions of their celestial spheres, their paths, and their details as well as their celestial and earthly mansions in all of their directions. That happened at the time of the king Jam son of Ūnjihān[5] the king.

The learned from among the people knew those things at that time and recorded them in their books and explained the contents of those books. They described, as well, the conditions of the surrounding universe in all its majesty and the causes of its foundation and its stars. They also knew the conditions of the drugs, medications and talismans, and such things that people used in their affairs that lead them to good and evil. They persisted as such for a while until the time of al-Ḍaḥḥāk b. Qayy. [From words other than those of Abū Sahl: *dāh āk,* means ten afflictions that the Arabs have transformed into al-Ḍaḥḥāk—we now return to the statement of Abū Sahl]. Ibn Qayy ruled during the time when the reign, allotment, sovereignty and governorship of Jupiter ruled over the years, in the land of Sawād [i.e. ancient Babylonia]. There he built a city whose name was derived from the name of Jupiter. He gathered in it the sciences and the scientists, and built in it twelve palaces, the same number as that of the zodiacal signs, and named (the palaces) according to the names (of the signs). There he stored the books of the learned and made the scientists live in those (palaces).

[From words other than those of Abū Sahl: he built seven houses, the same number as that of the seven planets and gave each of those houses to one man, thus giving the house of Mercury to Hermes, the house of Jupiter to Tīnkalūs,[6] the house of Mars to Ṭīnqarūs (Teukreus?).]

We return to the statement of Abū Sahl.

People obeyed them and followed their direction, since they knew how much more advanced they were from them in matters of knowledge and means of securing their wellbeing, until a prophet was then sent to them. As soon as the prophet appeared among them and once they knew about him, they forgot their sciences, and their minds became confused. As a result they became all dispersed and their opinions diversified according to their lusts and parties, so much so that each one of those learned men departed to a city in order to inhabit it and rule over its population.

Among them was a learned man called Hermes. He was the most perfect of all in terms of his intellect, and most precise in knowledge and most subtle in discerning. He came to the land of Egypt. He ruled over its population, enriched its land and improved the conditions of its residents, and manifested his learning in it.

But most of that learning and the best of it remained in Babylon, until the time when Alexander the king of the Greeks invaded the land of Persia, from a city the Greeks call Macedonia. That was at the time when [the Persian king Darius] denied the tribute that was imposed on Babylonia and Persia, and thus [Alexander] killed the king Dara the son of Dara and took over his dominion. He destroyed the cities and the towers that were built by devils and giants and demolished all the buildings with all the sciences that were engraved on their stones and woods. He shattered and burnt all that and scattered their contents everywhere. He copied what was gathered in the libraries and government offices of the city of Istakhr. He translated all that into Greek and Coptic. And after he had finished copying what he needed from it all, he burnt all that which had remained written in Persian, including a book called *al-Kushtaj*. He took all that he needed of the books of the science of the stars, of medicine, and of the natural sciences and sent them together with what he had gathered of the other sciences, treasures, and scientists to Egypt.

There remained few things that were sent by the kings of Persia for safekeeping to India and China at the time of their prophet Zaradasht (Zoroaster) and the Wise Jāmāsp. [This took place] when they were warned by that prophet and by Jāmāsp of the deeds of Alexander and his conquest of their land, his destruction of their books and sciences, and his transporting them to his own country.

At that time, learning in Iraq disappeared, was torn apart, and the scientists, few as they were, disputed among themselves and differed greatly. People split along partisan lines and scattered into schisms, so much so that each group of them took a king to itself, and were thus called *mulūk al-ṭawā'if* (kings of the sects).

Then the Greeks came under one dominion during the time of Alexander; and after having been all dispersed and engaged in war with one another, they were finally united with one hand. While the dominion of Babylonia remained weak, corrupt and fragmented, and its people were oppressed, and defeated, so much so that they could not defend their honor nor dispel any harm. [These conditions prevailed], until there came the reign of Ardashīr, the son of Bābak, of the dynasty of Sāsān. He healed their divisions, united their various sects, and conquered their enemy and took control of their country. He united them under his rule, cured their partisanship and assumed full reign over them. He then sent to India and China as well as to Greece for the books they had in their possession. He copied all that had fallen into their hands, and pursued the very few remaining sciences that had survived in Iraq. He gathered together all that had been dispersed, and collected all that which was scattered.

His son Shāpūr, who succeeded him, persisted in this policy, until all those books were copied into Persian: among them the books of Hermes the Babylonian who ruled over Egypt, Dorotheus the Syrian, Phaedrus the Greek from the city of Athens which was famous for its learning, Ptolemy the Alexandrian, Farmāsp the Indian. They then explained those books and taught them to the people, just as they have learned from those books, which were originally from Babylon.

Then Chosroes Anūshirvān succeeded those two [meaning Ardashīr and Shāpūr] and did his own gathering and collecting of books, and employed them on account of his love and passion for knowledge.

And so is the case of every people who had had such events befall them at times and their lots had changed. They would acquire new sciences in accordance with the decree that was meted to them by the planets and the zodiacal signs that control this world by the order of the almighty God. Here ends the speech of Abū Sahl.[7]

This story indeed exhibits the stuff of legends. But it is easy to detect its intent and the reason why Abū Sahl recounted it in the first place. Besides taking a jab at the possible conflicts between kings and prophets, Abū Sahl obviously sought to highlight two main issues: 1) the antiquity of the science of the stars, that is, astrology, which was his profession, and 2) he wished to relate the origins of all sciences to Babylonia, and by extension to Persia which ruled over Babylonia for long periods of time. He may have done that in order to boast of his Persian origins—and here one may detect a subtle racial boasting that formed part of the *shuʿūbīya* (anti-Arab) sentiment of the time—or of his control of his discipline of astrology, or both.

The *shuʿūbīya* sentiment, which may be lurking in Abū Sahl's account, does not preclude his attempt to explain other cultural phenomena at the same time, and to assert a special place for Persian culture to which he definitely belonged. By starting with the sciences originating in Babylonia, he followed that by taking a swing at Alexander the Great, the traditional enemy of Persia, for burning the Persian sciences. To Abū Sahl that could explain the disappearance of those sciences and the need to reclaim them later at the time of Shāpūr and Chosroes. In this manner Abū Sahl was in all likelihood also participating in the general literary traditions of his time as well: traditions best exemplified a few years later in the works of Jāḥiẓ (d. 869), who devoted special treatises to the virtues of various nations. Abū Sahl's insistence that the sciences were all gathered back into Persian under

Shāpūr and Chosroes is almost a transparent attempt to glorify the Persian role in the preservation and transmission of science. But we only have his word for that, despite the fact that it is admittedly true that some of the Greek and Indian sciences, especially the elementary astrological sciences, were in fact translated into Persian during that time. But to generalize that account to include all the sciences pushes the story into the realm of legend.

It is also true that legends can also contain a kernel of the truth. And Abū Sahl's account may indeed contain some unintended facts that I do not think were sought by Abū Sahl, or even known to him at the time. Only now, we know that many of the observations of ancient Babylonia were certainly used by such important Greek astronomers as Hipparchus (c. 150 B.C.) and Ptolemy (fl. 150 A.D.)[8] and indeed formed the observational foundations of their work, as we shall see. Some of those Babylonian parameters that were adopted by the Greek astronomers were too subtle for Abū Sahl to recognize.

Abū Sahl's true intention, however, was not to relate all that, for I believe he was trying to assert the validity of his own discipline of astrology, and that this validity rested partly on the antiquity of the discipline. His last statement about the decrees of the planets and the zodiacal signs and how they controlled the fate of nations reveals after all his true intentions as an astrologer. That was the kind of belief that was also expressed by the other Persian astrologer, Māshā'allāh in his own history of nations.[9] In sum, one could easily assume that Abū Sahl would make the next logical assertion that his discipline was valid, and that he was the one most knowledgeable about it. Furthermore, by tracing the discipline to a specific person in each period and place, sounds very much like the story that was recounted by al-Fārābī[10] about the history of philosophy, which we have already cited, and where besides having a philosopher associated with a particular king in each period (just as Aristotle was associated with Alexander) Fārābī ended up tracing the history of the philosophical discipline so that he would come out as the major beneficiary of that discipline. Here again, each sage had a country to rule, and a planet to empower him. All these legends have to be contextualized within the prevalent astrological framework that was very well established in the first part of the ninth century, and which stipulated that all people and all nations were subject to the decrees of the stars, as was so cogently attested in the work of Māshāʾallāh just cited.[11]

These legendary attempts, whether they were Fārābī's, or Māshāʾallāh's, and now Abū Sahl's, denote a desire at all times to seek origins, irrespective of whether those origins were origins of science, origins of cultures, or even origins of legends and epics. That is, Abū Sahl's story could also be seen as a creation story in its own right, except that it specifically targets the creation of culture.

What concerns us more at this juncture is the reason why al-Nadīm used this anecdote in the first place. Without any further evidence, the question is difficult to answer. But from the perspective of al-Nadīm, here posing as the intellectual historian of his time, one could argue that he used this particular anecdote in an effort to present the narrative regarding the rise of science that was probably prevalent among the Persian community members of the Abbasid Empire at the time of al-Nadīm. And as we shall come back to affirm once more, Abū Sahl's account may have also been used by al-Nadīm in order to share with his reader the prevalent stories about the transmission of science that were known in al-Nadīm's time. And since the story had some kernel of truth, as was just said, al-Nadīm may have felt that he could use it as a first plausible explanation of the transmission of science, an explanation that was obviously commonly adopted by the Persian community in early Abbasid times.

Furthermore, beginning with Abū Sahl's story also gave al-Nadīm the chance to start from the beginnings, that is, from the origins of science in Babylonia, which may have been his own belief as well. The later stories, that we shall soon see take up the rest of the narrative from the point where Abū Sahl left it (that is, from the time when all the sciences were gathered back in Persia). From then on, al-Nadīm could follow their progression until they reached Islamic civilization, which was his trajectory from the very beginning. In order to illustrate this intention, we should stress first that in this story Abū Sahl said nothing about the history of the sciences of his own time, and ended the story as if all the sciences of Persia were still there to be had, which we know was not historically true. For al-Nadīm, though, all he had to do was to string this story together with others in order to bring those same sciences from Persian into Islamic civilization. The following anecdotes achieve that purpose excellently well, as we shall see.

Abū Maʿshar's Story[12]

Al-Nadīm's second story also comes from an astrologer, this time the equally famous Abū Maʿshar al-Balkhī, a one-time *ḥadīth* scholar who was distracted from his pursuit of pilgrimage and *ḥadīth* scholarship by studying astrology.

According to Tannūkhī he continued to study astrology until he became an atheist (*ḥattā alḥada*).[13] The reason for his switch from *ḥadīth* to astrology was reportedly caused by his enmity with the famous philosopher al-Kindī (d. 870) on account of Kindī's attachment to the ancient sciences, a fact that was first frowned upon by Abū Maʿshar. It was al-Kindī who convinced him to study geometry and arithmetic, apparently on account of their utility to religious studies, and that entry into the ancient sciences led Abū Maʿshar to pursue astrology. The story is indicative of the relationship of astrology to the religious sciences at the time, and reflects an early attempt to attack the ancient sciences on account of their relationship to the religiously condemned discipline of astrology. We shall have a chance to return to this dynamic later on. But it is significant to note here as well that Kindī used arithmetic and geometry here as entries to the foreign sciences and that they were apparently condoned by the religious people. We shall also have occasion to return to this topic as well.

For now, the importance of Abū Maʿshar's story to al-Nadīm was that it began where Abū Sahl's story had left off, and thus Abū Maʿshar's story had the potential of tracing the transmission of the sciences one more step forward before they were finally brought into the Islamic civilization. Abū Maʿshar's book in which the story appears is called *Kitāb ikhtilāf al-zījāt* (Book on the variations among *zījes*), and like Abū Sahl's book this one too is apparently lost, except for this fragment which is still obviously preserved in the work of al-Nadīm. For al-Nadīm's purpose the story is of special interest for the following reason:

Abū Maʿshar says, in his book *Kitāb ikhtilāf al-zījāt,* that the Persian kings' love for the preservation of the sciences, their extreme care in perpetuating them across the ages, and their concern to protect them from natural disasters, both climatic as well as earthly mishaps, has led them to seek for them the most stable of writing material (*makātib*), most durable, and least likely to be affected by decay and effacement, namely, the bark of *khadank* (poplar tree), which is called *tūz*. The Indians, the Chinese and the other nations next to them imitated them in that respect. They also chose the same wood for their arrows on account of their stiffness and smoothness and their durability.

Once they have gathered the best writing material they could find, on which their sciences could be saved, they sought for them a building in a place on Earth that had the best soil and clay, least likely to cause decay, and farthest from earthquakes and mudslides. They searched the regions of the kingdom, and did not find any place under the sky that had most of those qualities other than Iṣfahān. Then they searched within that region and could not find better than the encampment (*rustāq*) of Jayy. Nor could they find within Jayy a place that had more of those qualities than the place where the city of Jayy was built much later on.

They came to Quhunduz, which was inside the city of Jayy, and deposited their sciences there. It is still standing till our own days, and is now called Sārūyah. And because of this building, people knew who first built it. And that is because many years ago, a section of that construction fell, which revealed a vault constructed of very hard clay (*Siftah).* There they found a great variety of the sciences of the ancients, preserved on bark of *tūz,* and written in ancient Persian. One of those books was brought to someone who could understand it. He read it and found in it a letter written by one of the ancient kings of Persia. In it he related[14] that when king Ṭahmūrath, who loved the sciences and the scientists, had learned about the climatic event that was to take place in the west, where rain would fall continuously and would become overabundant and bypass the usual limits, and that a period of two hundred and thirty two years and three hundred days would separate his first day of reign from the said event—according to the astrologers who warned him at the beginning of his reign that this event was going to pass from the west to the east—he ordered the engineers to find the best place in the kingdom in terms of soil and air. It was them who selected for him the structure known as Sārūyah, which is still standing inside the city of Jayy. He then ordered that this solid building be erected. When it was finished he transported to it many of the various sciences from his own library, and they were all transferred to the *tūz* bark. He placed them in one side of the house in order that they would be preserved after the passage of the climactic event.

Among the (treasure books) was a book attributed to one of the ancient sages, which contained specific years and revolutions from which one could extract the mean motions of the planets and the causes of their motions. The people who lived at the time of Ṭahmūrath, and those who came before them of the Persians, used to call those revolutions the cycles of *hazārāt* (Thousands). Most of the scientists of India, as well as all of their kings who ruled over the face of the Earth, and the ancient Persians and Chaldeans, who dwelt in tents in ancient Babylonia, used to extract the mean motions of the seven planets from these years and revolutions. People who lived at the time dated [the mean motions] according to [the *zīj*] which they found to be the most correct according to the test and the most concise of all the *zījes* that were known at the time. Astrolgers extracted from it then a *zīj* that they called the *Shahriyār,* which means the king of *zīj*es. This is the end of Abū Maʿshar's statement.[15]

At this point al-Nadīm inserts his own corroborating report from the stories that were obviously circulating in his own time. The text adds:

Muḥammad b. Isḥāq [i.e. al-Nadīm] says: A trustworthy person reported to me that in the year three hundred and fifty of the hijra [= 961 A.D.] another vault collapsed as well, whose location was not detected because its roof was thought to be solid until it collapsed, and thus revealed many books that no one could read. What I saw with my own eyes were the fragments of books which were found around the year forty [that is, around 950] in boxes laid in Isfahan's ramparts and were sent by Abū al-Faḍl Ibn al-ʿAmīd [Muḥammad b. al-Ḥusain d. 970]. The books were in Greek, and were therefore entrusted to people like Yūḥannā who could decipher them. They turned out to be names of soldiers and their salaries. But they were extremely putrefied, and smelled so horribly as if they had been just taken out of the tannery. After being held in Baghdad for a year or so, they dried out and smelled no more. Some of them are still held with our teacher Abū Sulaimān.

It is said that Sārūyah is one of the old and well executed marvelous buildings, in the east, compared, to the Pyramids in the land of Egypt, in the west, in terms of its glorious and marvelous construction.[16]

The intent of the story is to demonstrate the Persian Kings' love for learning and the efforts they spent to protect science. It was on account of them, that mean motions of the planets were preserved for the astrologers who could use such values for their casting of horoscopes and the like. The detailed account regarding the kind of material on which they were written and the places where they were kept, and the extra care spent to preserve them all indicate that such mean motions were trustworthy, and astrologers such as Abū Maʿshar himself should opt to use them. This obviously gave Abū Maʿshar an advantage over others, on account of his intimate knowledge of such parameters.

On the other hand, the story also stresses the fact that it was the astrologers who predicted the climactic disaster that was to come from the west, and it was they who urged the preservation of books, thereby acting as the guardians of the intellectual legacy. The morale of the story is that the astrologers' knowledge is to be trusted and appreciated on account of their ability to predict future events as they have done, apparently successfully according to the story, with the climactic disaster.

Other sources from a century later, as in the case of Bīrūnī (d. ca. 1050),[17] seem to corroborate the intention of the story: to highlight the care with which Persian kings attempted to preserve and treasure books in the land of Persia. The fact that such stories kept being repeated could only mean that they must have been circulated widely in the tenth and eleventh centuries. Their purpose however, as can be seen from this one, is to stress not only the antiquity of the science of astrology, but that its sources had been secured and well preserved across the ages, a major requirement for a discipline that had to depend for its validity on repeated events that by their very nature took centuries to recur.

I read these stories less as historical sources, than as desperate attempts by astrologers to validate their discipline in the face of the severe attacks they must have been facing at this time, as is so well documented in the major surviving work of the same Abū Maʿshar: his *Introduction to Astrology,* which comes from the same period.[18]

Such stories regarding the transmission of science could not have a historical validity of their own. Their only worth is that they symbolically signal the existence of books in the libraries of Persian kings, but they all concur that no one could know then what kind of books they really were. All this can be easily detected from the occasional treasure-hunt trope in which these stories were cast. The books that al-Nadīm came in contact with, like those reported by al-Bīrūnī a century later, were indeed very fragmentary, stinky as al-Nadīm spares no pain in telling us (and thus apparently not so well preserved as the story intended to imply), were above all still in Greek. Only specialists could read them, and when they were finally deciphered they turned out to contain only names of soldiers and their salaries. They may have contained, however, some tables of mean motions that one could use for the construction of a *zīj* such as the Shahriyār *zīj*. But that is all these stories could say.

These reports cannot be taken as serious accounts for the transmission of science from one culture to another, for it is not historically possible to have a viable transmission of scientific knowledge that depended on the hazards of finding treasures, and when those treasures were found they could not be read or used except by the very few. For science to flourish, there must be a general infrastructure for it, and a much larger number of people in the society must be able to participate in its production. Otherwise the story becomes a story of a secret magical, alchemical or talismanic science that even Ibn Khaldun, in the fourteenth century had already condemned as an unhealthy environment for the spread of science. While attacking the discipline of astrology Ibn Khaldun asserts that this discipline could not be valid because it could not be published and freely debated in public, and the astrologers had to practice their craft in secrecy and thus could not have a valid science, since all valid sciences have to be practiced out in the open and the full light of day, so to speak.[19]

With Abū Maʿshar's story we are once more confronted with a story similar to that of Abū Sahl in which the stress is on the antiquity of the discipline of astrology. But here Abū Maʿshar adds the twist that the astronomical parameters, upon which all astrological predictions must be based, should also be reliable and of secure authenticity. It brings to the front the importance of such astronomical values as the mean motions of the planets, and stresses the fact that astronomical tables recording these values were composed during the Persian period.

But, like the first legend of Abū Sahl, this one too has a kernel of the truth, for we know from independent sources that such Persian astronomical handbooks existed, and that they were translated into Arabic, or at least were widely used in early Abbāsid times. That there was a *zīj,* called the *Shariyār-i zīj,* or *zīj-i Shāh,* is no doubt true as it was attested and used by so many early Abbāsid sources.[20] There is no doubt, as well, that early Abbāsid astrologers used planetary mean motions that were already preserved in earlier Persian sources. But the question remains as to how these mean motions were obtained in the first place, and the story, I am afraid, does not shed much historical light on that account. Legends seem to indicate the direction of history, but are woefully deficient in explaining its important details. Abū Maʿshar's story is no different.

But it is legitimate to ask about al-Nadīm's intention in starting his own account of the history of Islamic science with these two stories. My contention is that he only wished to relate the opinions that were circulating in his time about the origins of Islamic science, legends as they were. The first story related the transmission of the Greek sciences back to their original home in Persia, and the second simply preserved them there, and gave hints as to how they were then transmitted to early Islamic civilization. In both instances, the discipline of astrology was used as the template for the general history of science, in a way similar to our own undertaking of the discipline of astronomy as a template for the later developments in Islamic science.

Taken together, and with their emphasis on what happened in ancient and more recent Persia, the two stories seem to reveal the eastern sources of the Islamic sciences, or at least point to the direction where such sources could be sought. In all likelihood, that was the intention of al-Nadīm in grouping the two stories together in this fashion. What we should then expect him to do is to move to the west, that is, to the land of Byzantium, in order to complete the western component of the sources of Islamic science. And that is exactly what he did.

The third story speaks directly to the issue of the transmission of the Greek sciences into Arabic. And that is the same story that we referred to before when we spoke of Fārābī's account of the origins of Islamic philosophy. In that account, the important issue was the emphasis laid on the conflict that existed between Christianity and Philosophy. In al-Nadīm's account he tried to confront the issue of the sciences in his own time, and there he posed as a true historian of science who wished to investigate the manner in which those sciences could have crossed from one culture to another. In it he raised very important issues that have to do with the societal factors that inhibited or promoted the transmission and practice of science. He went on to reflect, in a very insightful manner, on the relationship between Islamic civilization and the other civilizations it came in contact with. Al-Nadīm's account reads as follows:

The Third Story[21]

In times past, philosophy (*ḥikma*) was restricted only to those whose natures could accept it. Philosophers (*falāsifa*) used to consider the horoscopes (*mawālīd*) of those who sought to learn philosophy (*falsafa*) and wisdom (*ḥikma*). And if the horoscope of the person indicated that he was from among those who could accept it, then philosophy would be taught to that person, and if not it would not. Philosophy used to be openly studied among the Greeks and Romans before the coming of the creed of Christ, peace be upon him. When the Romans adopted Christianity, they prohibited it, and burned some of its books, while they locked up (treasured) the others. People were prohibited from indulging in philosophical discourse, as it was then perceived to be against the prophetic creeds. At one time the Romans (*Rūm,* meaning Greeks, but here Byzantines) apostatized and went back to the doctrines of the philosophers. The reason for that was that Julian, who resided in Antioch, was then the king of the Romans and appointed a vizier by the name of Themistius, the commentator on Aristotle's books. And when Shāpūr dhū al-aktāf (Shāpūr II) sought to conquer him he was caught by Julian, either through battle, or that Shāpūr was recognized when he went into the land of the Romans (i.e. Byzantium) to scout it, and was captured. Stories vary on this account.

Julian then marched on to the land of the Persians until he reached Jundīshāpūr, where up till now there is a trench known as the Byzantine trench, and where he laid siege to the chieftains and commanders of the Persians. He besieged it for a long time but could not take it. In the meantime, Shāpūr was still kept captive in the palace of Julian. There, Julian's daughter fell in love with him and rescued him. He then traveled across the country in secrecy until he reached Jundīshāpūr and entered into the city. His followers, from among the residents of the city, took heart when they saw him, and went out of their homes and engaged the Romans in battle, taking Shāpūr's escape as a good omen. He then captured Julian and killed him.

> As a result the Romans quarreled among themselves.[22] The Great Constantine was in the army and they disputed among themselves for so long as to who should lead them until they could no longer resist him. And because Shāpūr was fond of Constantine, he put him in command over the Romans, and was graceful to them on account of him. He facilitated their exit from his country, after making conditions on Constantine that he should plant next to each palm tree in his country (i.e. Persia) and the *sawād* (i.e Mesopotamia) an olive tree, and that he should send from Byzantium machinery and supplies in order to reconstruct that which had been destroyed by Julian. He lived up to his promise, and Christianity resumed as it was before. And there resumed as well the prohibition of the books of philosophy and their [i.e. Byzantines'] treasuring of them as is the custom up till this day.
>
> The Persians had translated in the past some of the books of logic and medicine into their language, and those were in turn translated into Arabic by Ibn al-Muqaffaʿ (d. 759).[23]

In this account we clearly note al-Nadīm's intention to demonstrate how philosophy was persecuted in Byzantium, and as an afterthought he seems to indicate that some of the elementary books on logic and medicine had already been translated into Persian in ancient times. He concluded his story with the translations of Ibn al-Muqaffaʿ in order to give due credit to all those who did in fact translate old Persian books into Arabic. He did this in order to emphasize the role those people played in the transmission of science to Islam. Nevertheless, he still insisted in the previous sentence that when Christianity returned to Byzantium the prohibition of philosophy also returned, and the situation continued to be so till al-Nadīm's own time, that is till the end of the tenth century: "There resumed as well the prohibition of the books of philosophy and their treasuring of them as is the custom up till this day."[24]

When he was writing his *Fihrist,* toward the end of the tenth century, al-Nadīm, then posing as a historian of science of his own time, asserted therefore, that Byzantium of the tenth century did not encourage philosophy, and apparently used the philosophical books as trading treasures. He was apparently convinced that Byzantium had no appreciation for philosophy as such, despite the independent, but controversial evidence from the Byzantine side, which speaks of the rise of Byzantium's "first humanism" at that time.[25] Here again, it seems that the primary sources corroborate al-Nadīm's story as they did Fārābī's story that was mentioned before. In fact there are several "legendary" (and legends almost always have a kernel of the truth) accounts of missions sent by Muslim rulers to the Byzantine

Emperors seeking from them those very treasured books.[26] Some of those accounts would describe missions ending up in old temples, with restricted access, and would have great difficulties in acquiring the books that were treasured there. Such a science that continued to be locked up well into the tenth century, and which was being fought by Christian dogma, could not possibly produce a viable scientific tradition that could be passed on to another culture, neither through contact, nor through isolated pockets as was already argued before.

More importantly, the stories relating the mission of al-Maʾmūn to the emperor of Byzantium to request Greek books assert that al-Maʾmūn could not find such books in his domain, nor could the Byzantine emperor find them at first, until he was led to them by a priest who knew about the locked temple which had such books.[27] This should not be surprising, in light of the several accounts of the dearth of scientific books in Byzantium at the time.

The fact that such conditions, as the ones that were described by Lemerle, seem to have been prevalent in the Byzantine domain, especially during Byzantium's "dark ages" can be also confirmed when one considers the contemporary Syriac scientific material, which in my opinion was directly inspired by the Byzantine sources. And when one considers the extant Syriac sources, like those of Sergius of Rasʿaina (d. 536),[28] Severus Sebokht (c. 661),[29] or George, the Bishop of the Arabs (c. 724),[30] or even the works of Job of Edessa (c. 817), especially in the latter's encyclopedic *Book of Treasures,*[31] on the Syriac sciences in the early Abbāsid period, at the time when the translation movement from Greek into Arabic was at its apogee, one could easily discern elementary scientific books, very similar to the elementary logical and medical books that continued to be used in Byzantium and were translated into ancient Persian as we were told by al-Nadīm. Why should one expect otherwise? When we know that most of those who wrote in Syriac lived under Byzantine dominion, and were arguably persecuted by their Greek overlords. Echoes of this persecution are evident in the spontaneous remark that was made by Severus Sebokht, and which was already published by Nau, in connection with Sebokht's reference to the Indian numerals as an argument against the Greek claims that they were the masters of all the sciences and of all times.[32] What these sources very clearly demonstrate is that we could not possibly expect such Byzantine subjects to outsmart their masters and create a new science that was itself supressed in Byzantium.

We shall have occasion to return to the role that was played by this community of Syriac speakers, who also mastered Greek at times for their liturgical needs, in the transmission of the Greek sciences into Arabic but not before the end of eighth and early ninth centuries. We shall see how important that role was, and attempt to pinpoint the causes that led to it.

But for now, I wish to return to the intentions of al-Nadīm and ask once more about the reasons he must have had in mind for recounting this third story about the transmission of science. In my opinion, he did not only wish to indicate that the position of the philosophical sciences were really endangered in Byzantium in his own time, but that the situation was as such for a long time, at least as far back as the time that preceded, and followed immediately after, the death of Julian (The Apostate). He wished to stress that Julian was the only one who allowed philosophy to be studied and pursued. But when we remember that Julian ruled for barely two years only, that is, between the years 361 and 363, then the picture, al-Nadīm was trying to paint, becomes very clear. That picture recounts the story of the continuous Christian persecution of philosophy, thus echoing the expression of al-Fārābī who claimed that philosophy was "liberated" only when it reached the lands of Islam.

Up to this point, the reader of al-Nadīm's text still cannot account fully for the transmission of the sciences from the ancient cultures into Islamic civilization. Such a reader is still entitled to ask: how could such sciences, that were persecuted in their original Byzantine domain, provided there were such scientific activities to be persecuted, be passed on to another Islamic culture that did not have any science of its own as well, as we are so often told?

Al-Nadīm had not reached this stage of his narrative yet. His preparatory anecdotes, which he used to introduce his treatise on the "ancient sciences," have not yet reached their conclusion. But we can almost begin to see where he was going. He had already indicated that there could not have been a direct transmission of science from Byzantium into Arabic, as the classical narrative so often asserted, if the conditions were in fact as they were described by al-Nadīm. And to answer the question of how could those sciences be brought into Islamic civilization, especially from Byzantium if the situation was as he described, al-Nadīm's answer would lie in the fourth anecdote that was apparently used as a climax for the earlier ones. Because of its importance, and as it will become the focus of the discussion that follows, I give here a close translation of this fourth anecdote, just as it appeared in al-Nadīm's *Fihrist.*

The Fourth Story[33]

Khālid b. Yazīd b. Muʿāwiya was known as the wise man of the family of Marwān (*ḥakīm āl marwān*). He was distinguished in his own right, and was enterprising and full of love for the sciences. At one point it occurred to him to pursue alchemy, for which he gathered a group of Greeks from Egypt who had mastered Arabic. He then ordered them to translate the books of alchemy from Greek and Coptic into Arabic. This was the first translation in Islam from one language to another.

Then there was the translation of the *diwān,* which was in Persian, into Arabic during the days of al-Ḥajjāj [b. Yūsuf (d. 714)]. The one who translated it was Ṣāliḥ b. - ʿAbd al-Raḥmān, the client of Banū Tamīm, who had been one of the captives (*saby*) of Sijistān, and who used to work as a secretary to Zādān Farrūkh[34] b. Pīrī the secretary of al-Ḥajjāj, both in Arabic and in Persian, and Al-Ḥajjāj used to favor him. One day Ṣāliḥ said to Zādān Farrūkh: 'you are the cause of my (livelihood)[35], with the commander, and I feel that he took a liking to me. I will not be sure that one day he would not promote me ahead of you and demote you', to which he [i.e. Zādān] responded: 'Don't be so sure, for I think he needs me more than I need him, as he cannot find anyone to accomplish his accounting for him (*yakfīhi ḥisābahu*) except me.' [Ṣāliḥ] then replied: 'By God, if I wished to convert the accounts (*uḥawwil al-ḥisāb*) into Arabic, I could certainly do it.' To which [Zādān] replied: 'convert a few lines of it so that I see,' which he did. He then commanded him: 'feign sickness,' which he did. Al-Ḥajjāj then sent him his own doctor Theodoros who could not find anything wrong with him, and the news reached Zādān Farrūkh, who then ordered him to return to work.

At that time, it happened that Zādān Farrūkh was killed during the uprising of Ibn al-Ashʿath (d. ca. 704) as he was on his way from some place to his own home. It was then that al-Ḥajjāj replaced him with Ṣāliḥ as his secretary, who in turn told him what had transpired between him and his master at the *dīwān.* As a result al-Ḥajjāj determined to do it (that is to translate the *dīwān*), and he put this Ṣāliḥ in charge of it.

Mardānshāh the son of Zādān Farrūkh asked him: 'what would you do with *dehwīh* and *sheshwīh*?' to which he replied: 'I would write one tenth and half a tenth'. When asked 'what would you do with *wīd*'? He said: 'I would write 'furthermore' (*ayḍan*).' He said: '*al-wīd, is al-nayyif,* and the increase is added.' He was then told: 'May God uproot your descendants from this world as you have uprooted Persian.'

The Persians offered him a hundred thousand *dirhams* in order that he would feign his inability to convert the *dīwān,* but he refused and persisted in converting it until he completed it. ʿAbd al-Ḥamīd b. Yaḥyā (d. 750) [the famous Umayyad secretary and teacher of Ibn al-Muqaffaʿ] used to say: 'How great was Ṣāliḥ, and how great were his favors to the secretaries (*al-kuttāb*)!' Al-Ḥajjāj had set for him (meaning Ṣāliḥ) a specific deadline for the conversion of the *dīwān.*

As for the Syrian *dīwān,* it was in Greek. The one who was in charge of it was Sarjūn b. Manṣūr, under Muʿāwiya b. Abī Sufyān (d. 680), who was then succeeded by Manṣūr b. Sarjūn. The *dīwān* was converted during the days of Hishām b. ʿAbd al-Malik (rl. 724–743). And it was converted by Abū Thābit Sulaimān b. Saʿd the client of Ḥusain, who used to be an epistolary secretary during the days of ʿAbd al-Malik (rl. 685–705). It is also said that the *dīwān* was converted during the days of ʿAbd al-Malik as well. [For] it happened that ʿAbd al-Malik had ordered Sarjūn, one day, to do something, and the latter procrastinated in the matter, which angered ʿAbd al-Malik. He then asked Sulaimān who replied: "I shall convert the *dīwān.* . . ."[36]

After making the crucial connection between the importation of science into Islamic civilization with the translation movement, al-Nadīm seems to have laid down a careful strategy for his own narrative. With the third story he had ruled out the possibility of science having come by sheer contact with Byzantium, once he had demonstrated the poor status of that science in the northern and western lands of Byzantium. The importation from the east was equally unlikely since the two stories he reports were more in the form of legendary astrological lore, rather than historical events. Besides, they were reported by two astrologers who had a vested interest in that kind of connection. Therefore, even al-Nadīm himself would probably judge them as historically equally unreliable. Al-Nadīm must have known that. And he must have also known that he still had to explain the origins of Islamic science.

At this point he could not escape from giving his own account of the origins of the scientific tradition in Islam. And it is then that he demonstrated his preferred methodology and thus allowed us to look through a small window at the thoughts he was entertaining. For that specific reason, his own narrative gains tremendous importance for our discussion. By starting the last story with the statement about Khālid b. Yazīd as the first translator, he obviously wanted the reader to re-orient himself and think of the introduction of science into Islamic civilization as a willful act of acquisition taken by some historical persons who had a vested interest in acquiring those sciences. With this introduction he was also saying that science did not come into Islamic civilization by some 'natural' process of contact with another civilization, as he seemed to demonstrate that there was no such civilization to come in contact with, nor through a mysterious legendary survival of books in vaults whose ceilings were falling apart, nor through some pockets of high learning of which he makes no such mention. Rather the whole phenomenon was the result of a direct willful acquisition process that he wanted the reader to consider.[37]

And as soon as he finished the first three sentences about Khālid's role in the acquisition of the sciences, and here, he did not seem to have had enough information about this Khālid other than that he had a personal interest in such an acquisition, he quickly ended that short introduction with the blunt statement that "this was the first translation in Islam from one language to another," as if to say that translation was itself the answer to the importation of the sciences. The problem remained as to which translation. He could have recounted the classical narrative, at this point, and told us that there were translations from Greek into Syriac, or that the Abbāsids brought with them the Persian ideology of re-claiming the Greek sciences back to their origins, which their legends had already claimed. Instead he went directly into what he thought was the crucial step in this translation process, the "translation of the *dīwān*," and quickly noted that this process was an Umayyad process and not an Abbāsid one. For that purpose he gave the great details that he did on how this *dīwān* translation came about. As if he was leading us step by step to painstakingly appreciate the social dynamics during the times of the Umayyads that required such a translation to be undertaken.

As soon as he finished recounting the intrigues and the social conditions that governed life in the *dīwān*, and after explaining how the problems of its translation were resolved, both in Iraq as well as in Syria, he quickly connected that to yet another account, this time raising the question about the spread of the sciences in Islamic civilization, rather than questioning their beginnings as he was trying to do in the previous four stories. When he came to explain the spread of science in Islamic civilization he duly titled this new account as such: "Recounting the reason for which the books on philosophy and other ancient sciences had increased in this country." Notice, he was not saying the reason these books 'came about in the first place' but the reason "they increased," thus their beginnings had, at that stage of the narrative, been taken for granted.

The account that followed gave only "one" of those reasons, as it was duly titled again: "one of the reasons for that [increase]." He then went on to recount the now most familiar and famous story about al-Ma'mūn's dream[38] in the following terms:

Al-Ma'mūn once saw in his dream a man who looked as if he was white in color, with some reddish complexion, wide forehead, connected eyebrows, bald headed, dark blue reddish-eyed (*ashhal*), and good looking, sitting on his bed. Al-Ma'mūn said: "I was in front of him as if was filled with awe." I said: "Who are you?' To which he responded: "I am Aristotle." I was very pleased with that, and said: "May I ask you a question?" He said: "Go ahead, ask." I said: "What is good?" To which he replied: "That which is considered good to reason (*mā ḥasuna fī al-ʿaql*)." When I asked: "Then what?" he said: "That which is considered good by law (*mā ḥasuna fī al-sharʿ*." Then I said: "Then what?" He said: " That which is considered good by the people (*mā ḥasuna ʿinda al-jumhūr*)." And when I pressed on with: "What next?" He replied: "There is no next (*thumma lā thumma*)."[39]

To make sure that the reader got the point, al-Nadīm gave an alternate report of the same dream:

And in another report, I said: "Go on," to which he replied: "Whoever advises you about gold, let him be to you like gold, and be sure to follow *tawḥīd* [the Muʿtazilite doctrine of insistence on the oneness of God]." This dream was the surest reason for the acquisition of books (*ikhrāj al-kutub*). For al-Maʾmūn was in correspondence with the king of Byzantium, and al-Maʾmūn had gained mastery over him. He then wrote to him requesting permission that he sends to him someone who would make a selection from the ancient sciences that were treasured (*al-makhzūna al-muddkhara*) in the land of the Byzantines. He yielded to his request after initial hesitation. Al-Maʾmūn then sent a group of people that included al-Ḥajjāj b. Maṭar, Ibn al-Biṭrīq, and Salm the master of the House of Wisdom (*Bait al-Ḥikma*), as well as others. They took what they wanted from the books they found. And when they brought them to him [that is, to al-Maʾmūn] he ordered that they be translated, and they were. It was said that Yūḥannā b. Māsawaih was among those who went to the land of Byzantium.[40]

Muḥammad b. Isḥāq [that is, al-Nadīm] said: of those who took special care to acquire books (*ikhrāj al-kutub*) from Byzantium were Muḥammad, Aḥmad, and al-Ḥasan, the sons of Shākir the astrologer, whose account will follow. They spent [in that regard] huge gifts (*raghāʾib*), and sent Ḥunain b. Isḥāq and others to the Byzantine land, who brought back for them the most fascinating books (*ṭarāʾif al-kutub*) and most intriguing compositions of philosophy, music, arithmetic and medicine. Qusṭā b. Lūqā al-Baʿalbakī had also brought along some books, which he translated, and others translated for him.

Abū Sulaimān al-Manṭiqī al-Sijistānī said: the sons of the astrologer [that is Banū Mūsā b. Shākir] used to compensate a group of translators, among whom were Ḥunain b. Isḥāq, Ḥubaish b. al-Ḥasan, Thābit b. Qurra, and others, in a month, the sum of five hundred dīnārs for translation and dedication.

Muḥammad b. Isḥāq [that is al-Nadīm continuing] said: I heard Abū Isḥāq b. Shahrām recount in a general gathering that there was an old temple in the land of Byzantium, with a two-sided door of iron, the likes of its size has never been seen before. The ancient Greeks used to venerate it, pray and offer sacrifices in it in the old times when they worshiped planets and idols. He said: 'I asked the king of the Byzantines to open it for me, but he refused for it was closed since the time when the Greeks converted to Christianity. I persisted in my request, kept on corresponding with him, and asking him directly every time I was in his presence.' He went on to say: 'He then opened it for me, and lo an behold, that house was made of marble and most colorful stones, containing so many beautiful inscriptions and writings the likes of which I had never seen or heard about before. In that temple, there were books that could be carried only by several camels, he even exaggerated and said 'a thousand camels.' Some of those [books] were worn out, others were still in their original conditions, while others had been eaten by worms. He went on to say: 'I saw in it all kinds of golden vessels for offerings and other fabulous things.' He said: 'he then closed the door after my departure, and told me that he had made me a favor. He said that those [events] took place during the days of Sayf al-Dawla (945–967). He claimed that the temple was a three-days journey from Constantinople, and those who lived around that place were a group of Ṣabaʾians and Chaldeans, who had been allowed by the Byzantines to keep their faith after paying a poll tax.[41]

This concludes the report of al-Nadīm as to why the books of philosophy and other sciences began to flourish in Islamic civilization. After this account, he went on to list the details of the process of translation, beginning with the names of the translators from the various languages.

Al-Nadīm's Alternative Narrative

The alternative to the classical narrative, which forms the core of this chapter and is here proposed for the first time, as far as I can tell, takes its inspiration from this very narrative of al-Nadīm. After we have seen him survey the stories of his day about the introduction of the ancient sciences into Islamic civilization, and capping them with his own narrative, it became imperative to re-read the text of al-Nadīm in light of the problems that the classical narrative had failed to resolve as we have stated above. We can now assert, that the Persian element in the Abbāsid empire that was held responsible for the reclamation of the Greek sciences was based on a legendary story that was first proposed by al-Nadīm, but whose origin was the work of the Persian astrologer who obviously had a great interest in promoting that ideology in order to secure a job for him and for his descendants after him. In fact his ploy seems to have worked, although not for the same reason as we shall soon see, and early Abbāsid times witnessed the continuous employment of one Nawbakht or another as an astrologer at the highest level of the caliphal court for a period of 100 years or more.

From the Byzantine side, al-Nadīm's several accounts about the persecution of the philosophers in that land, the treasuring of books of the ancients in closed temples and the like, all the way till the middle of the tenth century, as he reported, only reflected the actual historical circumstances, as we have already seen, and further affirmed that the contact theory could not have worked, for there were no knowledgeable Byzantines who could master the classical Greek sources themselves in order to pass them on to the neighboring Islamic civilization.[42]

When the time came for him to introduce his own narrative, al-Nadīm, did not produce another legend of his own. Rather, he went directly to the historical phenomenon of translation. And he properly started with the report about the earliest translations that were known to him (the translations of Khālid b. Yazīd), rather than the translations during his own Abbāsid period, as the classical narrative would have wanted to argue. Al-Nadīm definitely wanted to return to the historical facts, and had no intention of arguing from the ideology that came later to frame the interpretation of those facts. He certainly wanted to emphasize the fact that the translation activity had already started during the Umayyad period, and with Khālid b. Yazīd in particular. What he failed to report, though, was the real reason for Khālid's actual interest at that time in classical Greek texts dealing with alchemy. Instead of delving immediately into the social, the political, the economic and administrative history of the period, so that he could locate the motivating forces for that translation activity, he only prefaced all that with the frequently repeated description of Khālid that he "was enterprising and full of love for the sciences (*lah^u himmat^un wa-maḥabbat^un li-l-ʿulūm*). If one were to read history in essentialist terms, one could simply stop at this preface and attribute to Khālid all sorts of desires and intentions.[43] But not al-Nadīm, for as soon as he concluded the three sentences about Khālid with the phrase "this was the first translation in Islam from one language to another," he immediately went on to the subject of the translation of the *dīwān,* as if to say that, in his mind, those two activities were organically connected. What that connection meant to al-Nadīm was straightforward. He apparently understood the process of acquisition of the ancient sciences to have started with the attempts of Khālid b. Yazīd which was contemporaneous or immediately followed by the translation of the *dīwān.*

As for Khālid's interest in those ancient sciences, of whose motivation al-Nadīm remains silent as we said, we have other sources to fill that motivational gap. We are told by Abū Hilāl al-ʿAskarī (c. 1000) in his *kitāb al-awāʾil,* among others, that

> ʿAbd al-Malik b. Marwān started to write *sūrat al-ikhlāṣ* (Qurʾan, 112) and the mention of the prophet on the *dīnārs* and *dirhams,* when the king of Byzantium wrote to him

the following message: 'You have introduced in your official documents *(ṭawāmīr)* something referring to your prophet. Abandon it, otherwise you shall see on our *dīnārs* the mention of things you detest.' That angered ʿAbd al-Malik, so he sent for Khālid b. Yazīd b. Muʿāwiya, who was greatly learned and wise, in order to consult with him upon this matter. Khālid then told him, 'have no fear o commander of the faithful! Prohibit their *dīnārs* and strike for the people new mint with the mention of God on them, as well as the mention of the prophet, may prayers and peace be upon him, and do not absolve them of what they hate in the official documents. And so he did.[44]

If this anecdote is taken together with Khālid's expressed interest in alchemy we can see why such books on alchemy may have come very handy to someone who was interested in striking new mint of gold coins. Who but the alchemists would be better prepared to identify pure gold, from other metals? And who but the alchemists would be the expert who could judge alloys and the like? That is, they had the kind of knowledge that a new mint master would desperately need.

Once we also remember that ʿAbd al-Malik's reforms did not only include the arabization of the *dīwān,* that is the internal administrative reforms of the empire, but that he went beyond that to create the new currency of the nascent Arab empire, which was up till that time still using the Byzantine coins of the realm in the west, and Sasanian coins in the east. Under such historical circumstances, Khālid's interest in the rules of alloying gold, which could be gotten from alchemical books, was definitely not only an academic interest. The fact that ʿAbd al-Malik would consult with him on such matters further affirms his reliability and the kind of answers he was supposed to supply from his alchemical books.

Going back to al-Nadīm's last story about the reasons for the spread of philosophical and scientific books in Islamic civilization and the relationship of that spread to the dream of al-Maʾmūn, all we need to remember is that although the story was of the legendary type, it still spoke to the spread of those books, and not to their coming to being in the first place. Nevertheless, the orientalists who created and championed the classical narrative in the first place harped on to that account, and made the direct connection between the expressions that were enunciated by Aristotle in that story like "reason," "*tawḥīd,*" the two specific Muʿtazilite key words, to derive from it that feature of the classical narrative that connected the importation of the ancient sciences into Islamic civilization with the Muʿtazilite leanings of al-Maʾmūn, as we already saw before. And as we have also said before this connection is still frequently repeated in the sources dealing with Islamic science. Those who repeat the story neglect to stress al-Nadīm's reason for recounting the dream, namely to give the reason for the *spread* of books and *not* their coming to be.

The same orientalists also gave the story another twist. By connecting Aristotle to the Muʿtazilites through the dream, and then by connecting the whole movement of translation to the Greek philosophical and scientific thought, concluded that the Muʿtazilites, who were the archenemies of what was then called *ahl al-ḥadīth* (people of tradition who later became *ahl al-sunna wa-l-ḥadīth*), or what they called the traditionalists, were the ones who were responsible for the importation of the ancient sciences into Islamic civilization, much to the dislike of the traditionalist Muslims. In Rosenthal's words: "It is probably no accident that the Muʿtazilah should have flourished during the decisive years of Greco-Arabic translation activity, that is, from the last decades of the eighth century until the reign of Caliph al-Maʾmūn (813–833) and his immediate successors. Rather, Muʿtazilah influence on the ʿAbbāsid rulers ought to be regarded as the real cause of an official attitude toward the heritage of classical antiquity that made impressive provisions for its adoption in Islam."[45]

In this manner, the already established conflict model that had been propagated in Europe since the age of reason, as a conflict between science and religion, was now transferred to the Islamic civilization in the form of Muʿtazilites versus traditionalists. With this "spin" people forgot the reasons behind al-Nadīm's account of that dream.

Once we strip this dream of this facile interpretation, tempting as it is, and if we understand it in its right context, we can then go back to the preceding paragraph where al-Nadīm's historical scholarship is best demonstrated. There we see al-Nadīm giving his own opinion of the story of the appearance of the sciences in Islamic civilization as a result of the administrative needs of the empire at the time of ʿAbd al-Malik, and not as a result of legendary stories told by self-serving astrologers who were struggling to keep their position at the Abbāsid court. That's why al-Nadīm began his own account by the stories of Khālid and the *dīwān* translations, and not by another legend like the dream of al-Maʾmūn.

We still have to determine what was on al-Nadīm's mind when he connected Khālid's translations of alchemical books and the administrative translations of the *dīwān* to the spread of philosophical and scientific books in the Islamic civilization. What was the connection between the translation of the *dīwān* and the translation of books on philosophy and science? If we are to gain some insights regarding these questions we must pursue the subtle hints that were already supplied by al-Nadīm himself.

The reason why those hints do not readily seem to connect the dots for us between the translation of the *dīwān* and the philosophical and scientific texts, and thus have deprived us so far from appreciating the real input of al-Nadīm on this matter, is to be sought in al-Nadīm's particular use of the word "*dīwān*" in this account. The term itself was also used in several earlier and later sources without ever specifying what was meant by it. The word is still used in modern Arabic, but has now come to designate a completely different entity, such as a government office (e.g., *dīwān al-muḥāsaba*) or a personal royal office (e.g., *al-dīwān al-malakī*). In some sense the word is at times still used in the classical sense when it referred to administrative offices that handled the affairs of the army as in *dīwān al-jaysh,* taxation bureau as in *dīwān al-Kharāj,* chancellery as in *dīwān al-rasāʾil,* etc. If we restrict ourselves to those common meanings of the word, we then find it difficult to connect such government offices to the translation of philosophical and scientific books.

But when we return to the story of the arabization of the *dīwān* itself, we find that both al-Nadīm, in his *Fihrist,* and the earlier tenth-century author al-Jahshiyārī (d. 942) in his *kitāb al-wuzarāʾ wa-l-kuttāb,*[46] both tried to guide us to the correct meaning of the *dīwān* by giving us examples of the kind of activities they knew were taking place in it. The only example that they give, which has been slightly distorted in al-Nadīm's version that has come down to us, denotes that both authors intended the *dīwān* operations to mean the *dīwān* accounting procedures that Zādān Farrūkh was bragging about when he claimed that he was the only one who could carry them out. On the basis of that specialized knowledge he could assert that al-Ḥajjāj needed him more than he needed al-Ḥajjāj. The example of the kind of accounting both authors give obviously required handling arithmetical operations carried over fractions and the like, the kind of arithmetic that is still slightly complicated by our modern-day standards. Therefore the *dīwān* that needed translation was the *dīwān* in which such complicated operations were performed, and not as most people thought the government office in which records of personnel and their salaries were kept.

The second kind of *dīwān,* where salaries were meted out, did not need any translation for it was in Arabic in the first place. We are explicitly told by al-Jahshiyārī: "There were always two *dīwāns* in Kufa and Basra: one in Arabic, in which records of people and their grants were kept, that is the *dīwān* that was instituted by Umar [b. al-Khaṭṭāb], and the other was for the purposes of revenues (*li-wujūh al-amwāl*) which was in Persian. The situation was similar in Syria, where there was a *dīwān* in Greek and another one in Arabic. Matters persisted in this fashion till the days of ʿAbd al-Malik."[47] Therefore, the *dīwān* that al-Nadīm was talking about was the *dīwān* of revenues, and revenues were the backbone of any government then, as now.

Since operations dealing with revenues required arithmetical operations which in their turn necessitated at least other elementary operations such as the surveying of real estates, and the re-surveying when estates were passed on as inheritance, a *dīwān* officer, as a revenue collector should have the qualifications to carry out those procedures. Furthermore, the computation of time in solar years, when taxes should be paid, and as we know solar and lunar years are not always easy to coordinate without at least some elementary astronomical knowledge, that too must have forced the *dīwān* officer to learn some astronomy. Similarly, re-apportioning payments, especially after the distribution of inheritance, digging canals, trading, etc., all necessitated that the said officer acquire such operational skills for which Muḥammad b. Mūsā al-Khwārizmī had to compose a complete book on Algebra just for that same purpose.[48] Incidentally, that requirement seems to have led to the creation of the discipline of Algebra *qua* discipline,[49] which was not known to the Greeks in the fashion that was articulated by al-Khwārizmī.

All the operations a *dīwān* officer was supposed to perform were not easy, and there must have been some elementary texts or manuals that were used to train those who worked in the *dīwān.* It is rather unfortunate that no such documents seem to have survived from this early period, probably because they were thought of as simple enough to be learned and discarded, or because their contents were orally transmitted from father to son, and thus there was no need to publish them to the public. But we do have some slightly indirect information about their contents, and the kind of operations that were required in these *dīwāns.* For we do find in the work of Ibn Qutayba (d. 879), who preceded Jahshiyārī by a half a century and al-Nadīm by almost a full century, and who himself was a contemporary of the last period of translation that followed the translation of the *dīwān,* a short synopsis of the qualification of those who sought employment in the *dīwān,* or those who were then called *kuttāb.* Those *kuttāb* were undoubtedly the heirs of the *dīwān* employees whose functions we are now seeking.

In his book *Adab al-kātib,* he regrets in the introduction the neglect that had become the share of the Arabic sciences of his time. Ibn Qutayba went on to stress that the *kātib* must seek the following sciences, if he were to be worthy of the name *kātib,* and not be among those who are after the office of *kātib* in name only:

> He must—in addition to our books—investigate matters relating to land surveying, so that he would know the right angled triangle, the acute, and the obtuse angled triangle; the vertical plumb lines (*masāqiṭ al-aḥjār*), the various squares (*sic*), the arcs and the curves, and the vertical lines. His knowledge should be tested on the land and not in books, for the one who reports is not like the eye-witness. And the non-Arabs (*ʿajam*) used to say: 'whoever was not an expert in matters relating to water distribution (*ijrāʾ al-miyāh*), the digging of trenches for drinking water, the covering of ditches, and the succession of days in terms of length increase and decrease, the revolution of the sun, the rising of the stars, the conditions of the moon when it becomes a crescent as well as its other conditions, and the control of weights, and the surface measurement of the triangle, the square, and the polygons, the erection of arches and bridges as well as water lifting devices and the norias by water side, and the conditions of the artisans and the details of calculations, he would be defective in his craft.[50]

Working in the *dīwāns* of the non-Arabs, as far as Ibn Qutayba could ascertain, should include a mastery of all those sciences that were just quoted by Ibn Qutayba from the earlier sources. As we can readily tell, those sciences made no mention of army grants and the like. This must mean that the *dīwāns* that were translated must have included the elementary texts of those sciences. For it was quite unlikely that Ibn Qutayba would call on the *kuttāb* of his time to acquire these sciences if there were not any texts through which they could be acquired. After all, he was the one who participated in supplying such texts by composing his *kitāb al-anwāʾ* (Book of the Rising and Setting or Stars), which touches upon some of those sciences, and particularly the sciences that relate the rising and setting of the stars to agricultural (read revenue) needs.[51] I shall soon return to mention other books in this regard.

For now, the interest in Ibn Qutayba's statement is that it confirms the meaning of the *dīwān,* which I claim was the one intended by al-Nadīm and al-Jahshiyārī. If that meaning is accepted, then one could say that the translations of the Persian and Greek *dīwāns* into Arabic must have included a group of elementary scientific texts, which were in turn very much connected to the philosophical and scientific texts that were mentioned before. How could it be otherwise when we know that any government must acquire such elementary sciences in order for it to function in any sophisticated manner?

Another confirmation for this reading comes from another contemporary of al-Jahshiyārī and al-Nadīm who was also interested in the education of the *kuttāb* and government bureaucrats. Several of his books have reached us from about the middle of the tenth century. The author in question was the famous scientist, Abū al-Wafāʾ al-Būzjānī (d. 998), whose name was very closely associated with the Greek mathematical and astronomical works that were translated into Arabic. It was this Abū al-Wafāʾ who had left us two books which directly address the geometric and arithmetical needs of the artisans and workers (obviously including government employees), that were called: *What the Artisans need by way of Geometry,* and *What the workers and* kuttāb *need by way of Arithmetic.*[52] In both of these texts, Abū al-Wafāʾ takes up elementary mathematical problems, of the types that were obviously discussed in the *dīwāns* of his time, or among those who were employed in those government departments who were then learning how to carry out the new functions that required those new sciences.

Moreover, we need only take a glance at *Keys of the Sciences* (*mafātīḥ al-ʿulūm*), a book by al-Khwārizmī al-Kātib, who lived some ten years or so after al-Nadīm and who himself was a *dīwān* employee, to appreciate the encyclopedic knowledge such an employee of the time needed to know.[53] Here we also see a direct connection between the kind of sciences that were practiced in the *dīwān* and the philosophical sciences, starting with logic. Most of the remaining sciences that were listed by al-Khwārizmī were in fact at the very core of the ancient sciences we are now discussing.

Even in the relatively later period, we see that those sciences continued to be practiced in the government *dīwāns*. This should not be surprising as we already know that most administrative offices are usually very conservative and tend to preserve practices for centuries at a time, practices that are usually inherited from one employee to the next, if not from father to son. From that tradition, we see in the work, *kitāb qawānīn al-dawāwīn* (The Book of the Rules of the *Dīwāns*) of Ibn Mamātī (d. 1209) the many arithmetical and natural scientific material that the *dīwān* employee was supposed to know.[54] And Ibn Mamātī ought to know better, for he himself was the descendant of a family that worked in the Egyptian *dīwān* for centuries.

Similarly, later generations have left us several *ḥisba* (market overseeing) manuals which mention not only the scientific books that the market overseer himself ought to know, but the scientific books that he should use in order to test the various professionals and to control their products from forgeries and the like. These professionals included bonesetters, physicians, pharmacists, as well as others whose names have been summarily mentioned in the work of Ibn al-Ukhuwwa of Egypt (d. 1329) called *Maʿālim al-qurbā fī Aḥkām al-ḥisbā*.[55]

For those who may object and say that this book is very late, and its contents may not apply to the kind of knowledge that the Umayyad worker was supposed to know, and the kind that al-Nadīm was talking about, I can only say: was it possible that there would not be in early Islamic times someone who would oversee the affairs of the public, their public health, their protection from deception, etc., and that these functions entered the Islamic administration in later times only? Was it not part of the duties of the administrator of the public treasury (*bayt al-māl*) to see to it that the right proportion of gold is cast in the minted dīnārs, together with what all that implies by way of managing alloys, composition of metals, and exacting weights and measures? Wouldn't such functions include some alchemy, or at least overlap with it, or what was then called *al-ṣanʿa,* that was being sought by Khālid? Wasn't this *ṣanʿa* also connected to pharmaceutical sciences, and the knowledge of weights and measures, as well as others?

In summary, despite lack of actual manuals that preserve for us a description of the actual operations that took place in the early *dīwān,* or of the contents of those early manuals or the sciences that were translated, this despite all the evidence that we have reviewed so far about the existence of those operations and sciences, we still cannot ignore the arabization of the *dīwāns,* which was tied by al-Nadīm himself to the process of the transmission of the ancient sciences to Islamic civilization. The consequences that can be drown from it can help us resolve some of the problems that were left unresolved by the contact or continuing pocket theories usually deployed as corollaries of the classical narrative.

From al-Nadīm's account, we note that the arabization process, including the restructuring of the foundation of the Islamic government, took place during the days of ʿAbd al-Malik, the first caliph to mint Arabic dīnārs that were independent of the Byzantine ones, who also engraved on them Qurʾanic verses rather than pictures of emperors, as we have already seen.

He is also the one of whom the sources speak as being the first to reorganize the administration of Islamic government and to centralize its functions and streamline it, to use modern parlance anachronistically. He apparently did all that through the arabization of the *dīwān*. Weren't these administrative reforms of the government absolutely essential for the foundation of the new Islamic state, when we also see that the Abbāsids themselves, who came to power almost fifty years after ʿAbd al-Malik, did not change back any of the reforms that ʿAbd al-Malik had introduced? This despite the enmity that the Abbāsids harbored and demonstrated toward the Umayyads, and despite the claims made by the classical narrative and some orientalists that the main backbone of the Abbāsid Empire was the Persian "element." Had this racial categorization been true, wouldn't the Abbāsids have reverted the *dīwān* back to Persian? Wouldn't this mean that ʿAbd al-Malik's reforms were extremely significant and cannot be simply bypassed in favor of focusing on the Persian "element" of the Abbāsids?

The Consequences of the Dīwān Translation: Ascension to Power by Other Means

Now that we can better appreciate the importance of the administrative reforms of ʿAbd al-Malik, after having stressed the need for relating them to the general translation movement of the philosophical and scientific texts, just as al-Nadīm had already done in his *Fihrist,* we should, at this point, go back to discuss the social conditions that paved the way for the importation of the foreign sciences into Islamic civilization. An importation that proved over time to be the most remarkable and unique achievement that was performed by the Persian and Greek speaking communities of the early Abbāsid empire. And by focusing on the social conditions we would be in a better position to answer the larger questions about the actual historical needs that were being met by the transmission of those ancient sciences.

Reading the texts that describe the translation of the *dīwān,* especially those that had been preserved by Jahshiyārī and al-Nadīm, give very clear indications of the serious social consequences of that activity. From among those consequences, the arabization of the *dīwān* seems to have led to the loss of the administrative jobs that were held by Persian and Greek speakers of the empire, who were mostly either Zoroastrian or Christian. Previous to this arabization, those early classes of bureaucrats must have felt so secure about their positions in the administration that they could afford the bragging of Zādān Farrūkh and the arrogance of Sarjūn.

We also saw that the Persian community was willing to bribe Ṣāliḥ b. ʿAbd al-Raḥmān so that he would feign the failure of the *dīwān* arabization. We also saw in the report of Jahshiyārī a reference to a meeting that was held, at the time when al-Ḥajjāj had just come to Iraq, by the Persian notables (*dahāqīn*), at the house of a man called Jamīl, in order to discuss among themselves how to protect the community from al-Ḥajjāj. They were then told by Jamīl: "You will fair well with him if you are not afflicted by a *kātib* from among you, meaning someone from Babylon. And they were in fact afflicted by Zādān Farrūkh who was a one-eyed evil man."[56] It was in that context that Jamīl related his famous parable about the head of an axe that was cast in a forest. The trees then spoke among themselves saying it was not for a good reason that this axe was thrown here. "To which a simple tree responded, if one of your branches does not go into its end, then you have no reason to fear."[57]

Doesn't this anecdote of Jahshiyārī point to the sense of a collective anxiety on the part of a community, this time the Persian community, and the eventual attempt of its members to accuse each other of treason, as any sociologist could have predicted under such circumstances? Wouldn't it be natural for such things to happen in a community that suddenly found itself disenfranchised, after it had already happily monopolized the positions of power in the government for years, just because the members of that community could control one language or other, or some science or other? Doesn't the sentiment commonly referred to with the term *shuʿūbīya* (racial prejudices), which is so often repeated in the sources of the period, represent something of the sort as well? Didn't the translation of the *dīwān* produce such a group anxiety so that Zādān Farrūkh had to tell his friends, when Ṣāliḥ had succeeded in translating few lines of the *dīwān*, "go seek an abode other than this," as reported by Jahshiyārī?[58]

I am almost certain that all that took place. And that the often repeated references to the competition between those who were employed by the government with those who were seeking such employment only confirms this, especially when we all know that the government was always a flourishing market, as was already known to Ibn Qutayba and later on to Ibn Khaldūn,[59] as it was usually the main employer at all times and in all places.

What could those communities do in response to those events? How could they awake from their first shock and try to reclaim their previous positions in the corridors of government? I think they did what most communities would do under such circumstances: go back and try to monopolize the government positions by other means. One such mean was to acquire the more advanced specializations in the very sciences that the government badly needed so that they would become once more indispensable to the running of the government.

How could that acquisition of advanced sciences happen when I have argued that there were no teachers and no experts to teach those disciplines? But if we stop to think that science does not always progress by the steady instruction of teachers, but rather by the leaps that are taken by very bright individuals who are capable of going beyond where their teachers had taken them, and who are usually inspired by an urgent need to do so, then the answer to this question would become slightly easier to comprehend.

Consider the following circumstances. The bureaucrats, who worked in the *dīwān* before it was arabised, were the very persons who knew the elementary sciences and used their linguistic and scientific skills to monopolize their positions at the *dīwān,* as we have already argued. Those same bureaucrats also knew that the very sciences that they mastered for their limited purposes were only introductions to more advanced sciences that they did not need to acquire as long as their positions were secure through the monopoly. I say this as I can almost hear someone like Sergius of Rasʿaina, who died toward the middle of the sixth century, and Severus Sebokht of the seventh century, say in their introductory treatises on astronomy: "whoever wants to verify this or that problem, more accurately, he should seek the more advanced texts of Ptolemy called the *Almagest,* or the *Handy Tables.*"[60] And those were the most advanced Syriac scientists of the period just before Islam or in early Islamic times. We note that they still used that kind of language about the Greek sources. Wouldn't their coreligionist and their community members, who were employed in the government, a few centuries later, share the same expectations from the Greek sources, and have at least the kind of knowledge that was similar to theirs? It is most likely that they too used to find in their own administrative scientific texts references similar to those that we can still find in the extant works of Sergius and Sebokht.

In order to be able to compete with the new occupants of the *dīwāns,* and go back to monopolize the high positions of government, members of these communities of bureacrats had to make use of their knowledge of both the Greek language and the elementary sciences that they used in the *dīwān,* and try to educate themselves or their children in the more advanced sciences, to which their elementary sciences referred for higher precision and sophistication. They did all that in order to be able to deploy that new information and win their previous positions at the *dīwān.* Now that they had lost their jobs, they had an excellent motivation to go to Ptolemy's *Almagest,* that they knew only by name before, when they had no need for it, and to which they were referred by their co-religionists.

Under the new conditions, and with the pain of unemployment, these bureaucrat communities would go back to teach their children and their co-religionists and to urge them to acquire the more advanced sciences about which they were well informed by the Greek as well as the Persian classical sources. And since Arabic had by then become the language of competition they were obliged to demonstrate their competence both in the new bureaucratic language as well as in the sciences of the higher order. Again, all those difficulties had to be re-negotiated before they could reestablich the monopoly that they once had in the *dīwān.* Within one generation or two, the children of these two communities managed to achieve that, and under the severe competition they also managed to perform the unique and remarkable feat that they did. It is the children who surpassed their teachers in acquiring the new advanced sciences, because they were motivated to do so by the pressure of their mere survival.

That such a thing did in fact take place is also reported in the classical sources, when those sources report the return of whole families back to the highest positions at the Abbāsid court. Families whose members knew perfectly well both the languages and the sciences of the Greeks and the Persians. Those new families could now occupy positions that were much more sensitive than the old *dīwān* jobs; they could become the personal advisors to the caliph himself. Think about the Bakhtīshūʿ family, which produced several high-ranking physicians for the Abbāsid court and whose members passed those jobs from father to son for nearly 100 years. The same Nawbakht family, of whom we spoke before, also achieved the high status of court astrologers, and also for several generations of fathers and sons.

Think also of Ḥunain Ibn Isḥāq, who managed to recruit his son and nephew, among others, into the court of the caliph as translators and physicians at the highest level of government. And to have a glimpse of the deadly competitive environment those new aspirants had to go through, think also of the very tough competition Ḥunain himself had to face from his own co-religionists and the speakers of his language, as he himself laments in an account that is preserved in the work of the thirteenth-century bio-bibliographer Ibn Abī Uṣaybiʿa.[61]

What is being proposed here is that the translation movement that is under discussion was generated by the desire of two communities to re-acquire jobs that their parents and co-religionists had lost in the government offices. And in order to do that, at that particular time, that is during the early years of the Abbāsid empire, they aimed to become indispensable to the government by their sheer possession of highly specialized knowledge. From those new posts, they tried to re-establish a new monopoly that the lower *dīwān* workers could not even dream of having as long as they stayed with their elementary sciences.

The evidence that such things did take place come from all those sources that speak of the competition among the highest bureaucrats in the government and their various attempts to exclude others from the competition through casting doubt about their competence in the advanced sciences. Ḥunain's treatise which was just cited, and in which he recounts the attacks he had to suffer at the hands of the other Christian physicians, who would malign him by referring to him as "just a translator and not a physician," is a brilliant example of that activity. It also opens for us a small window at the court of the Abbāsid's of the early part of the ninth century, with the court bureaucrats attempting very earnestly to create the new monopoly that will secure their jobs.

The sources have also preserved for us the communal solidarities that began to appear among the Syriac and Persian communities, and at times even among the people of the same city. We know, for instance, that Yūḥannā b. Māsawayh refused to teach Ḥunain b. Isḥāq medicine because Ḥunain was from the people of ʿIbād (a group of eastern Arabian tribesmen) of Ḥīra, whose members made a living mostly from exchanging money. Yūḥannā, on the other hand, was from Jundīsāpūr that produced the famous Bakhtīshūʾ family of which we just spoke. According to Ibn Abī Uṣaybiʿa, "the people of Jundīshāpūr especially, and their physicians, shied away from the people of Ḥīra and abhorred introducing the children of merchants to their profession."[62]

We also read in the same classical sources about the new environment of arrogance that in the past used to characterize the life of the *dīwān,* as we saw in the cases of Zādān Farrūkh and Sarjūn in their respective *dīwāns.* In the new era, we now begin to see a new class of people, who managed to create, somehow, a new monopoly, at the highest levels of government. And those people seem to have been emboldened, as in the case of the same Yūḥannā b. Māsawayh, who, according to al-Nadīm, was "a venerable physician, given due respect by kings, a great scholar and author, who had served under al-Ma'mūn, al-Mu'taṣim, al-Wāthiq and al-Mutawakkil,"[63] and above all dared to behave in the following fashion in the presence of the caliph al-Mutawakkil himself:

Al-Nadīm says: "I read in the hand of al-Ḥakīmī, who said: 'Ibn Ḥamdūn the boon companion [of the caliph] teased Ibn Māsawayh, one day, in the presence of al-Mutawakkil, to whom Ibn Māsawayh responded: 'if you had had as much intelligence as you have ignorance, and if that intelligence were distributed over a hundred beetles, then each of those beetles would have more intelligence than Aristotle.'"[64] If this is true, and there is no reason to doubt the veracity of al-Nadīm in this account, then we can say this competitive environment produced for the Abbāsid bureaucracy the finest class of servants, who had a remarkable competence, and who also tried to exercise their newly-found power by showing off at the highest positions of government. Those new highly qualified bureaucrats must have felt quite secure in their new posts, when they sat next to the caliph, and within his most intimate circles. Otherwise why would someone like the caliph al-Mutawakkil tolerate the behavior of Ibn Māsawayh when the latter dared insult the caliph's own boon companion?

This anecdote simply illustrates that this class of new bureaucrats had in fact managed to accomplish one of the most important feats in the history of Islamic civilization. They motivated and produced a translation movement, which was primarily an administrative movement in the first place, in which various competences were fighting over government positions, and in which many accusations of treason and the like frequently took place. That should not be surprising, for any sociologist could easily predict that such competition and behavior would be quite natural in cases of extreme competition.

As a by-product of this movement and the competition it engendered, the Arabic language, which had become by then the language of the new sciences, also managed to widen the circle of competition, and to open the opportunity for the Arabs, now working in the *dīwāns,* to join in the competition in order that they too could acquire the new sciences and preserve their new positions. Those Arab or Arabic-speaking bureaucrats now had their own reasons to hold on to power, and thus had to join the competition as well, either by accumulating knowledge directly, or by securing the services of men who could acquire that advanced knowledge for them. That was the case with many bureaucrats of the time. And for that reason we see that most of the translations, which were produced during the ninth century, were themselves patronized by bureaucrats, who were close to the center of power. Those translations were rarely patronized by the caliph himself, if they ever were. The caliph only got the best competent class of employees, but the employees sorted themselves out by their own sifting and competition. We all know that political power usually remains distant from science itself and occasionally even devotes itself to the exploitation of the scientists. Why should it be any different during Abbāsid times? Only at very rare occasions does one find a learned potentate, and if that person ever existed his influence could not have spanned the vast period of scientific activity that was produced during Abbāsid times and thereafter. Something else must have been at work, and our model predicts a continuous competitive environment at the bureaucratic level that kept those sciences alive and prospering.

In another extant treatise of Ḥunain Ibn Isḥāq, about the medical books of Galen that were translated into Arabic, and whose account he was asked to give by one of those bureaucrats who was also close to the caliph, he related in great details the conditions that led to the translations of 129 books of Galen.[65] In it he tells us that most of those books were translated for the sons of Mūsā b. Shākir, and especially for Muḥammad and Aḥmad—the two brothers who had together patronized more than 80 books of the total—and not a single book had ever been translated for the caliph. This in addition to the fact that Ḥunain himself, who had the lion's share in those translations, was at the same time the caliph's physician.

Conclusion

Looking at this translation movement, which was responsible for the introduction of the ancient sciences into Islamic civilization, from this perspective, allows us to open new windows onto Islamic intellectual history, and to begin to discern the motivation that gave rise to this movement. We can then see how certain members of the society, whose livelihood was threatened by ʿAbd al-Malik's reforms, had to insure that livelihood by other means. They naturally resorted to higher specialization through the translation of the more advanced sciences. That in turn helped them gain an edge in the new competition. And as a result they could secure, as Ibn Māsawayh and some others tried to do, a new monopoly at the higher echelons of government. When we remember that those echelons were in fact the very top caliphal court itself, we can then appreciate the vast power those people managed to garner for themselves and often for their descendants. This "healthy" competition also led to a healthy increase in the acquisition of the more advanced sciences, only to produce further competition, and so on.

Therefore, the conditions that prevailed during the first century of Abbāsid rule seem to me to have been the healthiest conditions for competitive acquisition of science where caliphs had a whole group of highly qualified people who could compete for whatever projects the caliph dreamt of executing. Of course, the resulting spread of science on its own created the even healthier conditions for further developments in science. In fact, it may have been this very environment that created what came to be known later as the Golden Age of Islamic civilization, and which was celebrated by the classical narrative.

All this competitive activity apparently had nothing to do with the Persian "elements" of the Abbāsid Empire who were supposedly trying to recapture their own antiquity by reclaiming their sciences from the Greeks. On the contrary it apparently happened because the Abbāsids turned out to be the unwitting heirs to ʿAbd al-Malik's reforms that preceded them by about one full generation. It was those reforms that set the healthy competition in motion in the first place, and through this competition the ever-increasing desire to acquire more and more advanced scientific books to keep the competition going.

All these conditions need to be investigated much more thoroughly. Various historians of varied scientific and philosophical disciplines need to re-examine these activities, which have only been scarcely touched upon here, before any more definite conclusions could be drawn. But this revision itself should hopefully make room for a better understanding of the dream of al-Maʾmūn, the role of the Muʿtazilites, and the actual role of the Syriac and Persian-speaking communities. It was the members of those communities who needed to seek the Greek and Persian classical sources, which had been treasured in dark inaccessible temples for centuries, and to dust them off and re-deploy the information therein for their own needs in order to survive the deadly competition they were facing during the early Abbāsid times.

Most importantly, the revision, that this alternative narrative forces upon us, can now definitely demonstrate that this acquisition of the classical sciences, and especially those of classical Greek, was not simply an act of blind imitation, but had to be adjusted to the needs of the time as we shall see later on. But much needs to be done still before one can substantiate in a comprehensive manner the effects of all these activities in very concrete terms.

And yet, some preliminary results have already been reached by just applying the framework of this alternative narrative. We can now put some of those results on the table and hopefully use them to paint a slightly different picture from the one the classical narrative usually offered. As we can now see, the translation movement was not a movement to imitate a higher culture that was there standing in competition with one's own. Instead, the acquiring culture had to dig out texts, that is *really appropriate* those texts, which were practically forgotten in the source culture. For although the Byzantines still spoke and wrote in Greek, they kept the classical books in vaults for years until they were brought out as a result of the demand in Baghdad, where they were now better appreciated. In a sense those sources were given a new lease on life as a result of the dire needs of the Syriac- and Persian-speaking communities who needed to reclaim the government positions their forefathers had lost. But more importantly, the competitive environment forced those new seekers of knowledge to quickly bypass the scientific production of Byzantium at the time and to seek their interlocutors in the best of the classical sources. It is no wonder that the names of philosophers and scientists who were not even contemporaries, and who all produced their knowledge before the third century of the Christian era (Plato, Aristotle, Galen, Ptolemy, and Diophantus, among others) became household names in ninth-century Baghdad.

Another result that can now be seen much more clearly, and continues to become more obvious everyday, is that the translation movement of early Abbāsid times, since it was generated by social conditions of the Islamic government itself, did not simply translate the classical texts, digest them and then began to create a science of its own as the classical narrative continues to tell us. What seems to have happened is that the translation and creation were taking place at the same time, as we shall also see again. Or better yet, with the alternative narrative we can discern some creative activities to have preceded the translations of the advanced text, and that those creative activities by themselves required further translations in order to lead to more creative thinking and so on. In this manner we can understand why al-Ḥajjāj b. Maṭar had to read Ptolemy's text very carefully and to correct it whenever he thought it was in error.

Furthermore, these preliminary results also demonstrate that both the translators as well as the patrons of those translations were themselves, and in most cases, scientists in their own right. And although they were close to political power, they were digging niches of their own within the ruling bureaucracy that could outlast the caliphs themselves. In other words, those bureaucrats had their own needs for those sciences and for the scientists that sometimes accompanied them. In a good number of cases they were scientists themselves as well. To illustrate their own hold on power, all we have to do is to consider their relationship to the person of the caliph himself, only to realize how much more established they were in order to survive in some cases at the caliphal court even after the succession of three to four caliphs who would be at times violently removed from power. Yet the physicians, the astrologers, the engineers, etc., would survive and continue to exercise that indispensable role their parents had wished for them, when they set them out to reach for the more advanced classical sciences.

Modern historians of Islamic science have already begun to demonstrate the ingenious research that seems to have taken place in early Islamic times, just as the translations were being carried out. And if we come to realize, as we now hopefully do, that the *dīwān* translations had already opened the door for further more advanced translations, then it becomes only natural to expect such creative results once the door for creative activities had been swung wide open for all qualified people to compete. This would be an ideal dream for a society that was undergoing what we would now call nation building. And something of the sort seems to have happened.

Modern research has also begun to uncover that this creative activity included most and foremost a process of re-assessment of the Greek scientific legacy, as we shall see later on, which constituted an active program of correcting the Greek mistakes. It even went further than that to create new scientific disciplines, such as algebra and trigonometry, as we have already seen. It even reformulated older disciplines, as was the case with the discipline of astronomy when the new science of *hay'a* (theoretical astronomy) was created at the same period. All these results need to be fleshed out and their consequences pursued even further before we can come to grips with their full social and cultural implications.

But we can also say that the results that have been established so far can definitely demonstrate very clearly that the process of monopoly which was first exercised by the *dīwān* employees, and then attempted again by the more educated class of their descendants, as was clearly demonstrated by Ibn Māsawayh's treatment of Ḥunain and the group of physicians at al-Mutawakkil's court, who afflicted the same Ḥunain with all sorts of calamities and intrigues, came to no avail. The reason for its failure came from the very nature of science itself, which does not easily allow for the monopoly of such activities, especially when there is a desperate societal need to pursue them. We can also say that the resulting flourishing activities at the time of the early Abbāsids, who themselves simply inherited all those competing classes of very qualified people from the Umayyad reforms, created an unprecedented recovery of the sciences of antiquity with a deep desire to deploy them for the purposes of the time, a phenomenon that was not to be repeated until the time of the late European renaissance.

At this point, I would like to go back and raise the question about the actual benefits that could be derived from the adoption of this new alternative narrative. In my defense, all I can say is that this new narrative had to be adopted after I have been fully convinced by al-Nadīm's strategy in presenting his argument about the translation movement. It was in that argument that he made the direct connection between the Islamic Civilization's appropriation of the ancient sciences and 'Abd al-Malik's reforms which were mainly centered around the order to translate the *dīwān*. It was al-Nadīm as well who saw that appropriation as a consequence of the reform. One wonders if 'Abd al-Malik himself ever contemplated all the consequences that his order entailed. But for us, by adopting this new narrative, if it does not do us any good, at least it will certainly help us explain the behavior of the *dīwān* employees, and the social conditions that ensued by isolating them as a class whose children will from then on strive to come back to the government at the higher, more desirable and more indispensable positions.

But on the theoretical level, what would be the benefit from adopting this new narrative in preference to the classical narrative that was in fact the brain-child of some of the most distinguished orientalists? This, when we also know that this very classical narrative seems to have served the community of Islamic intellectual historians for more than a century now. The answer to this question must be sought on two levels: The practical level which touches directly on the process of narrating the internal history of science itself, where we could pursue the developments of scientific ideas from one concept to the next, and the methodological level, which touches upon the reasons for which the history of science is written in the first place. As a corollary the answer also touches upon the best way to write history in general.

On the practical level, by adopting the alternative narrative, we would be able to answer some of the questions that will be discussed later when we use the discipline of astronomy as a template for the remaining disciplines and as a direct application of the impact of the new narrative. This will serve us well when we undertake to explain the developments in that discipline once it came to be pursued and reformulated within the Islamic civilization. We will then see that many phenomena, which had remained as veritable enigmas under the classical narrative, could now become easily understandable with the alternative narrative. To give only one quick preview at this point, I point to the language of the translation itself and the manner in which this very language could resolve the scientific technical terms so that someone like al-Ḥajjāj b. Maṭar could produce the earliest surviving translation of the *Almagest* in a fluid, technical and highly readable Arabic. This, when we know that this book is probably one of the most densely-written technical books, if not *the* most, and in which such terms as "*awj*," "*ḥaḍīḍ*," "*ufuq*" for "apogee," "perigee," and "horizon" respectively, were freely used without having to transliterate the Greek as was done in other works from the same period or even from a later period, as in the works of Qusṭā and Isḥāq b. Ḥunain. How could al-Ḥajjāj, who was one of the earliest Abbāsid translators, create this technical language? How could he succeed when we know how difficult such an enterprise can be? To convince one's self of that difficulty, all that one has to do is to consider the heroic efforts that have been pursued in the modern Arab countries, over the last fifty years or so, and continue to be pursued, to create such a technical language? If the alternative narrative does not answer any question other than this one, it would indeed prove its worth over and above the older classical narrative that remained silent about it, or turned it into an irresolvable puzzle in the first place. That is, if we stick to the classical narrative, which assumed that there were no sciences to speak of before the Abbāsid translation period, the hegemony of the Muʿtazilites, the dream of al-Maʾmūn and the like, we will not be able to explain the rise of this technical language of Ḥajjāj at this early period.

But if we go along with al-Nadīm and affirm that the translation movement had already started with the translation of the elementary sciences of the *dīwāns*,[66] and then remember that this *dīwān* translation movement preceded the translation of al-Ḥajjāj by about a full century, then it would become easy to understand the benefit that these earlier translations must have produced at the level of coining technical terms for someone like al-Ḥajjāj to use so freely 100 years later. There is no doubt that al-Ḥajjāj must have introduced some of his own terms, as we can still see in the hesitation of people like Qusṭā and Isḥāq to follow him. At this point, I have no desire to underestimate the efforts that were definitely expended by al-Ḥajjāj himself in accomplishing this project, but I do wish to emphasize that the alternative narrative puts him in his historical context, which allows him to pick up from the newly available language of the *dīwān* translations, add some of his own, as could be understood in a normal historical process, and not force him to perform miracles by creating a whole new technical language from scratch, as the classical narrative would have asked us to believe.

Al-Ḥajjāj's technical language is only one of the many sources of difficulty that we shall encounter in the following chapters, and where we will have occasion again and again to harp back on the benefits that could be derived from the adoption of this alternative narrative.

On the theoretical level, why do I call for the adoption of this alternative narrative? In response I must point to the importance of connecting the history of science to the social conditions in which science is spawned. For although I do not think we will be able to pinpoint exactly why a certain science is supported in a specific society at a specific time, while other fields of knowledge were stifled, I am certain we cannot fully understand the inner workings of the interaction between scientific production and the social, economic and political conditions without paying attention to this dialectic relationship. Adopting the alternative narrative will allow us at least to understand why certain translations were done at specific times, and why the very act of translation became important when it did. This will surely save us from the confusion usually offered by the classical narrative that attributes the origins of the translation movement to essentialist features of Islamic religion itself at times, while at other times it focuses on the racial composition of early Islamic society, like attributing the interest in the translations to the Persian "elements" of the Abbāsid empire, as we are so often told.

With this alternative narrative, we can see for the first time, after the insight of al-Nadīm, the clear relationship between scientific production and the social factors that made that production essential on the first hand, and possible on the other. With that insight we can come close to understanding the early intellectual history of Islamic civilization. And from that perspective we may finally come to appreciate the role played by the government bureaucrats (the *kuttāb* and the viziers) in promoting the acquisition of the ancient sciences, by patronizing this acquisition for their own purposes of competition and advancement in their jobs.

We would no longer need to continue to attribute such interest to a caliph's dream or the like, as if history marches in tune with the dreams of a single ruler or other. Furthermore, this alternative narrative allows us to explain why the 129 Galenic books on medicine were all translated for bureaucrats and not a single one of them to a caliph, as we were told by Ḥunain himself in the aforementioned treatise. We can also understand why the third, or maybe the fourth, translation of the *Almagest* was also patronized by another bureaucrat by the name of Abū Ṣaqr b. Bulbul (d. ca. 892), who worked first as a *kātib* and later promoted to a vizier, and not by the caliph himself.

I shall have occasion to return to these issues in light of the history of the Arabic astronomical tradition, where, as I have already said, I will use that discipline as a template against which I will check the validity of this alternative narrative. I will continue to use every possible occasion to illustrate the advantages gained by adopting this alternative narrative over and against the classical one, hoping that we can come to understand better the development of Islamic scientific thought.

With all that I hope that I have stressed enough the need to go back to the primary sources, both historical and scientific, and to try to re-read them without the biases of any ideological narrative, as much as possible, in order to detect from the sources themselves the direction that was taken by the scientific production and why. That process will hopefully lead us to a better understanding of the real developments in Islamic science, the various periods it went through, and come to finally appreciate, as much as we can, the real social forces that made all that possible.

Now that I have explained the motivation for the acquisition of the ancient sciences, and hopefully explained the processes and the social factors that brought it about, it is time to turn to the social conditions once more and try to detect the impact that these new "ancient" sciences had on the nascent Islamic civilization and how they were themselves transformed by this civilization. I shall devote the lion's share of the discussion to the impact of the Greek sciences on early Islamic society, for no reason other than the fact that those sciences became the focus of great concern from the earliest centuries and continued to capture the imagination of later scientists in everything they did, much to the neglect of the Persian and Indian sciences whose impact apparently began to fade rather quickly around the middle of the ninth century.

3 Encounter with the Greek Scientific Tradition

Like all messages that suffer from the reputation of the messenger, the incoming translations of Greek and Sanskrit texts, that began to be produced toward the end of the Umayyad and beginning of the Abbāsid period, as a result of ʿAbd al-Malik's reforms, began to be naturally associated with those classes of people who were now considered outside the bureaucracy of the *dīwān,* thus foreign to the body politic of the government hierarchy itself. On the opposing side were those who acquired their new jobs by virtue of their mastery of the Arabic language, which was now the new language of the *dīwān.* The natural allies of the second group were those who had also staked a position for themselves that depended on the mastery of the Arabic language as well. But their dependency was for slightly different purposes. These allies who were mainly religious figures and jurists required the mastery of the Arabic language in order to use it as an authoritative tool that allowed them to master the Qurʾanic text, in the first place, as well as master the other ancillary sciences like the prophetic traditions (*ḥadīth*), grammar, lexicography, literature, poetry, as well as all disciplines that served the purpose of deriving juridical opinions from such texts. Those two groups: the religious scholars and jurists on one side and the new bureaucrats of the government on the other, whose claims to authority were based on their mastery of the Arabic language, began to be perceived together as one larger group only when they were contrasted with those whose main claim to power was based on their mastery of those "foreign sciences" that were being recently translated, and were naturally from outside the culture. In this context it is easy to see why the early epistemological division between "foreign sciences" and "Islamic sciences" could very quickly gain ground, in this early period and could persist throughout Islamic intellectual history.

Although the translations came from the two main cultural depositories of India and Persia, in the east, and from Hellenistic lands, in the west, the Greek classical tradition soon began to outshine the other competing traditions. We have already seen that some major Sanskrit texts began to be translated during the reign of the second Abbāsid caliph al-Manṣūr (754–775) if not before,[1] some texts on logic even before that,[2] and it has been generally accepted that the Persian and Sanskrit texts, few as they were, were indeed the first to be translated.[3] The fact that the Sanskrit and Persian translations seem to have come first, must mean that members of the Persian-speaking community were the first to arrive at the conscious realization of the need to import "foreign sciences," in order to compete in the new government market. It may also explain the proliferation of rebellions during the first half of the eighth century, all led by Persians who contested the authority of the then decaying Umayyad empire and whose rebellious efforts were finally crowned by the success of the Abbāsid "revolution" in the middle of that century. That revolution was mainly perceived, at first, as an alliance of various factions including several Persian ones, who were by then all dissatisfied with the Umayyads.[4]

It was after the initial successes of this Persian community that the Syriac-speaking community began to follow suite and to commence the translation of the Greek texts into Arabic. For the philosophical and scientific sources, the forerunners of the Syriac translations included, for example, the early attempts of Ibn al-Muqaffaʿ to translate the Persian texts on logic.[5] One has to assume that at least some other Sanskrit/Persian texts, dealing with medicine and pharmacology, quickly followed suit. And they would have obviously included the attempts of al-Fazārī and Ibn Ṭāriq, who were already mentioned before, to translate the Sanskrit astronomical texts into Arabic and to produce Arabic texts that were modeled after the Sanskrit ones. Those compositions may have also included especially modified Sanskrit texts that allowed their contents to fit the new Arab environment by adjusting, for example, the years of the mean motions into Arab years, meaning Hijra years that are smaller than the solar years by about 11 days each. This conversion task was not a trivial task as we have said before. But we are quite certain that it was in fact accomplished according to the report of al-Nadīm about al-Fazārī when he says that the latter had produced a "*zīj ʿalā sinīy al-ʿArab*" (an astronomical table according to Arab years).[6]

From a cultural perspective, and in contrast, the Arabic translations of Greek sources, which were mainly executed by the Syriac-speaking community, were much more comprehensive, and included, besides the pure sciences and medicine, very sophisticated texts on philosophy and logic. Taken as a whole, the Greek philosophical corpus, which was also understood to include such disciplines as medicine, astronomy and mathematics, appeared as a self contained and integrated body of knowledge that could explain many varied phenomena by resorting to an all encompassing philosophical system such as the Aristotelian system. And in all likelihood, this system was found particularly appealing, especially because of its general applicability to various phenomena and because of the interconnectedness of the various parts of the scientific principles embedded into the formulations of that system. Within a few years, that is, in just half a century or so, between 820 and 870, almost all translations shifted, for all practical purposes, from the Persian and Sanskrit, as source languages, to Greek as the preferred language to be tapped on all levels.

The success of the latter translation attempt was unparalleled. It included almost all serious philosophical and scientific Greek texts. And technically speaking, the translations themselves began to be more organized, more systematic, involving teamwork, and at times operated very much like workshops in their own right. When one thinks of someone like Ḥunain b. Isḥāq and his son Isḥāq as well as his nephew Ḥubaish,[7] all involved in similar activities or joint projects during that same period, one can begin to detect the family structure of that activity. One may also anticipate the possible abuse of these activities by monopolizing entrepreneurs, or by patrons who at times wished to control the information those translations were bringing into Islamic civilization. These conditions could also explain why Ḥunain b. Isḥāq almost devoted his full time translating for Banū Mūsā, while he also occupied the formal position of the Caliph's physician, especially during the reign of al-Mutawakkil (847–861).

The monopolizing entrepreneurs did at times include professionals who required translations of specific Greek texts into Syriac rather than Arabic, so that they would at least monopolize the information for a while before the text would eventually be translated into Arabic. We know from Ḥunain's account of the translations of the Galenic books that he had translated some into Syriac for physicians like Jibrāʾīl b. Bakhtīshūʿ.[8] The same may be true of all the Aristotelian books that were reported by al-Nadīm[9] to have been translated into Syriac as well during this period, or just before. As we already said, the so-called "old translations" (*naql qadīm*), may have also been part of this competitive attempt at monopolizing information by the Syriac-speaking community.

To those who were not involved in the translation activity themselves, the world looked like it was already governed by two main groups. On the one hand, there were those who possessed the information contained in the "foreign sciences" now understood to be mainly Greek. Those same people were either employed at the highest echelons of the government offices like advisors to the caliphs or were competing for those same jobs from outside the government. On the other hand, there were those who possessed the mastery of the Arabic language and who worked at the lower echelons of the government at the old *dīwān* jobs but now allied to religious figures mentioned before who also jointly claimed the same sources of power: Arabic linguistic sciences. This intellectual split continued to express itself, as we just saw, in various forms like "foreign" versus "indigenous," "ancient" versus "modern," "rational" versus "traditional," etc., all signifying those two main centers of the new power structures.

In such an environment, and with the affiliations of the people involved in the pursuit of those sciences, it is easy to explain the appearance of such movements as the *shuʿūbīya* movement, which was widespread during the first half of the ninth century, and which pitted Arab versus non-Arab in almost every field of life. By the ethnic term Arab at this period one should probably understand it to mean an Arabophile as well, or designate people who laid their claim to power through the use of the Arabic language. Anecdote after anecdote relates this sentiment, even when the purpose of the anecdote was purely entertainment. Al-Jāḥiẓ's story, for example, as reported in his book *al-Bukhalāʾ*,[10] about the Arab physician Asad b. Jānī (before 850) speaks directly to this widespread sentiment. Asad was once told that his medical business was expected to flourish during a plague year, to which he answered that it was no longer possible for someone like him to make a living. When asked for the reason he said: that he was a Muslim—and people always thought, even before he became a physician or he was even born, that Muslims would never succeed in medicine; his name was Asad when it should have been Saliba, Morayel (sic), Yūḥannā or Pīrī; his agnomen (*kunya*) was Abū al-Ḥārith when it should have been Abū ʿĪsā, Abū Zakarīyā, or Abū Ibrāhīm. Moreover, he said, he wore a white cotton cloth, when it should have been a black silk robe. And his diction was Arabic, when it should have been the tongue of the people of Jundīshāpūr.

The competition between Arabs and non-Arabs, and among Muslims, Christians, and Jews, could not have been expressed any better. Furthermore, the anecdote illustrates the clear separation between those who depended on the foreign languages to make a living, like the people of Jundīshāpūr, and those Arabs or Arabophiles who sought to establish their authority through the Arabic Language. The anecdote also illustrates why people like Asad would naturally ally themselves with their co-religionists who also sought their power through the Arabic language.

In that environment we can also understand why even non-Arabs, mainly those of Persian descent, like Sībawaih (765–796), would also attempt to master the Arabic language once they apparently realized that they could not compete in the realm of the "foreign sciences"; the latter were being quickly monopolized by those who knew Greek rather than Persian. Those linguists, although of non-Arab origin but were probably Arabophiles, would eventually ally themselves with the religious sciences as well, and with those who steered away from the proximity of political power, in opposition to those who kept on translating "foreign" sciences into Arabic and getting closer and closer to the person of the caliph himself through the patronage of high ranking bureaucrats. The areas that depended on the acquisition of the foreign sciences, at the highest echelons of government, were the areas that became indispensable to the government as was already mentioned before.

This does not mean that the competition was restricted to people of differing linguistic groups and ethnicities, if one could speak of ethnicities at that time. For we know that the deadly race definitely spread throughout the bureaucracy to include at times people of the same religion and profession as we have seen with the case of Ḥunain b. Isḥāq at the caliphal court of al-Mutawakkil.

In such an environment, it would be natural to expect that any intrusion from the outside would be welcomed by some and immediately rejected by others. And since any imbalance in the available fields of knowledge, now understood as tools to political power, would necessarily mean the loss of livelihood for some and boon to others, as had already happened with the *dīwān* translations, a fact that was still fresh in people's minds during the eighth and early ninth century. Under those conditions then, everyone concerned would quickly scrutinize the introduction of any new idea. And this scrutiny would be first conducted by the opponents of those who were importing the new idea, or science in this case, and then by its proponents in order to make sure that it could withstand the attacks from the opposing camps. It would not even be unusual to have some of those new incoming ideas also attacked by those who were in the same camp, but who were themselves also competing and jostling for a greater share of power. It was in such an environment that the newly imported Greek sciences were cast.

The importers of Greek astronomy had to make sure that their field was dissociated from astrology, which was religiously frowned upon. In order to accomplish that, it was those very astronomers who succeeded in re-casting their discipline and in creating the new astronomy, which came to be known as *ʿilm al-Hayʾa* and for which there was no Greek equivalent term as such. Once they could shun the discipline of astrology, then the importers of Greek astronomy, as well as the composers of the *Hayʾa* texts could simply pose as allies of the religious establishment as well, and could then flourish within that establishment as they brought their work to bear on the religious sciences themselves. This trend explains the creation of the new field of *mīqāt*[11] toward the beginning of the eleventh century, in addition to the creation of *ʿilm al-hayʾa* itself, as much as it also explains the creation of the mathematical discipline of *ʿilm Farāʾiḍ* around the same period.

In that polarized environment, which survived throughout Islamic history, we can explain the appearance of certain new disciplines and the disappearance of others. We can also detect the flexibility of scientific production when it acclimatized itself to new social conditions. These developments proved to be crucial to the lasting character of Islamic science in general, and were to cast a particular shadow on the developments that took place within the particular field of astronomy, where the brunt of this conflict was focused. And it is for that reason that the reception of the Greek astronomical tradition offers us the best illustrative glimpse of the general conditions the other disciplines must have encountered as well. The emergence of Islamic astronomy as a discipline on its own is very much conditioned by these early labor-pangs the discipline went through and continued to color its developments in the later centuries.

Reaction to the Greek Scientific Legacy

It is easy to see why the seekers of the Greek scientific texts were less vulnerable than those who sought the philosophical ones; or say that their battle was easier to win. In the case of the sciences, especially the exact ones, like mathematics and astronomy, it was easier to detect errors and to prove the superiority of one opinion against another. Only when such disciplines as astrology were included with those sciences, as was done in the Greek tradition, the situation became slightly more complex.

In the purely scientific texts, as we shall soon see, there were those astronomical values in the Greek tradition that could easily be proved wrong. And that in itself could constitute a danger to those who were bringing those sciences into Arabic. For if they were not extra careful to weed out the mistakes, their whole enterprise could easily be denounced. In the case of philosophical ideas the boundaries between true and false were not as sharp, and the domains they covered overlapped dangerously with some of the domains reserved for religious speculation.

While it would be of great scientific significance to find out that Ptolemy's measurement of this or that parameter was wrong and thus needed to be corrected, this very finding would not have any dangerous immediate social implications. But trying to uphold the philosophical idea that the world was eternal as some of the Greek philosophical texts would say would immediately run into problems with the religious circles that were definitely set in their belief in the doctrine of the creation of the world, by a unique God.

By paying attention to such social conditions we can then appreciate the circumstances under which certain ideas were accepted while others were rejected. Those conditions will also shed light on the very process of the importation of the "foreign" sciences and the battles those sciences had to endure. The fact that the proponents of the foreign sciences themselves were extra alert to the kind of science they were importing, and wanted to make sure that this science was free of any blemish, so that it can withstand the attacks we just described, this care could now explain the reason why someone like al-Ḥajjāj b. Maṭar would end up correcting the Ptolemaic text of the *Almagest* as he was translating it.

In the text of the *Almagest* [IV, 2], al-Ḥajjāj found Ptolemy's report about the length of the lunar month. In it Ptolemy says that he was simply following Hipparchus who had in turn taken two lunar eclipses that were separated by 126,007 days and 1 hour, during which the moon made 4,267 revolutions. Ptolemy went on to say that if one divided the number of days by the number of revolutions, that is, divided $126{,}007^d$ and 1^h by 4,267, one would get the length of the lunar month to be 29 days, 31 minutes, 50 seconds, 8 thirds, 20 fourths (or alternatively written as $29;31,50,8,20^d$). In fact if one were to carry out the division, as prescribed by Ptolemy, the answer would not be the one given in the Ptolemaic text, rather it would be 29;31,50,8,9,20^{d}, which is exactly the number found in the earliest surviving Arabic translation of the *Almagest* by al-Ḥajjāj.[12]

Remembering that al-Ḥajjāj was apparently conscious of the environment of competition that we just spoke about, he could not afford to have what looked like a mistake in the translation, and took it upon himself to correct the Greek text. Now whether this number was in fact a 'mistake' in the Ptolemaic text or not, a problem that was already confronted by Asger Aaboe almost half a century ago, is immaterial here.[13] The important point to make is that al-Ḥajjāj must have thought that it was a mistake, and thus felt that he had to correct it, so that another more competent translator or astronomer would not point to it and thus belittle his scientific abilities.

The classic narrative could not possibly explain such nuances in the translation process, for it did not pay any attention to the competition generated by ʿAbd al-Malik's reforms, nor did it acknowledge the experience gained by the translators of the elementary sciences of the *dīwān* for a generation or two that could give someone like al-Ḥajjāj the necessary skill to carry out the correction. But with the alternative narrative all such activities become quite natural, and historically understandable.

On the observational side, the competition was not any less severe. We know from early reports that astronomers were eager to correct the astronomical values of the classical Greek tradition, not only because they were probably driven by the same motives as al-Ḥajjāj, i. e. to make sure that the texts they were translating were free of mistakes, but that they could also benefit from the passage of time in order to double-check the Greek astronomical parameters that lend themselves to refinements over time.[14] For example, the Greek value for the motion of precession was taken to be 1°/100 years (or about 0;0,36°/year), as recorded in Ptolemy's *Almagest* [VII, 2, et passim]. If that were true, it would have meant that during the first half of the ninth century—that is, some 700 years later—the fixed stars, and particularly the star Regulus (α Liones) whose longitude was easy to observe due to its proximity to the ecliptic, would have been displaced in longitude from the position at which they were observed during Ptolemy's time by about 7°. Instead the longitudes varied by as much as 11° during that period, thus necessitating a new value of precession: about 1°/66 years (0;0,54,54°/year), or about 1°/70 years (0;0,51°/year), whereas the modern value for this parameter is around 0;0,50°/year.

Once the new value was found, astronomers working during the first half of the ninth century began to use it in their works as was actually done by the Ma'mūn astronomers.[15] Ibn Kathīr al-Farghānī (c. 861), who wrote his summary of astronomy around the same time, however, continued to use the old Ptolemaic value of 1°/100 years,[16] most likely in an attempt to be true to the Greek tradition. At other times, as in the case of the adoption of the new value for the inclination of the ecliptic, he did not hesitate to abandon the Ptolemaic value and to side with the new observations of his time.

As we have also seen before, the inclination of the ecliptic which was determined by Ptolemy to be 23;51,20° was found to be too large, and the new measurements that took place in Baghdad, sometime during the early part of the ninth century, concluded that it was closer to 23;33°, a value[17] that is still in use today.

These new values must have come as a result of refinements that were obviously applied to both the methods of observations, as we shall soon see, and the types of instruments used for the purpose, as well as the size of those instruments.

Then there was the solar apogee, which was taken to be fixed at 5:30° of Gemini by Ptolemy, and which was also seen to have moved considerably by the ninth century. In fact the motion of this apogee was found to correspond very closely to the precession motion of the fixed stars, and thus by the time it was observed in Baghdad it was found to have moved by some 11°.[18]

All these findings must have been definitely determined by very competent astronomers, at least as competent as Ptolemy if not more so. As a result they force us to raise the very same question that was raised before: Who trained those astronomers to conduct such refined observations and to determine such precise values that have obviously withstood the test of time as we still find them in current use today? The classical narrative would, at this point, fail dramatically to explain this phenomenon. But if we take the implications of ʿAbd al-Malik's reforms into consideration, and assume the competition we have been assuming, then it becomes plausible to suggest that this very competition may have generated enough care and seriousness so that each astronomer would try to outsmart the others and continuously keep trying to find better and better values for those basic astronomical parameters.

But the sheer accumulation of so many parameters that were at variance with those reported in the Greek texts must have led to more serious research in early Abbāsid times. For we find that sometime during the first half of the ninth century new methods of observations were suggested in order to avoid the pitfalls of the Greek ones, and in order to improve over the Greek results. People began to discuss the impact of the instruments themselves, as well as the strategies for the observations, all in an attempt to explain the reasons why they were finding results that were certainly different from those that were found by Ptolemy some 700 years earlier.

One of the early challenges to the Ptolemaic observational methods came in that same century, when someone suggested that the position of the solar apogee could be determined by a more refined technique. The new technique involved observing the daily declination of the sun at the midpoints of the seasons, rather than their beginnings as was done by Ptolemy. That is, those ninth-century astronomers understood that Ptolemy's observational strategy, which required that the observations be carried out at the time when the sun would cross the equinoxial and solsticial points, was woefully flawed. And they also understood that this strategy would necessarily lead to the difficulty of determining the daily declination of the sun, say, on a mural quadrant, no matter how big, particularly when the sun was around the solstices. In fact, at those points the sun would not exhibit any appreciable declination and the daily variation in declination would actually be very small. Thus it could not be observed accurately. Those astronomers must have reasoned, therefore, that one would be much better off if he were to observe that declination when the sun was crossing the midpoints of the seasons, that is, the fifteenth degrees of Taurus, Leo, Scorpio, and Aquarius, and where the declinations would be much more apparent. The new method that was used for these observations was dubbed the *Fuṣūl* method (method of the seasons) on account of its reliance on the midpoints of the seasons as points of observation, in clear contradistinction to the beginnings of the seasons as was done by Ptolemy.

With this shift in observational techniques, and in one full swoop, the new values for the solar apogee, the solar eccentricity, as well as the concomitant value of the maximum solar equation could all be determined at the same time, and to a much higher degree of precision. And so they were, as was reported in the so-called *mumtaḥan zīj* (Verified astronomical table) that was presumably composed during the reign of al-Maʾmūn (813–833).[19]

All that was happening during the early part of the ninth century, a feat that could not have been accomplished by novices who were just beginning to acquaint themselves with such sophisticated Greek texts that they were translating at the same time. Several generations of earlier translators of elementary sciences must have paved the way for such activities, as the alternative narrative now wishes to propose.

Such activities continued to be performed on regular basis, and methods of observations continued to be checked and double-checked. New research on the types and sizes of instruments must have also been undertaken, and must have continued in the later centuries to constitute a tradition by itself. We have echoes of all this in reports preserved from the tenth century. One copy of such reports has been quoted in the thirteenth-century bio-bibliographical dictionary, the *Taʾrīkh al-ḥukamāʾ* of Qifṭī.[20] In it we are told that during the early Buyid times, i.e. during the latter half of the tenth century, the famous astronomer Abū Sahl al-Kūhī (c. 988) was called upon to conduct fresh observations in order to double-check these same values for the solar apogee, the solar eccentricity as well as the maximum equation of the sun. The report went on to say that Abū Sahl preferred to determine the sun's entry into the summer solstice and the autumnal equinox, just as was done by Ptolemy before him. But, more importantly, we are also told that Abū Sahl had a whole group of people present at the time of the observations, including religious scholars, judges, mathematicians, astronomers, the famous bureaucrat Abū Hilāl al-Ṣābiʾ (d. 1010) as well as other officials. Abū Sahl had all those officials affix their signatures to the report of the observation. The sheer variety in the professions and ranks of the individuals involved can only emphasize the social significance of such activities at that time. But the question remains: Why would Abū Sahl choose the method of Ptolemy, when he should have known that it was already superceded by the *fuṣūl* method more than a century before? Was he trying to "outsmart" Ptolemy by carrying out the very same observation?

Other echoes of the research on better and larger instruments also come to us from the works of al-Khujandī (d. ca. 1000), in which we are told that he attempted to build very large instruments in a continuous bid to get more precise results.[21] Khujandī was supposed to have attempted to build a sextant whose radius was some 20 cubits, and graduated in such a way that it would allow the observer to measure down to minutes of arc rather than degrees.[22]

In later centuries, similar activities continued to be pursued, and instruments continued to be further refined. By the thirteenth century, the same *fuṣūl* method itself, first invented in the first half of the ninth century, was itself refined as well, and another new method was developed that required solar observations to be taken at only three points on the ecliptic instead of four, and only two of the observations had to be diametrically apart.[23]

Subtler Observations

Other mistakes that were found in the *Almagest* were slightly more sophisticated in nature, and were not apparently immediately noticed as the text of the *Almagest* was first translated into Arabic. Two examples of such mistakes should suffice at this point.

The first of these has to do with a statement made by Ptolemy in connection with the relative apparent sizes of the two luminaries as they affect eclipses.[24] At that point Ptolemy does not only say that the apparent size of the solar disk appears to the observer on the Earth to be just as large as the lunar disk when the moon was at its greatest distance from the Earth, but that it was always so and that it did not exhibit any change in size for the same observer. Of course when the moon was closer to the observer, there was no question of its relative size with respect to the solar disk for then the duration of solar eclipses would settle the point. But the occurrence of annular eclipses, a phenomenon not even mentioned by Ptolemy, would certainly provide a counter example to the Ptolemaic statement. Such annular eclipses could then demonstrate that when the moon was at its farthest distance, its apparent size was still smaller than that of the sun, otherwise the sun would not appear like a ring around the disk of the moon during such annular eclipses. In his *Taḥrīr*, Ṭūsī (d. 1274) singled that phenomenon out and supplied records of more recent observations that actually documented such annular eclipses.[25] He went on to say further that the apparent solar disk itself was not in fact fixed, as Ptolemy had maintained, but that it changed in size. And that change could be detected by calculations of the various durations of eclipses at various relative positions of the luminaries. The same conclusion was reached a century or so later by Ibn al-Shāṭir (d. 1375) of Damascus, who even went as far as calculating the variations in the apparent size of the same solar disk, and was forced to construct a mathematical model describing the motion of the sun in order to accommodate those fresh calculations that he probably based on his own detailed analysis of eclipses.[26] We shall have occasion to return to the analysis of this construction by Ibn al-Shāṭir when we discuss the alternative solutions that were given to such Ptolemaic problems during Islamic times.

The second example of sophisticated but subtle mistakes that were found in the text of the *Almagest* involved the mathematical configuration that was described by Ptolemy in connection with the movements of the moon. In that specific configuration, which gave Ptolemy a considerable amount of trouble before he settled on a final version of it [*Almagest* V, 5–10], Ptolemy had to concoct a crank-type mechanism that could account for the variation in the second equation of the moon from a value of about 5;1°, when the moon was in conjunction or opposition with the sun, to about 7;40°, when the moon was at quadrature from the sun (i.e. some 90° away from the solar mean position). The Ptolemaic mathematical model worked reasonably well when it came to predicting the position of the moon in longitude. But as it was correctly observed by the same Ibn al-Shāṭir the model also "required that the diameter of the Moon should be almost twice as large at quadrature than at the beginning, which is impossible, because it was not seen as such *lam yura kadhālika*."[27]

Ibn al-Shāṭir was absolutely right in affirming that such a variation in the apparent size of the moon would result from the Ptolemaic model for the lunar motion. And because of his apparent reliance on his own newly conducted observations of eclipses, Ibn al-Shāṭir had to construct an alternative model for the motion of the moon that will also be discussed in the context of the solutions that were developed during Islamic times in opposition to those of Ptolemy.

All of these corrections, new techniques, new solutions, and developed refinements would not have been generated had the astronomers who produced them not read the Ptolemaic astronomical text with a critical spirit. Nearly all of the astronomical parameters that they had encountered in the *Almagest*, proved fundamentally defective, and a basic program of observation was needed to correct them. What seems to have happened in this early period is exactly that, for we hear of one astronomer after another all attempting to negotiate a way out of the difficulties that the *Almagest* had confronted them with. The resulting body of literature that they produced in response, whether in treatises devoted to the subject of methods of observation, or in the production of new astronomical tables called *mumtaḥan* (verified), or the like, could all be regarded as the logical results of that critical approach with which those early astronomers received the Greek scientific masterpieces. At the same time, this new literature could also be seen as a by-product of the clear desire to establish more reliable parameters for the new field of astronomy that was then emerging; parameters that would eventually be far superior to the ones that gave rise to the problems embedded in the *Almagest*.

Critiques and correction of fundamental parameters and critiques of the methods that produced them were not the only things that led the receiving culture to negotiate the difficulties encountered in the Ptolemaic text. One section of the *Almagest,* Books VII and VIII, dealt specifically with constellations and descriptions of constituent stars that the receiving culture had some experience with; although it did not seem to have had the comparatively systematic tabulation of such stars. But in this domain we still lack substantial information about the events that took place during this early period. What can be asserted, however, is that some modifications of the Greek text did already take place on the occasion of the various translations themselves, where alternative names were given to constellations either in addition to the ones that were being translated from Greek or to replace them altogether.

By the tenth century, the literature on the fixed stars began to generate two competing traditions of its own: One was directly derivative from the Greek, and was thus recorded in astronomical handbooks and the like, and of course perpetuated in the various translations of the *Almagest* and the books that derived from them. While the other tradition was represented by a whole host of texts devoted to *anwāʾ* literature[28] that can best be characterized as being concerned with the utility of the risings and settings of constellations for agricultural purposes and for the general purposes of daily life. This latter tradition approached the subject from a native Arabic background by drawing on the native sciences and the native knowledge of the constellations known from the widely read Arabic literary sources themselves.

Here again one could detect the opposing camps splitting along lines similar to the ones discussed above: There were those who favored reliance on the non-Arab "ancient" sciences, and were themselves identified as higher government bureaucrats, and those who preferred to rely on the ways of the Arabs in the non-governmental or lower bureaucratic circles. As a result an enormous literature began to be written on the subject of the stars. And because of the various traditions it involved, the same body of literature begged for systematization.

It was ʿAbd al-Raḥmān al-Ṣūfī (d. 986) who undertook that job by producing a masterpiece on the constellations that was not surpassed until modern times. His book *Ṣuwar al-Kawākib al-thābita* (Figures of the Fixed Stars) did not only include a general background description of each constellation and its constituent stars in both the Greek and the Arabic traditions, identifying, whenever possible, the multiple names given to the same star or groups of stars, but included as well systematic tables of longitude, latitude and magnitude of the individual stars themselves. This text, which is available only in a preliminary edition from Hyderabad,[29] has never been studied in any detail.[30] But even a casual reading of it reveals that it contains lengthy dialogues with the Greek tradition, particularly expressed in terms of objections to the Ptolemaic received text. One cannot help but notice the many occasions when Ṣūfī would say that this or that star or constellation is such and such according to Ptolemy, but I say it ought to be this or that, and the Arabs, in contrast, say this about it.[31] On account of its deliberate comprehensiveness, and probably on account of its authoritative standing as the standard reference book on the constellations that it must have become, this text lent itself to royal patronage production, and copies of it were so beautifully illustrated that many of them are still considered among the chefs d'oeuvres of Islamic art.[32]

Mathematical Reconstruction of the *Almagest*

Two other types of criticism that were directed at Ptolemy's *Almagest,* also need to be mentioned in this context, although they touch on slightly different issues from the ones that have been discussed so far. This group of critical ideas did not touch the issues of mistakes in the *Almagest* per se, as was done before. Rather it touched upon two other areas of the text where it could stand some updating: First, there was the criticism that could be classified under the heading of attempts to update the text of the *Almagest,* i.e. bring the mathematical approaches deployed in the text into par with the current mathematical knowledge of the time. For example, the very famous mathematical theorems that were used at the beginning of the *Almagest* to set up a trigonometric system that was used throughout the text, employed the classical Greek spherical trigonometric theorem which used chord functions as was done, for example, by the Menelaos theorem.[33] To his exposition of the theorem, and his proof of it, Ptolemy attached a chord table in order to facilitate the following computations in the rest of the book. It was this material that became an obvious target of the various revisions in early Islamic times. And that should have been expected, since by then the astronomers who were reconstructing the discipline of astronomy had at their disposal an almost fully developed trigonometric system of sines, cosines, tangents, and the like. In addition, this system was already fully embedded in the receiving culture into which the *Almagest* was being translated, and at times could very comfortably co-exist with the inherited Greek chord system with its comparative clumsiness for everyone to see.

From the translators themselves we would not know of the existence of this other field of trigonometry, which was itself unknown to the Greek tradition. But the various writers who were producing their own astronomical works, as the *Almagest* was being translated, did not shy away from using the new trigonometric functions to describe the same phenomena that were described in the *Almagest.* Of the several examples that can be cited regarding the use of the new mathematics to update the text of the *Almagest,* by far the best one comes from a slightly later period, around the middle of the thirteenth century. In Ṭūsī's *Taḥrīr al-majisṭī* (Redaction of the *Almagest*), already mentioned before, that was written in 1247, Ṭūsī approached this section of the *Almagest* in the following fashion. After concluding his exposition of the *Almagest*'s table of chords, he went on to make the following remark: "I say, since the method of the moderns, which uses the sines at this point instead of the chords, is easier to use, as I will explain below, I wish to refer to it as well."[34] He then went on to give a spherical sine theorem equivalent to that of Menelaos and affixed to it another configuration using the tangent function instead of the sine. He concludes that section by producing tables for sines and tangents to complete the mathematical and trigonometric tools for the rest of the book.

This updating of the *Almagest,* although not stressed often enough in the literature, is of crucial importance to understanding the life of the *Almagest* in the Islamic domain. And when we juxtapose this treatment of the *Almagest* text in the later centuries with the independent works of someone like Ḥabash al-Ḥāsib from the ninth century, in which we see these trigonometric functions used so freely as we shall soon see, we can then clearly appreciate the immediacy of the *Almagest* text to the practicing astronomers, and clearly see their willingness to merge its contents with the kind of astronomy they were already practicing.

In a parallel development, and this was also to be expected, one finds those same astronomers, using the results of the *Almagest,* at times, whenever they thought that those results were still valid, while at other times they would reject them completely in favor of new ideas of their own. This multiplicity of approaches to the *Almagest* text can only signal the very vital reactions it must have created within the receiving culture of early Islamic times. But one should also remember that at all instances this very vitality produced an *Almagest* text that was considerably enriched by the process.

Returning to the earlier astronomers, such as Ḥabash al-Ḥāsib (fl. ca. 850) in particular, who produced their own independent *zījes* (Astronomical Handbooks), that were only constructed in the tradition of Ptolemy's *Handy Tables,* we find that they too have also used the most recent and fully developed trigonometric functions in those works.[35]

Looking at the complete picture of that period, and after examining the scientific sources themselves, one can begin to see a process in which one finds that as soon as the Greek scientific texts were being translated, they were also being immediately updated by the currently known material and put to use in new compositions, all in order to improve the kind of science that was then produced.[36]

The second type of intervention in the text of the *Almagest,* had less to do with updating it mathematically or correcting its errors as we have already seen. Instead it was more like reconstructing it or re-editing it so that it would become more useful for students of astronomy. In this regard great liberties were taken with the text, feeling completely free to add to it and delete material from it, all in order to make it a more up-to-date functional text.

The best illustration of this type of intervention is also exemplified by Ṭūsī's *Taḥrīr,* which was mentioned before and in which one finds a new treatment of some chapters of the *Almagest,* such as *Almagest* X,7, where Ptolemy used an iterative method to compute the eccentricity of one planet and then repeated it in great detail for each of the other planets.[37] Instead of the Ptolemaic approach, Ṭūsī adopted a new technique of explaining the method in great detail in the case of one planet and then generalizing it to the others without repeating it in every case.

I mentioned the corrections Ṭūsī had introduced in the same text regarding the apparent size of the solar disk, and the counterexample of the annular eclipses that were apparently unknown to Ptolemy. I also mentioned the correction of other factual errors, including errors in the rate of precession, the inclination of the ecliptic, and the motion of the solar apogee, as well as the development of the methods of observations like the introduction of the *fuṣūl* method. In all those instances, we find the text of the *Almagest* critically reviewed and updated before it could become useful to the receiving culture. Far from being a model to be followed, although one could argue that it was in some sense, it was more like a foundation to build upon, but only after making sure that it was a safe foundation and that its errors and contradictions had been already weeded out. There were other instances in which the text of the *Almagest* was also found wanting, but this time for much more fundamental considerations than the ones that have been discussed so far.

Cosmological Problems of the *Almagest*

By considering only the factual corrections that have been discussed so far, one could easily arrive at the conclusion that the text of the *Almagest* would have become functionally serviceable once those corrections were adopted. The text would have been sufficient, for example, for practicing astronomers and astrologers and no further elaborations of it would have been necessary. But with astrology and its practice facing a veritable resistance from the main intellectual centers of the society, especially the religious ones, and thus its relationship to astronomy being consciously severed by the theoretical astronomers who invented the discipline of *hay'a* as we have already seen, the purpose of the discipline of astronomy was apparently defined in a slightly more nuanced fashion. This purpose can best be seen if one reads the two most famous works of Ptolemy together. Those works are: the *Almagest,* where one finds a detailed account regarding the relationship between the observed phenomena and the construction of geometric predictive models that explained the behavior of the planets at all times, and the *Planetary Hypotheses,* where one would find a detailed account of the celestial spheres that were made, in a true Aristotelian fashion, responsible for the motion of those planets. By reading those two texts together, as most people did, once they became available in Arabic, some serious cosmological problems began to appear. Most of those problems focused on Ptolemy's violation of the most basic cosmological tenet of Greek astronomy: the uniform circular motion of the planets around a fixed Earth located at the center of the universe.

That the Earth was fixed at the center of the universe was undoubtedly at the core of that Aristotelian cosmology, so much so that if one did not have such an Earth one would have had to suppose the existence of such an Earth at the very center of heaviness around which everything else revolved.[38] The real challenge was to explain the apparent phenomena from within that cosmological vision, and still retain some predictability in the geometric models that described the planetary motions.

From that cosmological perspective, the *Almagest* failed at almost every count. While there is the Ptolemaic pretense that the universe, which was being described, was an Aristotelian universe, within which all the Aristotelian elements were to be found as the building blocks of that universe, yet at every juncture the *Almagest* described situations that were physically impossible when looked upon from the perspective of the *Planetary Hypothesis* that emphasized that Aristotelian cosmology. It is this inconsistency between the mathematical models constructed in the *Almagest* to account for the motion of the planets, and the physical objects those models were supposed to represent that I have so often referred to as the major problem of the Greek astronomical tradition.[39]

Because these inconsistencies are of a completely different nature than the ones that were touched upon before, and because they were a direct by-product of the application of Aristotelian cosmology, some people have referred to them as philosophical problems. As a result they tried to read the *Almagest* as divorced from that same cosmology that was wholeheartedly adopted in the *Planetary Hypotheses,* the text that was supposed to complement the *Almagest,* which it followed. And yet these inconsistencies were perceived as touching the very foundation of science; in the sense that science should not harbor contradictions, as Ptolemy seems to have allowed it to do, between the physical side of the science and the mathematical representation of the same physical universe that was being described.

One can only assert that such problems were philosophical problems, if one were to think of them only in the medieval sense of natural philosophy, where such cosmological issues were properly discussed. But they also did matter to the scientists who were also trying to make sense of the physical phenomena around them, and who would have demanded that their scientific disciplines did not contradict each other. In that sense, those problems became real scientific problems, and did not remain only in the domain of philosophical speculation.

Take, for example, the physical spheres that were supposed to constitute the Aristotelian universe, and which were simply represented by circles by Ptolemy in the text of the *Almagest.* If one were to limit himself to philosophical speculations only, then those spheres would pose no serious problem if they were understood as mere mathematical representations that had no connection to reality. But if at the end of the day one used those spheres to account for the motion of planets and used them to predict the positions of those planets for a specific time, then one had to face their reality in a much deeper sense than is already admitted. And when that reality was reiterated in the *Planetary Hypotheses,* the contradictions became much more serious. Again, there would be no problem if one were only using those models of spheres to compute positions of planets only. But when one says that those spheres were actually physical in nature, in the Aristotelian sense of physical, it would become then impossible to think of them, for example, as being able to move uniformly, in place, around an axis that did not pass through their centers.

This was the most important impossibility in the whole of Greek astronomy, or at least it was so perceived. Such glaring absurdities that were embedded in almost every model of the *Almagest* could not pass unnoticed by astronomers, who were not only being watched over by their opponents in the society, who in turn did not want them to bring those "ancient" sciences into the Islamic domain in the first place, but they were also being watched by their own fellow astronomers who definitely believed, as al-Ḥajjāj must have done, that they could outsmart their fellow astronomers if they could cleanse the imported system from those blemishes.

That people were really thinking along those lines is best illustrated by one of the earliest texts to address the sheer physicality of the spheres: the text of Muḥammad b. Mūsā b. Shākir (d. 873), who was not only one of the major patrons of the translation of Greek scientific and philosophical texts, but was also himself a scientist in his own right. In his capacity as a practicing scientist, he devoted a treatise to the absurdity of assuming the existence of a ninth sphere, as Ptolemy had done. According to Ptolemy that last ninth sphere was responsible for the motion of the eighth sphere, which, in turn, carried the fixed stars. And yet Ptolemy had both spheres share the same center of the universe. The problem was then reduced to the impossibility of having two concentric spheres move one another without assuming a phenomenon like friction, which could not be allowed in the celestial realm of the Aristotelian universe where celestial spheres, by their very ethereal nature, did not allow friction to take place.[40]

That Ptolemy himself was thinking along the same lines is evident from the preface of the *Almagest,* where he says that the celestial motions should not be compared to the motions that we observe around us, for they belonged instead to some form of a deity. To which one could respond: if that were the case, and if the deities were responsible for the motion of the planets, then there would be no need for the science of astronomy, nor would there be any need for scientific observations, for who are the humans who could predict the behavior of deities? The readers of the Ptolemaic texts in their Arabic translation saw a different world, and could not simply resort to such whimsical deities in the midst of a competitive society that was watching every step they took.

This incompatibility between the mathematics of the *Almagest* and the physics of the *Planetary Hypotheses* would not have been noticed had those two books not been read together. And in the tense environment in which they were thrust their coming into conflict with one another was simply unavoidable. In addition, if one were to remember that those problems were being raised by Muḥammad b. Mūsā b. Shākir toward the middle of the ninth century, when the first translation of the *Almagest* by al-Ḥajjāj was barely two decades old, and the translation of Isḥāq b. Ḥunain (d. 911) had not yet taken place, one can then begin to appreciate the sophistication with which the Greek astronomical tradition was being received as it was being translated, a sophistication that could not be explained by the classical narrative. Furthermore, it is a kind of sophistication that could only come from this comprehensive understanding of the Greek philosophical tradition, where cosmology was read together with observational science, a reading that was nowhere to be found in any other civilization up till that time.

In later centuries, as other contradictions began to appear, further sophistication began to be necessary. But all the basic problems still focused around this major issue of the lack of consistency in the imported Greek astronomical tradition. In a word they still dealt with these foundational issues of science.

Once those issues were widely recognized by the various sectors of the society, they began to develop a tradition of their own. The various treatises that began to appear in the later centuries, and in which those issues were recounted, began to constitute a scientific genre of their own normally referred to with such titles as *Shukūk* (doubts). And because of the social dynamics within which those doubts were expressed, they were by no means restricted to the field of astronomy alone.

The similar text by Abū Bakr al-Rāzī (Latin Rhazes, d. 925) called *al-Shukūk ʿalā Jālīnūs* (Doubts Against Galen), falls in this category as well, and with it we can easily detect a general cultural trend that has yet to be elaborated. Only skeletal sketches of these developments and of the major issues that were raised in those texts could be attempted at this time. Of course the astronomical tradition still received the lion's share of those discussions, and can only be used here as a representative model of the other discussions that were obviously taking place in the other disciplines.

The Astronomical *Shukūk* Tradition

If one were to disregard the earlier objections to the Ptolemaic observational parameters, or even the cosmological questions by Muḥammad b. Mūsā b. Shākir, just mentioned, as early expressions of doubts that had not yet developed into a genre of their own, then one will have to say that the genre was born with Rāzī's book, which was expressly called *Shukūk,* despite the fact that in the case of Rāzī his book was restricted to medical and philosophical doubts. Astronomical *Shukūk* were soon to follow, even though they seem to have taken a slightly different route.

During the eleventh century, possibly in the latter half of that century, an Andalusian astronomer, whose name is yet to be identified, has left us a treatise called *Kitāb al-Hayʾa* (A Book on Astronomy), which is still preserved in an apparently unique copy at the Osmania library at Hyderabad (Deccan, India). In this treatise the author had several comments to make about the problems in the received Greek astronomy. But almost every time he made such comments he would quickly say that he had gathered those problems in a book that he called *al-Istidrāk* [*ʿalā Baṭlamyūs*] (which could be freely translated as: Recapitulation Regarding Ptolemy). This book has yet to be located. But from the context in which it is mentioned, and the problems it seems to refer to, it sounds like it was of the same nature as the other *shukūk* texts under discussion.[41]

In the east, and around the same time, Abū ʿUbayd al-Jūzjānī (d. ca. 1070), the student of Ibn Sīnā (Latin Avicenna d. 1037), also left us a small treatise *On the Construction of the Spheres*. In this treatise he mentioned that he had discussed with his teacher, Ibn Sīnā, the famous Ptolemaic absurdity, which was by then known as the problem of *muʿaddil al-masīr* (Equator of Motion, or *equant* for short).[42] The fact that both such texts existed, one from Al-Andalus, the farthest western reaches of the Islamic world at the time, and one from Bukhara, the farthest east, and the fact that the second text comes from the philosophical circle of Ibn Sīnā and not from the circle of astronomers and mathematicians, could only mean that the cosmological issues that were perceived to have plagued Ptolemaic astronomy were by then circulating in widespread intellectual and geographical circles; they were no longer restricted to the elite of astronomical theoreticians. The equant problem itself, which had the longest staying record, is none other than the physical absurdity of proposing that a physical sphere could move uniformly, in place, around an axis which did not pass through its center. This absurdity permeated almost all of the models, which were proposed in Ptolemy's *Almagest*. What the texts of al-Andalus and Bukhara suggest is that by the eleventh century that proposition was apparently widely recognized as a physical impossibility.

In his own rather humorous story Abū ʿUbayd informs us that when he discussed the proposed solution for this Ptolemaic absurdity of the equant, with his teacher Ibn Sīnā, he was told by Ibn Sīnā himself that he had also resolved it, but refrained from giving out the solution in order to urge the student to find it for himself. In the very next sentence the student went on to say that he did not believe that his teacher had ever resolved that problem.

The anecdote, legendary as it may be, is still indicative of the kind of problems those custodians of the "foreign sciences," the philosophers in particular, were competing to solve, and the challenges they were facing, as well as the fame they hoped to acquire if they could rid the Greek astronomical tradition of its absurdities. The anecdote also indicates that if the philosophers were already aware of this joint reading of the Ptolemaic texts (the

Almagest and the *Planetary Hypotheses*) in which such problems would arise, this must mean that the astronomers were obviously much more deeply entrenched in that perspective. And the debates of the latter must have informed the former.

For the good fortune of the astronomers, it also appears that their debates over such kinds of issues were socially condoned. They did not only transcend their circles to reach the circles of the philosophers, but they probably also gave rise to the likelihood of rebutting the incoming Greek tradition altogether, since they were being critical of it.

Such discussions had no direct bearing on the other more socially controversial perception of the Greek tradition: its willingness to harbor those astrological sciences that were not as widely accepted as the theoretical critiques seem to have been. For our immediate purposes, however, it is important to document the tradition of the critiques themselves in order to demonstrate the sophistication of that tradition, and its wider implication on the very formation of Islamic science.

Again in the same century, and still in the east, we find the prolific polymath and famous astronomer Abū al-Raiḥān al-Bīrūnī (d. ca. 1048) who also had something to say about the physical absurdities of the Ptolemaic system. This, despite the fact that Bīrūnī's main astronomical production was really geared toward the mathematical observational part of astronomy and paid much less attention to the cosmological aspects of the discipline. In his *Ibṭāl al-buhtān bi-īrād al-burhān* (Disqualifying Falsehood by Expounding Proof), which seems to have been lost but which was quoted by the astronomer Quṭb al-Dīn al-Shīrāzī (d. 1311), Bīrūnī had this to say about the Ptolemaic description of the latitudinal motion of the planets: "As for the motions of the five epicyclic apogees in inclination, as it is commonly known, and is mentioned in the *Almagest,* those would require motions that were appropriate for the mechanical devices of Banū Mūsā, and they do not belong to the principles of Astronomy."[43] That was Bīrūnī's polite way of saying that Ptolemy's discussion of the planetary latitudes was not astronomy proper, and that it amounted to nothing. Such was the extent of criticism of Ptolemy, even by people who had a vested interest in defending him against his detractors. And yet they could not remain silent about the Ptolemaic absurdities, probably because they apparently felt that they had a greater interest in competing amongst themselves by demonstrating that they could outsmart Ptolemy.

The best-preserved and most elaborate text in the genre of *shukūk* was a criticism of Ptolemy that was leveled by another polymath, by the name of Ibn al-Haitham (d. ca. 1040 Latin Alhazen), who was also a contemporary of the astronomers mentioned above, and whose work on Optics was the only work that was known in the Latin West and which earned him his well deserved fame. His critique of Ptolemaic astronomy is contained in an Arabic text which has survived, but which was apparently never translated into Latin. The text in question is his extensive *al-Shukūk ʿalā Baṭlamyūs* (*Dubitationes in Ptolemaeum*) [*Shukūk*],[44] in which he took issue with several of Ptolemy's works in which he found fault.

The three Ptolemaic works in question included the *Almagest,* the *Planetary Hypotheses,* and the *Optics*. Their mere grouping is a clear indication that those works were read together, in a comprehensive manner, and not in isolation as is sometimes claimed.[45] For Ibn al-Haitham, the common thread that connected the three books together is that they all contained problems or doubts (*shukūk*) that revealed contradictions that could not be explained away (*lā taʾawwul fīhā*).[46] This phraseology also indicates that every effort was already made to give Ptolemy the benefit of the doubt. Problems were obviously explained away wherever that was possible,[47] and only those absurdities that could not be justified were attacked. Ptolemy's books were taken up in the following order: the *Almagest,* which had the lion's share, to be followed by the *Planetary Hypotheses,* and then the *Optics*. In the sequel I will take a few select examples from this treatise in order to illustrate the kind of issues that attracted the attention of Ibn al-Haitham.

In his critique of the *Almagest,* Ibn al-Haitham passes very quickly over the early chapters of that book, and commences the real critique with the Ptolemaic description of the model for the lunar motion. In it Ptolemy assumes that the motion of the moon, on its own epicycle, is measured from a line that passes through the center of the epicycle, but is directed, not to the center of the world, around which the motion of the epicycle itself is measured, nor to the center of the sphere that carries the epicycle, called the deferent, but to a point, called the prosneusis point (*nuqṭat al-muḥādhāt*) by Ptolemy. In the Ptolemaic model, this point falls diametrically opposite to the center of the deferent from the center of the world. In his overall assessment of this model Ibn al-Haitham clearly said that it was basically fictitious and that it had no connection to the real world it was supposed to describe. He singled out the soft spot in the model with the following remark: "The epicyclic diameter is an imaginary line, and an imaginary line does not move by itself in any perceptible fashion that produces an existing entity in this world."[48] Furthermore: "Nothing moves in any perceptible motion that produces an existing entity in this world except the body which [really] exists in this world."[49] Later on, he went on to affirm once more: "no motion exists in this world in any perceptible fashion except the motion of [real] bodies." And then he concluded this section by stating that a single epicycle could not possibly move the moon by its own anomalistic motion and at the same time move in such a way that its diameter will always be directed toward the prosneusis point. That would entail that a single sphere was supposed to move in two separate motions by itself, which was impossible.

Books VI–VIII of the *Almagest* did not bother Ibn al-Haitham very much. Instead he moved very quickly to Book IX where the issue of the equant is discussed. In *Almagest* IX, 2, Ptolemy made the explicit statement that the upper planets moved in a uniform circular motion, just like the other planets he had discussed before. But by *Almagest* IX, 5, Ptolemy had laid the foundation for the equant problem when he insisted that "we find, too, that the epicycle centre is carried on an eccentre which, though equal in size to the eccentre which produces the anomaly, is not described about the same centre as the latter."[50]

The point that Ptolemy was trying to make at that occasion was that the two spheres, whose combined motion was responsible for the motion of the planet, were distinct spheres: one, the deferent, simply carried the epicycle of the planet, and the second, taken to be equal to the deferent in size, was responsible for the uniform motion of the planet's epicycle, but explicitly stating that the motion of the last sphere did not take place around the same center as the deferent. It was the center of the latter fictitious sphere, the sphere of uniform motion, that was later called the equant. In chapter IX, 6 of the *Almagest,* Ptolemy went on to describe much more clearly the equant center. There he defined it as a point along the line of apsides such that its distance above the center of the deferent was equal to the distance of the deferent's center from the center of the world.

Moreover, the line connecting this equant point to the center of the epicycle, when extended, constituted the line from which the mean motion of the epicycle was measured. In effect, this said that the deferent sphere, which carried the epicycle, was forced to move uniformly around a center, now called the equant, other than its own center, which was physically impossible.

By then, Ibn al-Haitham seems to have obviously realized the seriousness of the problem, as his following statement indicates: "What we have reported is the truth of what Ptolemy had established for the motion of the upper planets; and that is a notion that necessitates a contradiction."[51] This was in fact the contradiction between the physical reality of the celestial spheres and the mathematical model that was supposed to represent them. For as Ptolemy had accepted the uniform motion of the upper planets, the epicyclic centers of those planets were carried by deferents, which were supposed to move in this uniform motion. But with the equant proposition, one was told that the epicyclic center described equal arcs in equal times, i.e. moved uniformly, around a center that was not the center of the deferent that carried it.

But by Ptolemy's own proof in *Almagest* III, if a body moved uniformly around one point it could not move uniformly around any other point. Therefore the epicyclic center, as stipulated by Ptolemy, must move non-uniformly around the center of its own carrier, the deferent. And since the equant sphere was a fictitious sphere, and thus could not produce any perceptible motion of its own, as was often repeated by Ibn al-Haitham, the only sphere that could produce a real motion was that of the deferent, and that was now proved to be moving non-uniformly around its own center. This contradicts the assumption of uniform motion that was accepted by Ptolemy in the first place, hence the contradiction that was realized by Ibn al-Haitham. The other alternative was to assume that the same physical sphere, the deferent, could move uniformly around an axis that did not pass through its own center which was physically impossible, for it was exactly the physical absurdity mentioned before.

All the other models of the *Almagest,* except the model of the sun, which had problems of its own, shared this absurd feature of the equant. In the case of the moon, its epicycle too was also supposed to be carried on a deferent that moved in such a way that the epicyclic center of the moon did not describe equal arcs around its own deferent center, in equal times, but rather around the center of the world. That was in essence requiring a sphere to move uniformly around an axis that did not pass through its own center as well, which was exactly the point of the equant problem.

Mercury's model, which was considerably more complicated than the other planetary models, shared this feature as well. There too, the deferent that carried the epicycle of Mercury moved in such a way that its motion was not uniform around the deferent's center but around a point that was along the line of apsides half way between the center of the world and the center of another director sphere that carried the deferent sphere of Mercury.

Furthermore, both in the case of Mercury as well as the case of the upper planets, Ptolemy did not even attempt to demonstrate how he arrived at the location of the equant. It was simply stated to occupy such and such a position without any further discussion as to why, or any proof, as would have been expected in a mathematical science such as astronomy. It was this issue in particular that gave rise to the question, which was raised by another Andalusian astronomer by the name of Jābir b. Aflaḥ (middle of the twelfth century) and was singled out in his own research.[52]

From all those Ptolemaic configurations, Ibn al-Haitham could draw only one conclusion: that they were all extraneous to the field of astronomy. This much was even admitted by Ptolemy himself in *Almagest* IX, 2, where he had stated, in no ambiguous terms, that he was using a configuration that was contrary to accepted principle (*khārija ʿan al-qiyās* as in the Arabic translation of the text, or "not from some readily accepted principle" as in Toomer's translation of the *Almagest*). From that admission, Ibn al-Haitham could then only conclude with a rebellious voice against the whole of Ptolemaic astronomy, articulated in the following terms:

> [Since Ptolemy] had already admitted that his assumption of motions along imaginary circles was contrary to [the accepted] principles, then it would be more so for imaginary lines to move around assumed points. And if the motion of the epicyclic diameter around the distant center [i.e. the equant] was also contrary to [the accepted] principles, and if the assumption of a body that moved this diameter around this center was also contrary to [the accepted] principles, for it contradicted the premises, then the arrangement, which Ptolemy had composed for the motions of the five planets, was also contrary to [the accepted] principle. And it is impossible for the motion of the planets, which was perpetual, uniform, and unchanging to be contrary to [the accepted] principles. Nor should it be permissible to attribute a uniform, perpetual, and unchanging motion to anything other than correct principles, which are necessarily due to accepted assumptions that allowed no doubt. Then it became clear, from all that was demonstrated so far, that the configuration, which Ptolemy had established for the motion of the five planets, was a false configuration (*hayʾa bāṭila*), and that the motions of these planets must have a correct configuration, which included bodies moving in a uniform, perpetual, and continuous motion, without having to suffer any contradiction, or be blemished by any doubt. That configuration must be other than the one established by Ptolemy.[53]

This was not a criticism of Ptolemy. Rather, it was an extremely well-articulated condemnation of the very foundation of Ptolemaic astronomy and an open call for its toppling in favor of an alternative astronomy that did not suffer from such contradictions. It did not only expose the fatal mistakes and contradictions in Ptolemaic astronomy, but rose to the occasion of articulating a new set of principles upon which an alternative new astronomy had to be based.

Such attacks, articulated by various astronomers working in the Islamic tradition, did in fact constitute an essential shift in the very conceptualization of the new Islamic science that was being articulated. The new conceptualization did not only condemn the Greek legacy, but laid the foundation for the new consistent science. In the new science, which was then born out of those attacks during Islamic times, physical objects would be, from then on, mathematically represented by models that did not deprive them of their physicality as was done by Ptolemy.

Ptolemy's latitude theory, as expounded in the *Almagest,* did not fair any better. In it Ptolemy himself had expressed doubts about its exact workings, an admission that only encouraged Ibn al-Haitham to conclude:

> This is an absurd impossibility (*muḥāl fāḥish*), in direct contradiction with his [meaning Ptolemy's] earlier statement about the celestial motions—being continuous, uniform and perpetual—because this motion has to belong to a body that moves in this manner, and there is no perceptible motion except that which belongs to an existing body.[54]

What Ibn al-Haitham was referring to was the seesawing motion of the inclined planes, which carried the epicycles of the lower planets of Mercury and Venus. That motion was also another impossibility that could not be tolerated by Ibn al-Haitham, and was simply dismissed as another grave error on the part of Ptolemy. Ibn al-Haitham's argument can be summarized as such: with such motions Ptolemy was forcing physical bodies to move in opposite motions, which was in itself physically impossible.

Over and over again, Ibn al-Haitham returned to the vision of the new astronomy he would like to see—an astronomy based on the new principles of consistency between the physical reality of the universe we live in and the mathematics one uses to represent that reality. In the new astronomy, those two fields of science had to be constantly consistent, otherwise we would end up talking about imaginary motions as was done by Ptolemy:

> The contradiction in the configuration of the upper planets that is taken against him [meaning Ptolemy] was due to the fact that he assumed the motions to take place in imaginary lines and circles and not in existent bodies. Once those (motions) were assumed in existent bodies contradiction followed.[55]

Furthermore, Ptolemy knew very well that he was embracing such contradictions as he was quoted by Ibn al-Haitham to have said: "We know that the use of such things is not detrimental to our purpose, as long as no significant excesses are introduced on account of them."[56] To which Ibn al-Haitham could only say:

He means that the configuration that he had posited necessitates no excesses in the motion of the planets. This statement, however, should not be an excuse for assuming false configurations (*hay'āt bāṭila*) that could not possibly exist. For if he assumed a configuration that could not possibly exist, and if that configuration anticipated the actual motions of the planets as he had imagined, that would not release him from the fault of having erroneously assumed such a configuration. For it is not permissible to stipulate the actual motions of the planets by a configuration that could not possibly exist. Neither is his statement regarding the assumption of things that are contrary to the accepted principles, that they are only hypothetical and not real and thus are not detrimental to the motions of the planets, an excuse that would allow him to commit such absurdities (*muḥālāt*) that should not exist in the configurations of the celestial bodies. Moreover, when he says that 'things that are posited without proof could have only been reached through some scientific mean, once they are shown to agree with the observable phenomena, even though it is difficult to describe the method by which they were reached' is a valid statement. By that I mean that he had indeed followed some scientific mean when he assumed what he assumed by way of configurations. Except that the mean that he had followed had led him to admit that he had assumed things that were contrary to (the accepted) principles. Once he knew that it was contrary to the principles, he had no excuse to assume it, saying that it was not detrimental to the motions of the planets, unless if he were prepared to admit that the real configuration was different from what he had assumed, and that he could not reach its essence. Only then would he be excused to do what he did, and it would be known that the configurations that he had assumed were not the real ones.[57]

In this long passage, Ibn al-Haitham leaves no doubt as to his real intentions. He obviously means that real physical bodies do exit in the universe and once that was assumed those bodies must be represented by mathematical models that did not violate their true physical nature, as was done by Ptolemy when he assumed the existence of an equant that would force a physical sphere to move uniformly, in place, on an axis that did not pass through its center. That was physically absurd in Ibn al-Haitham's new astronomy.

In the larger cultural context, this passage also demonstrates the extent to which these cosmological debates began to influence the very foundation of science; they allowed for the new requirement of consistency to be clearly demonstrated with such vivid examples from the field of astronomy.

The timing of those remarks is also important, for they allow us to conclude that the eleventh century, which has produced so many critiques of Ptolemaic astronomy as we have already seen, seems to have been the time when new research projects were launched, and new re-organization of the sciences on new conceptual grounds must have begun to take place. The appearance of the new disciplines of *mīqāt,* and *farā'iḍ,* soon after that or very close to that time, are only few of the features that must have characterized this period. Similar results can be derived from an analysis of the developments in the mathematical and medical disciplines, and those who work in those fields may also reach similar conclusions. For astronomy, this vigorous discussion of the foundations of science seems to have given rise to long-term developments whose repercussions eventually led to truly revolutionary results. Those results, in turn, led to the final overthrow of the Greek astronomical edifice.

Returning to Ibn al-Haitham's critique of Ptolemy's *Almagest,* I will quote his conclusion at some length, not only because it draws the real demarcation lines of the new astronomy Ibn al-Haitham was calling for, but also because it demonstrates the utter dissatisfaction that was obviously felt with Greek astronomy. No one could possibly chart the contours of the new astronomy, or express the sentiments of dissatisfaction with the old, better than Ibn al-Haitham himself. In his own words:

> We must elucidate the method that was followed by Ptolemy for determining the configurations of the planets. That is, he had gathered together all the motions of the individual planets that he could verify with his own observations, or the observations of those who had preceded him. He then sought a configuration that could possibly exist for real bodies that moved with those motions, and was not able to achieve it. He then assumed an imaginary configuration with imaginary lines and circles that could move in those motions, even though only some of those motions could indeed take place in [real] bodies that moved in those motions. He was obliged to follow that route for he could not devise another.
>
> But if one were to assume an imaginary line, and made that line move in his imagination, it would not follow that there should be a corresponding line that would move in the heavens with that motion. Nor would it be true that if one imagined a circle in the heavens, and imagined a planet to move on that circle, that the [real] planet would [in fact] move along that imaginary circle. And if that were so, then the configurations that were assumed by Ptolemy for the five planets were

false configurations, and that he had established them after he knew that they were false, for he was unable to obtain others. The motions of the planets, however, have correct configurations in [real] existent bodies that Ptolemy did not come to understand nor could he achieve. For it is not admissible that a perceptible, perpetual and uniform motion be found without it having a correct configuration in [real] existent bodies. This is all we have regarding the book of the *Almagest.*[58]

With this summary condemnation of Ptolemaic astronomy, Ibn al-Haitham was obviously setting the field of Arabic astronomy on completely new footing. He could not stress any more forcefully the need for the consistency between the assumptions about the nature of the bodies that constitute the universe, and the construction of mathematical models for planetary motions that could represent those bodies without violating the very physical reality of the spheres of which the world was supposed to be made. That is the most succinct statement of the principle of consistency that was to characterize the new astronomy from that time on.

Put briefly, it should be clear that one does not accept a set of principles regarding the physical formation of the universe, and then develop mathematical models to illustrate the behavior of that universe in such terms that would contradict the very physicality of the objects that were originally accepted, or transform them so that they would no longer be recognizable. It is like assuming the world is made of a sphere and then for purposes of demonstrating how it moves one ends up representing the world with the mathematical figure of a triangle.

Similar criticisms were also directed at the Ptolemaic texts in the earlier centuries, as was documented before, and some of them had hinted to this new approach of consistency between the physical world and its presumed behavior. But at no time before Ibn al-Haitham was this new understanding of the fundamentals of new astronomy so well articulated.

The text of the *Planetary Hypotheses* did not fare much better in Ibn al-Haitham's estimation, and most certainly did not advance the new ways of thinking about astronomy. In contrast to the *Almagest,* where one could find excuses for Ptolemy and claim that he was talking about imaginary circles and lines, i.e. abstract mathematical models, and not about real physical bodies whose motions would entail the absurdities enumerated, in the case of the *Planetary Hypotheses* Ptolemy spoke of physical bodies explicitly. Thus the type of criticism advanced by Ibn al-Haitham became much more pertinent with respect to that book. In addition, since the *Planetary Hypotheses* was written after the *Almagest,* Ibn al-Haitham then took advantage of that chronology, and seized the opportunity to compare Ptolemy's thinking about the subject at two different stages of his scientific career and in two different works. He combed the second work, the *Planetary Hypotheses,* in order to determine if the absurdities of the *Almagest* had by then been resolved.

Surprisingly, he found out that the problems became much worse. Instead of resolving some of the outstanding problems of the *Almagest,* Ptolemy added some new ones in the *Planetary Hypotheses.*

Ibn al-Haitham went through both texts and produced a comparative list of spheres and motions that were described in the *Almagest,* and were now changed in the *Planetary Hypotheses.* While the configuration that was drawn for the sun remained the same in the two texts, and while the motions of the moon were nominally also the same, the motion that was described in the *Almagest* as producing the correction for the prosneusis phenomenon was not mentioned in the *Planetary Hypotheses.* In the case of Mercury only five of its motions that were mentioned in the *Almagest* were retained, and three were dropped. Similarly in the case of Venus, four motions were retained and three were dropped. The upper planets retained all the motions that were described in the *Almagest* and only the latitude motion around the small circles was dropped. But there were some more drastic changes made in the rest of the arrangement that Ptolemy had stipulated for the motion of the planets in latitude.

After going through this comparative survey in some detail, Ibn al-Haitham reached the preliminary conclusion that the configurations that were described in the *Planetary Hypotheses* were different from those described in the *Almagest,* if for no other reason except that some ten motions were no longer mentioned in the new text and the motion in latitude was overhauled. To which Ibn al-Haitham says:

> This arrangement, which was detailed in the first treatise of the *Planetary Hypotheses* is contrary to the one that was proposed in the *Almagest,* and it is also contrary to the observed latitudinal motions of the planets to the north or to the south when they were close to their epicyclic apogee. Then it becomes evident that the configuration that is described in the first treatise of the *Planetary Hypotheses* is not only contrary to observation, but that it was also contrary to what he had established in the *Almagest.*[59]

After a thorough study of the various motions that were described in the *Planetary Hypotheses,* and their causes, Ibn al-Haitham found himself quoting Ptolemy, in several passages, where Ptolemy would be caught saying that all those motions should be accounted for by real spherical bodies that were responsible for them. That left Ibn al-Haitham with one conclusion: that Ptolemy had explicitly committed himself to "finding for every motion that was mentioned in the *Almagest* a corresponding body that moved by that motion."[60]

As for obvious contradictions, even in the same book, those were definitely used as further fodder to support Ibn al-Haitham's thesis. To give just one example of the kind of issues Ibn al-Haitham emphasized, he noted that in the second treatise of the *Planetary Hypotheses* Ptolemy had said that motion by compulsion was not permissible in the celestial spheres, when he had already said in the first treatise that each one of those spheres would have a motion of its own and another one that was forced upon it.[61]

As for the new physical bodies that were introduced by Ptolemy in the *Planetary Hypotheses,* namely the slices of spheres (*manshūrāt*) instead of the full spheres that were assumed in the *Almagest,* Ibn al-Haitham thought that the *manshūrāt* were a step in the wrong direction. For those slices in turn entailed "absurd impossibilities (*muḥālāt fāḥisha*), which are of two kinds: One takes place when the body empties one space to fill another, and the second when the body had to move in different and contrary motions."[62]

In the case of the full spheres that were assumed in the *Almagest,* they at least entailed "only one kind of impossibilities, and that is the different and contrary motions, and did not entail the other, namely, the emptying of one space and filling the other."[63] The example of the spheres, which had to move in different and contrary motions, is mentioned once more in connection with the equant problem that was already faced in the *Almagest.*[64]

Ibn al-Haitham's attitude toward those spherical slices of the *Planetary Hypotheses* were echoed two centuries later, in the work of Mu'ayyad al-Dīn al-ʿUrḍī (d. 1266), who also said that, as far as those spherical slices were concerned,

> the impossibility that they would entail is even uglier (*aqbaḥ*) than that of the full spheres and more uncomely. For they would produce the same impossibilities mentioned before, like their moving non-uniformly around their own centers, and in addition they would entail orbs that were not spherical, but rather disconnected dissimilar surfaces, which is an impossibility in the natural sciences.[65]

Ibn al-Haitham had this to say about the motion in latitude, which Ptolemy had described in the *Almagest* by using a device of two small circles that would move the epicyclic radii, a feature that was dropped in the *Planetary Hypotheses:*

Then it becomes clear that Ptolemy was either in error when he disregarded the description of this configuration, or that he was wrong to establish this motion for the planets when he determined the latitudinal motion in the *Almagest.*[66]

Similarly, in the case of the inferior planets, Mercury and Venus, the small circles that were described in the *Almagest* to account for the motion of their epicycles in latitude, and which were now dropped in the *Planetary Hypotheses* had to lead to the conclusion that Ptolemy was either wrong in dropping them now, or in mentioning them in the *Almagest* in the first place. In whichever case, the treatment in the two books was contradictory, and that was one more obvious sign that the two books were read together.

Toward the end of the second treatise of the *Planetary Hypotheses,* Ptolemy seemed to lean toward the belief that it was possible to think of planets that would move by themselves, i.e. not to require a sphere that would move them. Ibn al-Haitham documented such statements very carefully only to conclude that not even the motion of rolling (*tadaḥruj*) should be permitted. For then

If Ptolemy could find it permissible that a planet could move by itself, without any body moving it, then that permissibility would make all the spherical slices as well as the spheres [themselves] invalid.[67]

In essence, Ibn al-Haitham was saying if planets could exhibit all those motions on their own, without any bodies moving them, then all of those assumptions of spheres and slices of spheres and the like would be completely superfluous. And here again ʿUrḍī adopted a similar attitude in his own critique of Ptolemy in a slightly different context:

If one were to accept such impossibilities in this discipline (*ṣināʿa*), it would have been all in vain, and one would have found it sufficient to take only one concentric sphere for each planet, thus rendering eccentric and epicyclic spheres superfluous.[68]

Ibn al-Haitham concluded his critique of Ptolemy's *Planetary Hypotheses* with the following statement:

He [meaning Ptolemy] either knew of the impossibilities that would result from the conditions that he assumed and established, or he did not know. If he had accepted them without knowing of the resulting impossibilities, then he would be incompetent in his craft, misled in his attempt to imagine it and to devise configurations for it. And he would never be accused of that. But if he had established what he established while he knew the necessary results—which may be the case befitting him—with the reason being that he was obliged to do so for he could not devise a better solution, and [yet] he went ahead and knowingly delved into these contradictions, then he would have erred twice: once by establishing these notions that produce these impossibilities, and the second time by committing an error when he knew that it was an error.

To be fair, had all this been considered, Ptolemy would have established a configuration for the planets that would have been free from all these impossibilities, and he would not have resorted to what he had established—with all the resulting grave impossibilities—nor would he have accepted that had he been able to produce something better.

The truth that leaves no room for doubt is that there are correct configurations for the movements of the planets, which exist, are consistent, and entail none of these impossibilities and contradictions. But they are different from the ones that were established by Ptolemy. And Ptolemy could not comprehend them. Nor could his imagination attain their true nature.[69]

By moving from the critique of the *Planetary Hypotheses* directly to the *Optics* of Ptolemy, Ibn al-Haitham did not only demonstrate that the Greek astronomical tradition was essentially flawed but that the other sciences, like optics, suffered from the same inconsistencies as well. This is a clear indication of the pervasiveness of this critical spirit in Islamic times, and confirms what was stated before about the social motivations for such critiques that were by no means restricted to astronomy alone. Furthermore, it also demonstrates the extent to which the Greek scientific tradition itself was taken as a whole; and as a whole was criticized from various perspectives. But the concentrated critiques of the astronomical tradition, as was amply illustrated so far, must convince us of the need to consider this stage of Islamic astronomy as a new beginning for astronomy, in the sense that the need for a new astronomy had by then been clearly demonstrated.

In order to illustrate the fecundity of our new historiographic approach, and appreciate the repercussions of such critiques of the Greek scientific tradition as the ones that were leveled by someone like Ibn al-Haitham, and against the expectations of the classical narrative that marks this period after the eleventh century as a period of steady decline, I now turn to some later critiques in order to demonstrate the longevity of that tradition, and to indicate the direction it continued to follow. Rather than preserve the Greek scientific tradition, these later critiques, to which we now turn, illustrate the pervasiveness of the attacks against it.

The astronomical developments that took place after the time of Ibn al-Haitham have special significance for another reason. Not only do they illustrate the continuity of the earlier critical tradition, but also demonstrate the kind of new questions that began to emerge, and the similarity between those questions and the ones that were raised later on during the European Renaissance.

Naṣīr al-Dīn al-Ṭūsī (d. 1274), who was mentioned before in connection with the various critiques of the Ptolemaic text of the *Almagest,* had his own doubts about the cosmological issues that have been raised so far. In his *Taḥrīr al-majisṭī* (completed in 1247), he only criticized Ptolemy sporadically. But in his later work, the *Tadhkira* (completed in 1260), he devoted much longer sections to the cosmological questions and proceeded to formulate his own mathematical models to replace those of Ptolemy. We shall have occasion to return to Ṭūsī's reformed models later. For now, and in the context of the encounter with the Greek tradition, the remarks he made in his *Taḥrīr* should give us an idea of his thoughts on the subject around the middle of the thirteenth century.

By comparing Ṭūsī's various works it becomes apparent that he began to ponder the importance of the cosmological issues for the first time when he was composing the *Taḥrīr,* a book that was devoted to the production of a useful, updated version of the *Almagest,* thus naturally offering an ideal occasion to voice his own reservations about the book he was re-editing.

In the *Taḥrīr,* and while discussing the lunar model of Ptolemy, *Almagest* [V, 2], Ṭūsī concluded that section with the following remark: "As for the possibility of a simple motion on a circumference of a circle, which is uniform around a point other than the center, it is a subtle point that should be verified."[70] Doubtless, this is the same irregularity that was mentioned before in the context of the equant problem, i.e. the absurdity arising from the situation when a sphere is forced to move uniformly, in place, around an axis that did not pass through its center.

Furthermore, in the case of the prosneusis point of the lunar model, Ṭūsī simply said: "This motion is similar to the motion of the five [planets] in the inclination and the slanting, as will be shown later on, except that that is a motion in latitude while this one is in longitude. One must look into the possibility of the existence of complete circular motions that would produce such observable motions [i.e. similar to the oscillating prosneusis motion of the epicyclic diameter.] Let that be verified."[71] One could easily see how this perplexity could have been at the origin of Ṭūsī's thinking, and which later led him to invent his famous mathematical theorem, now called the Ṭūsī Couple in the literature. The theorem itself achieved just that: an oscillatory motion produced by a combination of two circular motions.

In fact the latitudinal motion of the planets shares many features with the motions of the lunar spheres, and in particular with the slanting of the prosneusis point. That is, the oscillation of the axis, which marks the beginning of the epicyclic motion of the moon, is similar to the oscillation of the inclined planes of the lower planets. And it was this particular Ptolemaic theory of planetary latitudes that exhausted Ṭūsī's patience. He saved his sharpest criticism just for that notion. In a nutshell, Ptolemy accounted for the motion of the planetary inclined planes in latitude by suggesting that one could affix the tips of the diameters of those planes to a pair of small circles along which the tips of the diameters of the said planes would move. And as soon as he suggested those small circles he knew that he was not abiding by the accepted principles, as we hinted before, and thus felt that he had to justify his solution in the following manner: "Let no one, considering the complicated nature of our devices, judge such hypotheses to be overelaborated. For it is not appropriate to compare human [constructions] with divine, nor to form one's beliefs about such great things on the basis of very dissimilar analogies."[72] To this, Ṭūsī could only say:

> This statement is, at this point, extraneous to the art [of astronomy] (*khārij ʿan al-ṣināʿa*). For it is the duty of those who work in this art to posit circles and parts that move uniformly in such a way that all the varied observed motions would result as a combination of these regular motions. Moreover, since the diameters of the epicycles had to be carried by small circles so that they could be moved northward and southward, which also entailed that they would be moved as well from the plane of the eccentric [i.e. the deferent] so that they would no longer point to the direction of the ecliptic center, nor would they be parallel to the specific diameters in the plane of the ecliptic, but they would rather be swayed back and forth in longitude by an amount equal to their latitude, that, is contrary to reality. One could not even say that this variation is only felt in the case of the latitude, and not in the longitude, because they are equal in magnitude and equally distant from the center of the ecliptic.[73]

In the context of this very criticism of Ptolemy, Ṭūsī did not only redefine the function of the astronomer with respect to the observations and the mathematical methods with which these observations should be explained, but he went on to propose a new theorem that could resolve this specific predicament of Ptolemy. The new theorem, which was expressed in the

Taḥrīr in a preliminary fashion only to be developed further into the just-mentioned Ṭūsī Couple later on in the *Tadhkira,* will be revisited in the sequel when we return to the context of the non-Ptolemaic models that were constructed for the specific purpose of formulating alternatives to Ptolemaic astronomy.

Returning to the *shukūk* tradition, we note that about three centuries later, by the end of the fifteenth century, when the classical narrative had already preached the death of Islamic science, the problems (*shukūk/ishkālāt*) of Ptolemaic astronomy continued to attract the attention of the working astronomer. In fact those very problems became so famous, and so widespread by then, that they were taken up on their own and made into subjects of individual works, in a manner reminiscent of the specialized *Shukūk* of Rāzī and Ibn al-Haitham almost half a millennium before.

One such fifteenth-century work (of about forty folios in one manuscript) was composed by Muḥyī al-Dīn Muḥammad b. Qāsim, known as al-Akhawayn (d. ca. 1500). The title of the work is simply *al-Ishkālāt fī ʿilm al-hayʾa* (Problems in the Science of Astronomy), and seems to have been taken from the first sentence of the book which followed the usual introduction. The sentence began immediately with the enumeration of the famous problems of astronomy. By al-Akhawayn's count, those problems were reducible to seven, and they were all to be found in the received Ptolemaic astronomy.

Al-Akhawayn's treatise began thus:

> Know that the famous problems relating to the science of astronomy (*al-ishkālāt fī ʿilm al-hayʾa*) in regard to the configurations of the spheres are seven. The first is (the problem) of speeding up, slowing down, and mean motion. . . . The second (concerns the appearance) of planetary bodies being sometimes small, and sometimes large. The third (concerns) the stations, retrograde, and direct motion. . . . The fourth (concerns) uniform motion around a point different from the center of the mover. That is, when a mover moves another body in circular motion and the second body covers equal angles in equal times around a point other than the center of its mover. The fifth (concerns) a motion that is uniform around a specific point as it draws near to that point and moves away from it. The sixth (concerns) the slanting of the direction of the diameter of one sphere that is moved by another sphere from the center of that sphere (meaning the moving sphere). . . . The seventh (concerns) the lack of complete revolutions among the celestial motions as will be explained in detail.[74]

Al-Akhawayn's advantage over Ibn al-Haitham, with whose *Shukūk* al-Akhawayn's treatise could be easily compared, lied in the fact that

Al-Akhawayn could not only enumerate the famous problems of Ptolemaic astronomy, but by his time he could also offer solutions to them. Some were simple straightforward solutions already suggested in the Ptolemaic texts themselves. Others required much more ingenuity and were developed by later astronomers working in the Islamic civilization. Al-Akhawayn quoted both solutions when he could, but remained very brief, as if intending his treatise to be an introductory text for an advanced course on astronomy where the student's appetite would be only whetted by such problems and solutions and students would be urged to delve further in the more advanced texts.

As a result, the treatise managed to summarize not only the status of the problems in Ptolemaic astronomy at this relatively late date, but gave account of the many solutions that had become famous on their own. Al-Akhawayn did not offer all the known solutions to every problem, but restricted himself only to a few, well chosen ones. He only selected from the enormous corpus of solutions that had already accumulated during the few centuries before his time. Modern research has already documented those solutions in some detail. But the ones that were preferred by al-Akhawayn clearly carried the earmarks of a personal touch that is usually encountered whenever an anthology is attempted. Without going into great details, as anthologies are prone to do, al-Akhawayn simply stated, but explicitly so, that some of those problems were particular to specific planets, and that one should not expect each of those problems to be found in all the planets for which Ptolemy had suggested a mathematical model.

After a short introduction, al-Akhawayn devoted the rest of the treatise to a systematic exposition of the configurations of each planet, as given in the famous Ptolemaic astronomy, enumerated the number of problems that the specific configuration suffered from, and proceeded to give the solutions that he knew of. As such his treatise can be thought of as an interesting, simplified anthology of the kind of research that was done for almost half a millennium, and which was also focused on the shortcomings of Ptolemaic astronomy. As a result, one can simply say that by the sixteenth century there had accumulated a large corpus of critiques of, as well as alternative solutions to, almost all the major problems that plagued Ptolemaic astronomy. By the beginning of the sixteenth century, no self-respecting astronomer would have continued to uphold the long-discarded and obsolete astronomy of Ptolemy.

Nevertheless, for astronomers working at later dates, this astronomy was not completely forgotten. They continued to mention its major problems. But that should be read as a sign not of their intention to criticize Ptolemy in specific, but as an indication that the knowledge of such problems had become so widespread in the later centuries, as was stated before. By these later times, the discipline of astronomy itself, as it was reconstituted by the successive generations of critics, could no longer be pursued by seriously-minded astronomers without at least mentioning that such problems existed.

In hindsight, it appears that the so-called age of decline, after the twelfth century, could be characterized as an age during which theoretical astronomy, i.e. the pursuit of planetary theories, began to fork into two separate traditions. There were those who pursued the subject of the critiques themselves, which formed by then a well established genre of astronomical writing. And there were those who attempted to remedy the problems of Ptolemaic astronomy and who constituted a tradition of their own: the tradition of reconstructing Ptolemaic astronomy rather than just satisfying themselves with its criticism. A good representative of the former group was Ibn al-Haitham himself who offered his elaborate and scathing critique of Ptolemaic astronomy and offered no alternatives of his own. And for that failure he was severely criticized, in turn, by the later astronomer ʿUrḍī.

It was not unusual to find astronomers attempting to resolve these problems one at a time, rather than undertaking a whole reconstruction of Ptolemaic astronomy as was done by other astronomers such as ʿUrḍī and Ibn al-Shāṭir. In the fifteenth century, we find, for example, a good representative of the former group in the famous astronomer ʿAlāʾ al-Dīn al-Qushjī (d. 1474). He singled out one of the most notorious problems in Ptolemaic astronomy: the problem of the equant of Mercury, which could not be solved even by Ṭūsī, as he himself had already expressly confessed in his own *Tadhkira*. In contrast, and as a step in the right direction, Qushjī confidently expounded the problem in great detail, and immediately followed that by offering one of its most elegant solutions, all in a short treatise of few pages.[75] We will also have occasion to return to this solution in connection with the long tradition of alternatives that were proposed for the reformation of Ptolemaic astronomy.

But as far as criticism was concerned, such specific attempts at isolating individual problems for treatment eloquently express the continued dissatisfaction with at least some aspects of the Ptolemaic tradition. And as isolated problems, they should be best understood as advanced research topics quite similar to our modern practice of devoting individual articles to the treatment of particular issues in advanced journals.

Qushjī's grandson, Mīram Çelebī (d. 1524), who was an astronomer in his own right, and who was also the grandson of another distinguished fifteenth-century astronomer by the name of Qāḍīzādeh al-Rūmī (fl. 1440), left several astronomical works; some of them were direct commentaries on the more general works of his grandfather Qushjī. In one of those commentaries, he stated explicitly that he was going to devote an elaborate separate treatise to the problems of Ptolemaic astronomy, which he would call *Dhayl al-Fatḥīya* (Appendix to the *Fatḥīya*), where the *Fatḥīya* itself, his grandfather's work, did not mention any such problems. Instead, it was a rather straightforward exposition of Ptolemaic astronomy. The occasions at which Mīram mentioned the *Dhayl* were in connection with the problems of the Ptolemaic configurations for the Moon and Mercury. But until the text of the *Dhayl* is located and studied its full contents still remain unknown.[76]

The sixteenth century witnessed similar efforts by astronomers who mostly came from Persia. One of them was Ghiyāth al-Dīn Manṣūr b. Muḥammad al-Ḥusainī al-Dashtaghī al-Shīrāzī (d. 1542/3), who produced at least two works on planetary astronomy: *al-Hay'a al-manṣūrīya* (The Manṣūrī Astronomy), and *al-Lawāmiʿ wa-l-Maʿārij* (The Sparkles and the Ascensions), the second of which has not yet been identified. But in a third extant work, *al-Safīr,* he stated explicitly that he did not only criticize Ptolemy in those two earlier works, but that he even proposed new solutions for the Ptolemaic problems detailed therein, and spoke very flatteringly of the ones he produced in the *Lawāmiʿ*. Discussing the configuration for the Moon in *al-Safīr,* he said:

> The (fact that the) motion is uniform around the center of the world, rather than around its own center [meaning the center of the deferent], that is one of the problems (*ishkālāt*) in this discipline. . . . I have various other methods (for solving it), which I have explained in (the book) *al-Hay'a al-manṣūrīya,* and have also referred to (still) other marvelous methods in (the book) *al-Lawāmiʿ wa-l-maʿārij.*[77]

And while explaining the prosneusis problem in the *safīr,* he went on to say: "This prosneusis is also among the problems (*ishkālāt*). . . . The truth (concerning) it is what I have established in *al-Hay'a al-manṣūrīya,* which shines with the sparkles (*Lawāmiʿ*) of light."[78]

And in the course of discussing the equant problem in the configuration for the upper planets, he went on to say: "This too is among the problems (*ishkālāt*), which al-*Hay'a al-manṣūrīya* is capable of solving."[79]

With such explicit references, there is no doubt that this sixteenth-century astronomer was interested in pursuing the critical tradition that had already grown around the problems of Ptolemaic astronomy. But here too, unless his other two works are identified and studied in some depth, their actual contents and their real import still remain enigmatic and only speculative at this point.

Similarly, the Syrian astronomer Ghars al-Dīn Aḥmad b. Khalīl al-Ḥalabī (d. 1563) voiced similar concerns in his telling treatise that he called *Tanbīh al-nuqqād ʿalā mā fī al-hay'a al-mashhūra min al-fasād* (Warning the Critics About the Faults of the Generally Accepted Astronomy). In it he even raised an issue that has not been raised so far in this discussion, but which will be taken up in the section dealing with the relationship between astronomy and philosophy. For our current purposes, it is enough to signal that the issue itself expressed doubts regarding the permissibility of the eccentrics that were used in the Ptolemaic configurations. In that context, Ghars al-Dīn proclaimed:

> Since the generally accepted astronomy is not free from doubts (*shukūk*), especially those regarding the eccentrics, I have confronted them in this treatise, not in order to belittle the principles of this craft (i.e astronomy), but (to point to) slips where the intention did not match (the results), and to have that as a proof of what we have written (elsewhere). . . .[80]

The fourth chapter of that treatise was devoted to the problems of the lunar configuration, and the treatise itself was dated to 1551 A.D.

The same century also witnessed the most extensive, ingenious and unparalleled works of Shams al-Dīn al-Khafrī (d. 1550), which combined both the critical tradition as well as the tradition of alternative constructions to Ptolemaic astronomy. Some of those works have already been subjected to some analysis by the present author, and we shall have occasion to return to them in the section dealing with alternatives to Ptolemaic astronomy.[81]

The next century witnessed the production of the prolific scientist Bahā' al-Dīn al-ʿĀmilī (d. 1622), who did not seem to have confronted the Ptolemaic problems directly, as they do not seem to be especially mentioned in his treatise *Tashrīḥ al-aflāk*. But his commentators did not observe such reserve. Instead they composed full texts of their own, or added marginalia to ʿAmilī's text which was by then heavily read and widely distributed in schools, and thus continued to expose and indirectly popularize the faults of Ptolemaic astronomy. One of those commentators on al-ʿĀmilī's text added a marginal note in which he gave a quasi history of those faults and the people who had addressed them before. In one manuscript, the note reads:

The first among the moderns who spoke about the solution of the insoluble (problems) was al-Waḥīd al-Jurjānī, the student of al-Raʾīs Abū ʿAlī Ibn Sīnā [sic., meaning ʿAbd al-Wāḥid al-Jūzjānī]. He wrote a treatise, which he called *Tarkīb al-aflāk* (The Structure of the [Celestial] Spheres], and mentioned in it the models with which these problems (*ishkālāt*) could be solved. After him came Abū ʿAlī b. al-Haitham, then the inquirer Ṭūsī, and then the learned Shīrāzī, who collected from his contemporaries such as Muḥyī al-Dīn al-Maghribī—because the Principle of the inclined (*al-mumayyila/al-mumīla*) is copied from him—, and then the excellent master Shams al-Dīn Muḥammad b. ʿAlī b. Muḥammad al-Ḥammādī (?). You should note that the statements of Abū ʿUbayd were very weak, and nothing could be solved with Ibn al-Haitham's words, as it was already stated in the *Tadhkira* by the inquirer Ṭūsī. With the words of the inquirer (Ṭūsī) himself, as we have copied their gist, the problems of the porsneusis, Mercury's equant, and the latitudes of the cinctures (*manāṭiq*) of the epicycles and the deferents could not be solved. As for the author of the *Tuḥfa* [i.e. Quṭb al-Dīn al-Shīrāzī], he had elaborated too much. The master Muḥammad al-Munajjim al-Ḥammādī composed a treatise, in which he claimed that these problems (*ishkālāt*) could all be solved with one hundred and forty spheres. He indeed established three principles, which were, in reality, erroneous. Anyone requiring (more information about) them he should seek them in *al-Maʿārij* [part] of the *Lawāmiʿ of al-Manṣūrīya*.[82]

This quasi-historical synopsis, despite its historical shortcomings, at least reveals two important trends: First, it signaled that there were people who were themselves interested in the history of astronomy, and second that the problems of Ptolemaic astronomy continued to be discussed after the middle of the seventeenth century when this note was probably written. In addition, it also reveals that the works of Dashtaghī had by then become the standard references, at least as far as the author of this marginal note was concerned.

Historians of Arabic astronomy have not yet made any forays in the centuries that followed in order to determine the extent of criticism, if there was any, or to find out if the later astronomers continued to construct alternatives to Ptolemaic astronomy. This particular research would be of the utmost importance, especially in light of the fact that in these later centuries one would want to know how those astronomers dealt with the reception of modern post-Copernican astronomy in Islamic countries. Or whether the old Ptolemaic astronomy could still survive the onslaught of post Copernican astronomy. What little research has been done in this domain, i.e. in the domain of criticism of the natural philosophical underpinnings, reveals that during the latter part of the nineteenth century there were still those who defended Ptolemaic astronomy against its detractors, when the detractors had by then adopted the alternative Copernican and more modern astronomy.[83]

Theoretical Objections

Finally, there were objections of another kind: more theoretical in nature, in the sense that they addressed such theoretical issues that touched upon the very foundations of all scientific activities and were not only restricted to astronomy. And there were those who proposed new mathematical models to account for the same observations of Ptolemy, without explaining their motivations. But their alternative works can only mean that they were dissatisfied with the existing Ptolemaic models, and thus their activity must be perceived as an objection in itself. So when we come to survey the various alternative models that were proposed to replace the Ptolemaic ones, one could read that survey as an elaborate statement of theoretical objections to Ptolemaic astronomy as well.

Others raised further theoretical questions that could be read together with the philosophical questions that will be touched upon later on as we discuss the relationship of science to philosophy in the case of astronomy. In the present context of the encounter with the Greek tradition those questions gain a special significance as they touched upon the philosophy of science in a more focused sense. That is they tried to determine the domain in which one astronomer was justified in raising objections to the work of another. What was it exactly that one was allowed to object to, and what kind of evidence one was required to bring to the argument in order to make the case? What was the role of the observations in astronomy and what was an acceptable account of them? For this type of questioning the best representative was the Damascene astronomer Mu'ayyad al-Dīn al-ʿUrḍī, whose name was mentioned several times already. In his extensive treatise, *Kitāb al-Hay'a,*[84] which could be read in its entirety as a comprehensive statement of objections to Ptolemaic astronomy, he isolated such issues in particular when he attempted to reform, for example, the proposed Ptolemaic configurations for the planet Mercury. After enumerating the various spheres, their motions, and their relative positions with respect to one another, he went on to say:

The conditions resulting from the observations just mentioned—I mean the ones from which these conditions are known—are only the motions of the deferent's apogee and perigee. As for the directions of these motions, these were not necessitated (by the observations), rather they were simply given by Ptolemy.

Had these motions been in the manner which he had adopted, and hadn't they contradicted the principles, then he would have achieved his purpose.[85]

By questioning the relationship between the observations and the kind of results one was allowed to deduce from them, ʿUrḍī wished to direct the attention of his reader to the activity of model building itself. What part of it was dictated by the observations and what part was left to the astronomer himself to construct? And in the case of the model for the planet Mercury, ʿUrḍī had this to say:

This total (configuration) resulted from many factors: The observations, the proofs which are based on observations, the periodic motions, the configuration (*hay'a*) that he [meaning Ptolemy] had conjectured, and the directions of the (various) motions (involved). In regard to the observations, the proofs, and the periodic motions, nothing of them could be criticized, for nothing had come to light, which would contradict them.

As for the method of conjecture (*ḥads*), he (i.e. Ptolemy) has no priority in it, (especially) after his mistakes have been clearly exposed. If anyone else ever finds something, which agrees with the principles, as well as the particular motions of the planet which were found by observation, then that person would have a greater claim to the truth.

When we saw the error of this opinion, and sought to rectify it, as we did in the case of the remaining planets, we found out that we could perfect it if we reversed the directions of the two previously mentioned motions—I mean the motions of the director and the deferent orb.[86]

In very clear terms, this demonstrates the type of engagement that ʿUrḍī had sought to achieve with the Ptolemaic tradition. To some of its parts, especially the observational aspects, he had no objections to make because he had no observations of his own to bring to bear. The periods of the planets, he also had to accept as he also did not have better ancient sources to deduce his own. Ptolemy's mathematics was also superb, especially after it had been already updated by the generations of astronomers who worked on it since the ninth century. But when it came to conjecture (*ḥads*), ʿUrḍī's term for theorizing, there was no reason to prefer Ptolemy's theories over others, especially when those theories could not account for the observations and yet remain faithful to the Aristotelian cosmology that was already accepted by Ptolemy. It is not that astronomers like ʿUrḍī were blaming Ptolemy for abandoning Aristotle, and that they were so enamored by the latter to wish to re-install him through the adoption of his cosmology, but what they found objectionable was the theoretical contradiction between Ptolemy's acceptance of a set of principles on the one hand, no matter whose principles, and his contradicting those very principles when he came to describe the mathematical constructs that represented the same principles. On that level of theorization they felt that Ptolemy should have no priority over them. In fact they felt better qualified to theorize simply because they avoided the contradictions that plagued Ptolemy's astronomy. And yet their alternative models accounted for the observations just as well as Ptolemy's models could. Virtuous as he was, Ptolemy's authority could not overcome his inability to theorize properly.

At this stage, the Islamic astronomical tradition had obviously reached such maturity that it could profitably raise issues that were not raised before. It could contemplate problems, relationships, theoretical strategies, that were not dreamt by Ptolemy. The confidence in the new foundations of science gave these astronomers the ability to go beyond the criticism of Ptolemy, and to dare to oppose his models with their own, by either redeploying the same mathematics that he used, or by devising some of their own to replace it. This confidence also allowed them to look at the universe from a different perspective and to lay down new rules for the science that would eventually describe it. Such matters that were being explored in that manner touched the foundation of every science and were no longer restricted to astronomy alone. Again they should be borne in mind when we discuss the relationship of science to philosophy later on.

In the present context, and even at the expense of some overlapping with the later chapter, these theoretical issues should be highlighted here as well. The theoretical lines that were developed in response to the Greek astronomical tradition also gave rise to the debate over the admissibility of eccentrics and epicycles among the celestial spheres, a debate that was not in essence a debate over the violation of the physicality of the spheres as was discussed so far, but a discussion over whether in principle the celestial realm admitted such configurations at all. The origin of the problem was already locatable in several of the Aristotelian works, but most notably in the *De Caelo,* where Aristotle proved, with impeccable philosophical rigor, not only that the whole universe was spherical, but also that the Earth was at its center. And if one did not have an Earth there, one had to assume an Earth as the fixed point of any moving sphere, besides being the ultimate point of heaviness of the universe.[87] The argument was therefore of a necessary nature and not a merely convenient option to place the Earth at the center of the universe or not. Astronomers could debate as much as they wanted whether the observable phenomena could be explained by the assumption of a fixed Earth at the center of the universe, or by a revolving Earth around its own axis or around the sun. Some of them did raise these very possibilities in pre-Aristotelian and post Aristotelian times as well as during Islamic times, as was done by Aristarchus of Samos (c. 230 B.C.), for example, and by Bīrūnī (1048) centuries after him. They did acknowledge that the same phenomena could either be explained by a fixed Earth at the center or by a moving one. But that did not change the Aristotelian cosmological conditions one bit. According to Aristotle that "theoretical" Earth had to be motionless at the very center of the universe in the same way every moving sphere must have a motionless point at its very center.

The discussion that revolved around the admissibility of eccentrics and epicycles lied at the core of this theoretical discussion, and those who would not allow such concepts took the position that such eccentrics and epicycles would then introduce a center of heaviness, other than the Earth, around which celestial simple bodies would then move. Ptolemy tried to resolve the debate by introducing the Apollonian theorem, which allowed for the replacement of an eccentric with a simple concentric sphere carrying an epicycle. But the problem could not be resolved so easily as the epicycle itself was also found objectionable for it also introduced a center of heaviness around which the epicycle itself revolved, and worse yet the epicycle had to be placed out there in the world of the Aristotelian ether which was defined as the ultimate simple element par excellence.

Andalusian astronomers such as Ibn Bāja (Avempace d. 1138/9), Ibn Ṭufayl (d. 1185/6), Ibn Rushd (Averroes d. 1198), and al-Biṭrūjī (c. 1200), each in his own style, expressed their dissatisfaction with Ptolemaic astronomy specifically because it harbored such appalling non-Aristotelian bodies as eccentrics and epicycles. Al-Biṭrūjī went farther than all of them by undertaking to construct a complete alternative configuration that avoided these eccentrics.[88]

Al-Biṭrūjī's success was spoiled by the inability of his configuration to account for the observations in a quantitative manner that would allow for the predictability of the planetary positions for any time and any place. That fundamental test of any astronomical proposition was the main hurdle against which al-Biṭrūjī's configuration collapsed. It was only an attempt to resuscitate the old Eudoxian spheres that had at one point enchanted Aristotle himself, but could not even then predict the position of any planet at any time, despite the fact that they could give a rather naïve description of a planet's general behavior. And so was the case with Al-Biṭrūjī's construction, which also failed to account for the observable motions of the planets. For that reason alone Biṭrūjī's account remained to be a curious proposition that was not pursued any further by later astronomers. I suppose no practicing astronomer or astrologer, who needed to compute positions of planets, could take it seriously.

On more serious grounds, and for all those who wished to uphold the Aristotelian universe, at some point they had to admit that the Arsitotelian universe was not all that consistent anyway. Again we shall return to this philosophical issue later on. But in the context of this chapter, where we are focusing on the reception of the Greek scientific tradition into the Islamic civilization, let us complete the picture by indicating the range of objections the astronomers who worked in Islamic times were prepared to raise. According to Aristotle, all celestial bodies, spheres, stars and planets, were all supposed to have been made of the same Aristotelian simple element, ether. That element was supposed to be divine, thus the simplest of all elements, capable only of one motion: the circular motion that had no beginning, nor end. As a result the simple element ether did not partake of any composition or any generation or corruption, as was the case with the other sublunar elements that experienced linear contrary motions. If that Aristotelian proposition were to be taken literally, and there were some who did take it so, then one would wonder how could a sphere, say, that carried the sun, in the same fashion a ring carried a crown, emit such a bright light as the light of the sun, from only one part of it, where the sun is located, while the rest of its body acted like a crystalline transparent spherical substance that did not emit any light? This, when the sun and its carrying sphere were both assumedly made of the same element ether.

Ibn al-Shāṭir of Damascus confronted the Aristotelian universe along these very lines and with this exact understanding. In his own way though, he posed the question in the following manner: He said that since the stars, and the planets, were themselves different from the spheres that carry them, as in the case with the sun that emits light while the sphere that carries it does not, then Aristotle would have to admit that the celestial world was not all that simple and must admit of some type of composition. Now, since astronomers, Aristotelian ones included would know that some of the fixed stars were in fact considerably bigger than the largest epicycles of the planets, then if a composition is allowed for the fixed stars, the same composition must also be admitted for the much smaller epicycles as well. Ibn al-Shāṭir would then conclude that he was entitled to as much composition in the celestial spheres that would allow for the epicycles as Aristotle would allow for the fixed stars. He then went on to say, that while Aristotle and those who followed him could be right about the inadmissibility of the eccentrics, they were all wrong on the inadmissibility of the epicycles. The immediate consequence of this position led Ibn al-Shāṭir to construct very complicated mathematical models that would replace the Ptolemaic models, but at the same time they were all constructed without a single eccentric sphere. In his defense he simply changed the Aristotelian assumption to stipulate that the universe was not as simple and consistent as Aristotle had thought, but according to Ibn al-Shāṭir, that it admitted of some form of composition. In a real sense, Ibn al-Shāṭir's novel assumption was the only one I know of where an astronomer actually confronted the Aristotelian assumptions with a set of his own. This should have serious philosophical repercussions when taken in the context of the gradual collapse of the Aristotelian universe that culminated with the Newtonian final coup de grace.

In the context of the encounter with the Greek scientific tradition, and in the context of the relationship of the science of astronomy to the other sciences, a particular case should be made for mathematics. Not only because the astronomers used this discipline so profusely, nor because it was the demonstrative science par excellence, but because those same astronomers who went on to criticize Ptolemaic astronomy, and with their extensive proposals for alternative constructions, began to unravel the nature of the discipline of mathematics as well, by noticing that there were so many mathematical constructions that could explain the same observational results. The standard case in this regard was that of the Apollonius theorem which was used by Ptolemy himself to account for the same observations either by an eccentric construction or by an epicyclic one. Ptolemy was conscious of the fact that those two mathematical constructions depicted the same observational results, and opted to use the eccentric construction on account of its simplicity since it involved only one motion as he put it.

What Ptolemy did not say was that both constructions, the eccentric as well as the epicyclic, violated the Aristotelian cosmology. For in the first case, the eccentric assumed a fixed center of heaviness other than the Earth, which was inadmissible, and in the second case of the epicycle it assumed a center of heaviness out in the celestial realm as we just described.

Later astronomers who had no vested interest to defend the Aristotelian universe one way or another were at times ambivalent about those constructions and went along with the Ptolemaic choice of the eccentric constructions. As we have just said, only Ibn al-Shāṭir objected to the eccentrics and avoided using them in his reformulation of astronomy.

But what was also left unsaid was the validity of the discipline of mathematics itself in relationship to astronomical theory. How was one to assess which mathematical construction was to be preferred and which one was not, especially when both constructions could explain the observational data just as the Apollonius theorem could? We have already seen how ʿUrḍī, for example, already succumbed to the Ptolemaic use of mathematics, and did not raise any doubts in that regard. Only when he had to reformulate Ptolemy's mathematical model for the upper planets, he felt obliged to introduce a mathematical theorem that was not found in the Greek texts, and used that theorem only to account for the observations in a much better model than that of Ptolemy. But he went no further than that.

Not until later did astronomers stop to think about the connection between mathematics and astronomy, and as we have just said, did so only when they began to notice that there were many mathematical constructions that could lead to the same results, that is, account for the observations equally well. By the sixteenth century, the astronomer Shams al-Dīn al-Khafrī (d. 1550), who was already mentioned, employed this very same new understanding of mathematics to its fullest, where in his own description of the new models that he and others had developed he would supply several models for the same planetary motions. That is, he would give several mathematical alternatives to interpret the same observations in exactly the same fashion. In the case of the motions of the planet Mercury, for example, he gave in one of his works four different mathematical models all yielding exactly the same mathematical results, and thus all accounting for the observations in the same fashion. And in his own words he offered these models one after the other simply as different ways (which he called *wujūh*) of looking at the same physical reality. This new understanding, that mathematics was only a language that allowed the astronomer to describe the same physical reality in so many different ways, is nowhere better exhibited than in the works of Khafrī.[89]

Conclusion

This brief overview should have made it very clear that the Greek astronomical tradition, especially that which was represented by the most important texts of that tradition, the Ptolemaic texts, was not simply preserved in the Islamic culture, as is so often asserted, but that it had received a very critical assessment from the very beginning. From correcting the perceived mistakes in the Greek texts by the translators themselves, to the critical reevaluation of the observational results that led to changing the most fundamental astronomical parameters of that tradition, to raising objections against that tradition for its detected disregard of its own natural philosophical premises that were firmly grounded in the Aristotelian tradition, to the theoretical objections against that tradition for its lack of systematic consistency, and finally to the theoretical objections that were raised in regard to the actual foundations of astronomy itself, how the science itself was structured and which components of it were subservient to which other components, and which of the other sciences were deployed in it and in what capacity, all of those aspects of the Greek astronomical tradition were subjects of great dispute.

First, the most important aspect of the ensuing debate was that it was carried out with the most classical Greek authors, an observation that confirms our earlier assumptions about the lack of scientific sophistication of the contemporary Byzantine and Sasanian cultures. None of the critiques that we have signaled so far were of current Byzantine or Sasanian doctrines, rather they were directed against Ptolemy, Galen, Aristotle and the like. The most important feature of this encounter with the Greek tradition is that it was a confrontation with the classical authors, and thus in a round about way the confrontation itself brought those classical ideas back into currency, at the same time as they were being refuted and modified. It was not a polemic against contemporary Byzantine authors, which affirms once more that there was no such advanced civilization in Byzantium to come in contact with, as we have already repeated again and again.

Second, this confrontation took place in the context of very complex social forces that were at odds with each other, for political and social reasons, and were only secondarily directed against the very science itself. As we have already seen, the science and the philosophy that were being brought into the Islamic civilization were directly connected to the social position of the persons who were bringing them, usually political and economic positions, and in a sense their final chances of acceptance or rejection in the target Islamic civilization were conditioned by the success or failure of the groups that sponsored those activities.

Third, the Greek scientific and philosophical sources were being sought for reasons connected to the on-going debate that was taking place within Islamic civilization, a debate that was generated by the reforms of ʿAbd al-Malik, as I have maintained all along, and were not sources that were encountered by chance through innocent contacts between two civilizations. This conscious and willful selection of texts to be translated, which Sabra and before him Lemerle would want to call appropriation, because they served a specific purpose in the debate, also colored the manner in which those texts were accepted or rejected in the acquiring civilization. They were not ill-directed chance encounters that could bring their own momentum from the outside. Thus the Greek texts that were translated into Arabic simply enforced certain pre-selected directions and did not create their own directions, except in very tangential ways when they began to generate philosophical schools of their own in later centuries.

Fourth, because the texts that were being sought during the eighth and ninth centuries were already written some 700 years earlier, and in some cases even more, their scientific contents were already obsolete in the sense that their mistakes were already exaggerated with the passage of time. For example, Ptolemy's small mistakes, which resulted from his comparison of his own observations with those of Hipparchus who observed some two centuries earlier, were now quite exaggerated after the passage of some seven centuries before they were re-examined again in ninth-century Baghdad. Only from that perspective we can understand why it was easy to note the differences between the ninth-century results, such as precession, position of solar apogee and the like, and the results that were determined by Ptolemy some seven centuries earlier.

Fifth, these results, whether they were directly acquired from the Greek sources, or were modified by the fresh observations, were always used within the ongoing struggle between the proponents of the "ancient sciences," whose claim to power depended directly on those new results, and the proponents of the more Islamic classical orientation whose claim to power depended on their knowledge of the Arabic language. And because this main competition between those two groups also generated another competition with fellow scientists who were also trying to prove their relevance to political authority which employed them in the final analysis, then every scientist who was engaged in acquiring Greek texts had to worry about the two sets of opponents who were looking over his shoulder: the fellow scientists who wanted to claim greater authority to the texts that they had acquired, and thus compete for the same government jobs, and the opponents whose authority rested on their knowledge of the Arabic language that was already affiliated with the religious sciences where it was desperately needed. As we have already stated this phenomenon itself can explain why someone like al-Ḥajjāj b. Maṭar had to make sure that his translation was written in the best Arabic, in order to compete against those who possessed the Arabic language, and its contents had to be corrected from the scientific point of view so that his work would be better than the work of the other translators who at times simply translated the text mistakes and all. This may also explain why Ḥajjāj's translation was not the first, and that it was already an improvement over an older translation, as we are told by al-Nadīm. Furthermore, it had less transliterated words from Greek than the translation that was completed some fifty years later by Isḥāq b. Ḥunain a clear indication that this linguistic competition had already faded by Isḥāq's time and was then transformed into another competition based on ethnic and religious affiliation, which was highlighted by the *Shuʿūbīya* movement.

Ghazālī's later attack on the essentialism of the Greek causal philosophy could also be read as a continuation of the debate against the Aristotelian essentialism that required a fixed Earth at the center of the universe, and a strict adherence to the Aristotelian cosmological universe that was not always followed by Ptolemy. In other words, those within Islamic civilization who saw in the works of Aristotle a strict essentialist philosophy could not tolerate the divergences of Ptolemy and thus initiated a whole series of attacks against his deviations. But it was those same persons with that same strict Aristotelian essentialist interpretation who were perceived by the religious camp represented by Ghazālī as going too far in their essentialism on issues of causality for example. Thus in an ironic turn of events one can say that the objections that were raised by astronomers working in the Islamic domain against Ptolemy's astronomy were motivated by Aristotelian purist astronomers who were at the same time fighting their own battle with religious people who wanted to understand Aristotle in a much more relaxed sense, almost in the same relaxed manner in which Ptolemy understood Aristotle.

Moreover, the debate that expressed itself in the *Shukūk* literature, of which we have seen several examples, can be perceived as one feature of a much larger phenomenon that included religious attacks against Greek astrology, observational mistakes, factual errors in medicine, etc., where such disciplines also developed under the double watchful eyes of enemies from without, who competed over the sources of authority and who had the right to claim the possession of that source, and enemies from within who competed over who was the better scientist who could qualify for the government job.

It is within this complex environment that new disciplines such as *Hay'a, Mīqāt* and *Farā'iḍ,* came into being in order to satisfy the outside competition with the religiously inclined opponents, but at the same time to carve purely independent disciplines that could compete against the traditional ones, which were being promoted by fellow scientists from within so to speak. In that environment a science like *Hay'a* became at once a religiously acceptable science, and at the same time a more rigorous science that carried the brunt of the attacks against Greek astronomy in order to prove its rigor and its good religious standing.

Within the same environment we can better understand this insistence on scientific rigor as the motivation behind the constant emphasis by *Hay'a* authors on the inner consistency of science mentioned above, an emphasis that characterized the long history of the *Hay'a* tradition. *Hay'a* authors were obviously trying to keep this double edge advantage over their fellow astronomers, such as authors of *zījes* for example, by remaining more stringent in their scientific consistency requirements and by remaining religiously acceptable to the society at large. In that regard they scored a tremendous success as their discipline continued to be taught till recent times, and sometimes well within the religious educational institutions themselves.

Once we can see the double motivation to attack the Greek tradition, we expect this phenomenon to have had similar effects in other fields as well. And when we consider the field of medicine for example, we arrive at very similar results. We already had a chance to refer to the text of Abū Bakr al-Rāzī in which he objected to Galen's theories but also went ahead and composed his own rigorous scientific book on the difference between smallpox and measles, a difference that was apparently unknown to Galen, despite Rāzī's protestations.

This same critical spirit was also exhibited in the work of ʿAbd al-Laṭīf al-Baghdādī (d. 1231)[90] who visited Egypt toward the beginning of the thirteenth century and who also found himself on a collision course with Galen at certain medical points. In his account of his trip to Egypt, he mentioned that he had found certain anatomical points very difficult to explain to his students, and for them to understand those points, because in theory the written word was always less evident than the observation, or that "observation is always much stronger than words" as he put it.[91] This should not be surprising since dissection was not commonly practiced in premodern times. But Baghdādī went on to say that he profited from a recent plague that had befallen Egypt, and visited with his students the piles of skeletons, which were still lying on the outskirts of Cairo. During their investigation, Baghdādī related that he noted the jawbone of those skeletons, and that he found it to have been a single bone rather than two, as Galen had asserted. He went on to relate that he repeated the observation several times, in many skeletons, and that he always found it to be one bone. He then asked several other people who had also observed it in his presence and on their own and they all agreed that it was one bone. He then promised to write a treatise in which he would describe the differences between what he saw and what he read in the books of Galen. But he continued to say that he kept on investigating this issue in graveyards of various ages in order to see if a seam or a split would be observed in that bone with age. And as much as he wished to save the Galenic text, he found none.

In another instance, the same Baghdādī also reported that his original investigation was contrary to the teachings of Galen, but that later repeated observations confirmed the Galenic texts.

The works of Ibn al-Nafīs of Damascus (d. 1288), already mentioned before, fall in the same category, in that they exhibit this tendency to try to save the Greek texts from their own folly, so to speak, but having to object to them when there was better evidence of their error. Ibn al-Nafīs's discovery of the smaller pulmonary circulation of the blood comes from the same tradition as that of Rāzī and Baghdādī, and represents the empowerment that the scientists of the Islamic domain must have felt once they started noticing the exposure of a whole sequence of mistakes in the classical Greek scientific texts, and once they started believing in what they saw with their own eyes.

Other disciplines witnessed similar transformations in that they managed to cleanse the mistakes of the Greek tradition, whenever possible, but also went beyond to forge their own new terrain that the Greek authors did not know about. In particular the discipline of mathematics seems to have received a very interesting boost toward the sixteenth century when its relationship to astronomy was finally correctly understood at the hand of someone like Khafrī (d. 1550) who could finally see that mathematics was just a tool that could be used to describe physical phenomena, and that it did not retain the *Truth* itself.

The only astronomical criticism that was not touched upon in any detail in this overview was the criticism that was implicit in the various attempts of generations of astronomers who sought to reform Ptolemaic astronomy by constructing new mathematical models that could render the reality of observations, and the theoretical natural philosophical foundations in a much more coherent and consistent fashion. These will be explored in the chapter, which will survey the non-Ptolemaic models as was already promised.

The Islamic civilization did not seem to have produced a rigorous astronomical criticism of the type that would have questioned the natural philosophical foundations of Greek astronomy themselves. Although some religiously inspired cosmologies did in fact speak to that point, yet there were no astronomers that I know of who adopted these views or sought to interpret the astronomical implications of such cosmologies. The final rejection of Aristotelian cosmology had to come late in the history of astronomy, and only after a long and arduous struggle that was initiated by modern science under conditions that were completely different from those that prevailed in the Islamic civilization.

4 Islamic Astronomy Defines Itself: The Critical Innovations

Now that we have seen the kind of reactions the encounter with Greek science has produced in Islamic civilization, we can better appreciate the context for the astronomical developments that took place, as we continue to use astronomy as a template for the other disciplines that must have experienced similar transformations. In astronomy, the reactions expressed, at all levels, ranged from simple corrections of what was thought to be a mistake in the text, as was done by al-Ḥajjāj in the case of the *Almagest,* to correcting the basic parameters by fresh observations, as in the case of redetermining the better values of precession and the inclination of the ecliptic among others, to critiquing the methods of observation, as was done in the case of the *fuṣūl* method, and finally to casting doubt on the reliability of the very foundations of the Greek astronomical tradition itself when it seemed to violate the principles upon which it was based in the first place.

All these developments, when coupled with the very watchful eyes of the competing groups we spoke about earlier, from inside the profession as well as from outside, generated a skeptic attitude toward the incoming tradition. In itself this attitude emboldened astronomers to raise deeper and deeper questions as they continued to examine this Greek tradition in light of their own research. In this environment, it becomes easy to understand why good competent astronomers could not continue to practice astronomy by simply taking the Greek astronomical tradition at its face value. They had to compete by proving that they could achieve better results than those that were achieved by the Greeks and which were being continuously criticized at the time.

This did not mean that the Greek astronomical sources were yielding such dramatically erroneous mistakes that they could no longer be used to answer simple mundane questions like casting a horoscope or the like. But it did mean that the professional astronomer, from early Abbāsid times on, could no longer survive the competition if he limited himself to such simple questions in the first place. The serious astronomers had to answer more complex questions regarding the suitability of the proposed Ptolemaic astronomical configurations in accounting for the observations on the one hand, and in embodying the prevailing cosmological system of Aristotle on the other. For them, it was no longer sufficient to find the positions of the planets at any time for purposes of casting a horoscope or some such things, but they had to know how the planets moved, what caused their motion, why do they appear to go through all sorts of irregular behavior, and how does one account for that, all within the assumption of a universe made up of spheres all moving in place at uniform speeds as Aristotle had stipulated. At this degree of seriousness, Ptolemaic astronomy was found to be desperately wanting.

With the work of Muḥammad b. Mūsā b. Shākir, during the first half of the ninth century, regarding the properties and the admissibility of the existence of the ninth sphere, the stage was set for undertaking a total overhaul of the entire Greek astronomical edifice. When it was found later on, as we have already seen with the critiques of Ibn al-Haitham, that the physical foundations of the Ptolemaic configurations did not match the mathematical representations that were offered by Ptolemy, the motivation for the overall reform of that astronomy became a matter of necessity rather than choice. Only practicing astrologers could satisfy themselves, if they so pleased, with the use of the Ptolemaic *Handy Tables,* for example, to calculate the planetary positions that they needed for their horoscope casting. But those astrologers themselves fell under the censoring eye of the society at large, despite their ability to continue to function, and still try to make themselves useful to that society. Even then, they too had to require better and better astronomical tables (*zījes*) for their craft, as the old parameters of the Ptolemaic tables were continuously corrected as time went on. Socially though, no self-respecting astronomer would admittedly want to be cast in an astrologer's garb, if he could help it. This despite the fact that some of them did. While the best of them would want to associate themselves with the critical tradition that was beginning to pick up steam from the earliest decades of the ninth century. The latter had to cast a new name for the discipline they practiced (the discipline of *ʿilm al-hay'a*), because they did not wish to be associated with the lesser figure of the practicing astrologer.[1]

It was this environment that motivated the research of the new Islamic astronomy. Its main mission, as was enunciated later by Mu'ayyad al-Dīn al-ʿUrḍī (d. 1266) of Damascus,[2] one of the most distinguished astronomers of that tradition, was to create an astronomy that did not suffer from the cosmological shortcomings of Ptolemaic astronomy, that could account for the observations just as well as Ptolemaic astronomy could do if not better, and that did not limit itself to criticizing Ptolemy only, despite all the benefits that one derived from the detailed critique of Ptolemy's mistakes. This urgent need for a higher form of scientific astronomy was almost felt by all serious astronomers whose works we have come to know only recently, and who formed a continuous tradition inaugurated toward the beginnings of the ninth century and continued well into the sixteenth century as far as we can now tell. One astronomer after another would take very seriously Ibn al-Haitham's declaration which stated that there must be an astronomical theory, or in his language astronomical configuration (*hay'a*), that could account for the observations conducted in the real physical world without having to represent that world with a set of imaginary lines and circles as was done by Ptolemy.[3] One could hear them all repeat: If the world was real, made up of real spheres, as was argued by Aristotle, then let it be represented by mathematical models that did not contradict that physical reality.

On the more mundane level, when it was a matter of double-checking the observations that would account for the behavior of the real physical world, or that would help establish the very observations that were to be used as the building blocks of the theoretical representation, those had to be taken seriously as well. That is why one can document several attempts to double-check the values of the basic parameters, as we have already seen, or to initiate a whole discussion about the optimal methods of observation as we have also seen, or to initiate whole new fields of refining observational instruments or inventing whole new ones when there was a need for that. These activities continued to take place as the astronomical tradition continued to grow. The studies of Khujandī's surviving works on larger, and thus more precise, instruments, or those of ʿUrḍī regarding the construction of the Marāgha observatory instruments, among others, speak exactly to such concepts.[4] But what was really wrong with Ptolemaic astronomy that generated all those discussions?

The Problems with Ptolemaic Astronomy[5]

It is possible to say that Ptolemy saw the astronomical universe in four different ways: It was either a universe completely composed of Aristotelian spheres that could be described with the same language that was used in the *Planetary Hypotheses,* or a complementary world that was formed of those same spheres and represented by more precise predictive mathematical models as was done in the *Almagest,* or that it was a world that was already determined and its behavior tabulated as was done in the *Handy Tables,* or a world that was constantly at the mercy of the revolving celestial spheres that governed the world of change in the sublunar region in which we live, as was done in the *Tetrabiblos.*

For astronomers working in the Islamic civilization, the universe of the *Handy Tables* did not present much challenge, as it was a matter of fixing the mistakes of those *Handy Tables* by fresh observations whenever it was a matter of determining the positions of planets for any time and place. New parameters could do that, as was in fact done by generations of *zīj* writers who simply continued to update the *Handy Tables.* At times they added to them newer concepts that were not known in the Greek tradition that required tables of their own, such as tables for the visibility of the moon, or tables for prayer times, or *qibla* directions, etc., that were necessitated by the new religion of Islam and would not even occur to someone like Ptolemy. In such cases the newly established Islamic requirement of finding the best time and location for lunar visibility owed its inspiration to a religious practice rather than a scientific curiosity or astronomical need. And it is in such instances that religious thought would give rise to scientific thought and science could become a handmaiden of religion, as we shall see below.

The second Ptolemaic description of the world, that which was reflected in the *Tetrabiblos,* was quickly found to be way too general for use by the practicing astrologers. For although the *Tetrabiblos* gave a fairly sophisticated analysis of the manner in which the Aristotelian spheres and planets exerted their influence on the sublunar region, it did not offer detailed instructions on how to translate that theoretical analysis into practical horoscopes that could answer particular questions at specific times. For that reason more specific books had to be developed in order to make up for those shortcomings. Bīrūnī's book on the *Elements of Astrology* is a masterpiece in that regard,[6] as are the books of the various astrologers who attempted a more direct approach to the subject like the *Introduction to Astrology* of Abū Maʿshar.[7]

But the Ptolemaic books that caused the greatest amount of problems for the astronomers of the Islamic civilization were the *Planetary Hypotheses* and the *Almagest*. For although those two books were complementary to one another, yet they were mutually exclusive when it came to accounting for the Aristotelian cosmology in a more systematic fashion. From that perspective the *Planetary Hypotheses* spoke directly to a system of physical spheres, more or less in closer agreement with the Aristotelian spheres, while the *Almagest* spoke of circles representing spheres, and thus only implicitly acknowledged the Aristotelian spheres. Yet, both books spoke of physical impossibilities such as equants and the like. It was those impossibilities and absurdities that contradicted the Aristotelian cosmology that were found most objectionable.

It is not that Astronomers working in the Islamic civilization were enamored by Aristotelian cosmology and wanted to save it at any cost.[8] Rather it was that they saw in those two books clear indications of the Aristotelian assumptions about the composition of the universe and its constituent parts, and yet could not see the descriptive representations of that universe, as was done in the *Almagest,* really doing justice to the science of astronomy itself. When people read those two books, and they obviously read them together, as we have already mentioned at various occasions before, or when they read carefully the underlying assumptions as expressed in the *Almagest* too seriously, what they saw was a field that had accepted a set of Aristotelian cosmological premises, but went ahead and spoke about those premises in a language that contradicted their very essence. For instance, they saw Ptolemy speak about Aristotelian spheres as the constituent elements of the universe, and then turned around and represented those spheres with mathematical spheres whose properties would deprive them of their very sphericity. It was these kinds of fundamental contradictions that were thought of as detracting from the scientific basis of astronomy, and under no condition could serious astronomers accept those contradictions.

In what follows I only highlight the main features of these absurdities, and follow that with a description of innovative approaches that were taken by the astronomers of the Islamic civilization in order to emend, whenever possible, or create alternatives to the imported Ptolemaic astronomy.[9]

The Motion of the Sun

In the case of the sun, Ptolemy noted that if the observer were really located at the center of the Aristotelian universe, as the Aristotelian cosmology would require, then we would have routinely equal days the whole year round, we would have no seasons, and the sun would repeat its path around us day after day. But the observed reality was not like that. To account for that reality, Ptolemy first determined the basic parameters, like the length of a solar year, and then went ahead and proposed one of two solutions for the actual motion of the sun. He stipulated that the sun was either carried by an eccentric sphere, whose center did not coincide with the center of the Earth, as Aristotle would have wanted to insist, or that it was carried by another much smaller sphere—called epicycle—which itself was in turn carried by another sphere that was concentric with the Earth (figure 4.1).

In Book III of the *Almagest,* Ptolemy made sure first that both alternatives could still account for the observations well enough, and quickly resorted to the previous work of Apollonius (c. 200 B.C.) which in fact proved that both of these descriptions of motion could be represented by configurations that were mathematically equivalent in every respect. One did not have to chose, therefore, between them, if the purpose was only that of accounting for the observations.[10] In his own words "the mathematician's task and goal ought to be to show all the heavenly phenomena being reproduced by *uniform circular motions,*[11] and that the tabular form most appropriate and suited to this task is one which separates the individual uniform motions from the non-uniform [anomalistic] motion which [only] seems to take place, and is [in fact] due to the circular models; the apparent places of the bodies are then displayed by the combination of these two motions into one."[12]

Despite the fact that the actual motion of the sun could be represented in tabular form, both in terms of mean motion as well as in anomalisitic one, the problem still resided in the type of motions of the spheres that accounted for the observed motions. Obviously, Ptolemy's insistence on uniform motion here is undeniable. For he could not at the same time adhere to Aristotelian cosmology and yet allow any of the spheres to move at a varying speed as it pleased. And in case one forgets, in Book III, 3 of the *Almagest,* he made this uniform motion the general guiding principle of his astronomy in the following terms: "But first we must make the general point that the rearward displacements of the planets with respect to the heavens

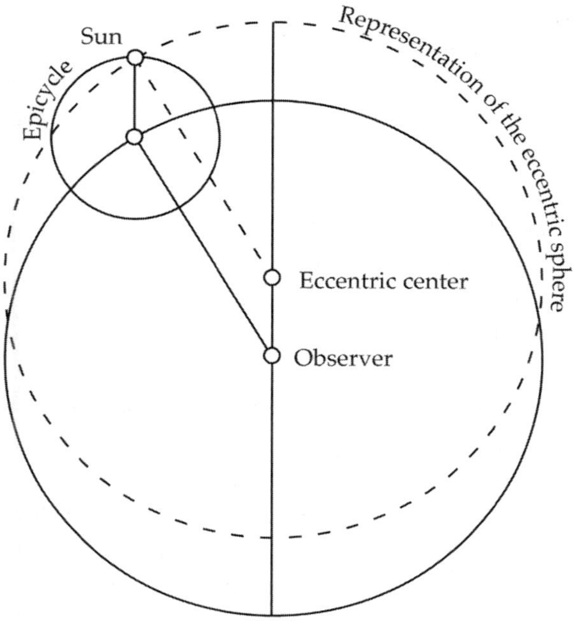

Figure 4.1
The equivalence of the eccentric and epicyclic models for the sun.

are, in every case, just like the motion of the universe in advance, by nature uniform and circular."[13]

If Ptolemy was talking, as he seemed to be doing so clearly, about Aristotelian spheres that moved uniformly in place, then both alternatives that he proposed for the motion of the sun suffered from other Aristotelian considerations as we have seen before: First, the eccentric model of Ptolemy, would propose that there is a center of heaviness in the universe around which the most obvious luminary, the sun, would move, which was different from the Earth that was taken by Aristotle to be the center of heaviness par excellence. That is contrary to all the Aristotelian assumptions about the composition of the universe, and about the need to have an Earth at the center of the universe, not only as a center of heaviness to which all other elements would "naturally" gravitate or recede from, but as the fixed center of the sphere of the universe which is as essential to it as any fixed center is to a rotating mathematical sphere.

The second alternative, the epicyclic model, also assumed the existence of a sphere, out in the celestial realm, which had its own relatively fixed center of heaviness, different from that of the universe, and thus would make the element ether, of which all the celestial bodies are made, a complex element rather than the simple element Aristotle defined it to be.

These are the obvious contradictions that gave rise to the medieval discussions about eccentrics and epicycles, and about their undesirability in general, which we have already mentioned. And as we have already seen, they irked Ibn al-Shāṭir (d. 1375) of Damascus enough so that he would try to resolve them, as we shall repeat below. As for Ptolemy, and despite his insistence on the Aristotelian feature of uniform motion, he remained absolutely silent on these other Aristotelian considerations. In fact, he proceeded as if nothing was wrong and went ahead to assess the merits of each of the two models with respect to the criteria of simplicity. From that perspective he judged the eccentric model to be simpler on account of the fact that it involved one motion instead of two.[14] As far as he was concerned this instrumental reason was enough to allow him this temporary lapse of memory that there were other Aristotelian conditions to be met.

In fact, the situation became even worse as he proceeded. For although he could offer two options for the case of the solar motions, either an eccentric or an epicyclic one, when it came to the other planets he knew quite well that he would no longer have these options. He would have to use both eccentrics as well as epicycles in order to account for their more complex motions. Without any reference to Aristotelian cosmology, or any recollection that it was his guiding cosmology from the beginning, he went on to say: ". . . for bodies which exhibit a double anomaly both of the above hypotheses [meaning the eccentric and the epicyclic] may be combined, as we shall prove in our discussion of such bodies. . . ."[15]

Although he had his own pedagogical reasons to do so, Ptolemy moved on to discuss the motion of the moon before discussing the motion of the other planets. It may be worth mentioning at this point that Ptolemy grouped the planets together in terms of the predictive mathematical model that he devised for their motions, of course with total disregard for Aristotelian cosmology as we just saw. As a result, he treated the motions of the sun, the moon, and mercury, separately and with separate models for each, and then grouped the other "upper" planets, Saturn, Jupiter, Mars, and Venus together and described their motion with one model. Furthermore, in the final arrangement of his presentation of the models, he also had to abandon the principle of simplicity and presented the models in the following order: sun, moon, upper planets, and mercury.

For the purpose of illustrating the underlying Aristotelian cosmological problems that these models entailed, I shall readopt the principle of simplicity and proceed to expose the problems with the model for the upper planets next, before I pass on to the models of the moon and mercury.

The Motion of the Planets

The motions of the upper planets Saturn, Jupiter, Mars, and Venus, described in Book IX of the *Almagest* and grouped together in IX, 6, can be briefly summarized in the following manner (figure 4.2): Each of the upper planets was supposed to be carried by an epicycle, attached to it in the same fashion a crown is attached to and carried by the ring, to use medieval descriptions. The epicycle was itself carried within the shell of and by an

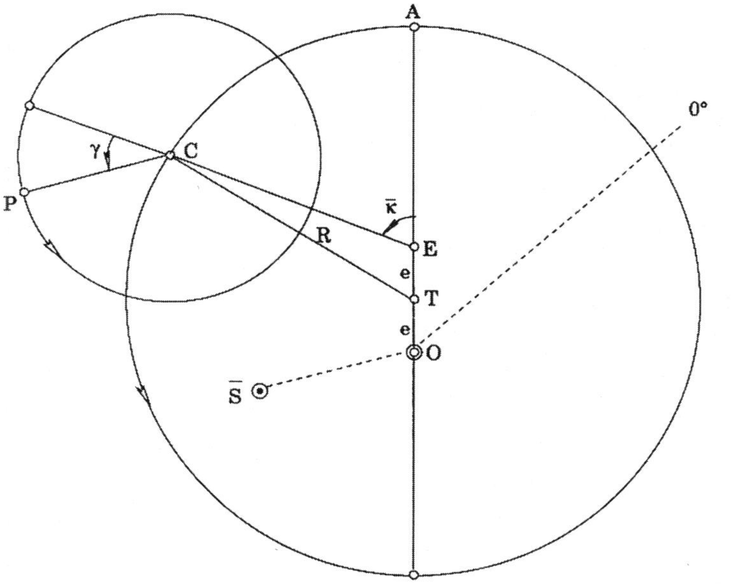

Figure 4.2
Ptolemy's model for the upper planets. The observer is at point *O*. The planet *P* is carried on an epicycle with center *C*, which is in turn carried by the deferent with center *T*. Note that the deferent rotates uniformly around the equant *E* and not around its own center *T*.

eccentric sphere called the deferent, here represented by a simple circle with center *T*. Each of those spheres moved uniformly in order to account for the anomalistic and mean motions respectively. But in order to account for the observations properly Ptolemy had to assume that the carrying deferent sphere did not move uniformly around its center *T*, nor around the Earth which was at the center *O* of the universe still, but around another point *E*, called later the *equant* point, or the center of the equalizer of motion when described by Ptolemy as a sphere. Without any proof of any sort, Ptolemy went on to stipulate that the equant point was located away from the deferent's center, by the same distance as the center of the deferent itself was removed from the center of the universe, and on the opposite side. That is the eccentricity *OT* was equal to *TE*.[16] With this arrangement, the deferent's uniform motion around its *equant* accounted for the mean motion of the planet, and the epcicyle's motion around its own center accounted for the anomalistic motion, and thus the phenomena were sufficiently saved.

But from the Aristotelian perspective, this new predictive model for the positions of the planets did not only violate the Aristotelian presuppositions once with the adoption of the eccentric sphere, but twice with the adoption of the epicyclic one as well. It even went beyond that to violate, in a very serious fashion, the very mathematical property of a sphere. With Ptolemy's new assumption that there could be a physical sphere that could move uniformly, in place, around an axis that did not pass through its center, it became very clear that the same assumption would require us to abandon completely the very concept of the mathematical sphere and its defining properties. The only axis around which a physical sphere could move uniformly in place was the one that had to pass through the fixed center of the sphere; otherwise it could not stay in place.

Even if Ptolemy could have satisfied the Aristotelian conditions by avoiding the eccentric sphere and accounted for it by another epicycle, as he did with the case of the sun, and even if he had allowed himself the license to use epicycles, arguably against the Aristotelian conception of the simplicity of the element ether, as was done later by Ibn al-Shāṭir of Damascus (1375), for example, still his requirement that any physical sphere could move uniformly around an axis that did not pass through its center would make the existence of such a sphere physically impossible. And as was clearly stated by Ibn al-Haitham later on, and already quoted before, we do not live in an imaginary world where such spheres may only exist in the mind, but in a very real one whose motions had to be accounted for.

On the positive side, Ptolemy's configuration, physically absurd as it was, could still account rather well for the longitudinal motions of the planets. It could explain the daily progression of the planets from west to east, i.e. contrary to the direction of the primary daily motion of the heavens, and in the direction of the ascending order of the zodiacal signs Aries, Taurus, Gemini, etc., as a result of the planet's own mean motion. The planet's particular motion, said anomalistic motion, took place uniformly on its own epicycle, and could account for the forward and backward motions of the planet as well as account for the stations in between. These positive conditions, when coupled with the ability of the model to predict the positions of the planets rather accurately, for its time, could satisfy the astrological predictions and the like.

On the negative side, the sheer absurdity of the equant concept, in the realm of physical spheres, turned this model into a point of contention to be taken up by every serious astronomer up to and including Copericus.[17]

The Motion of the Moon

In the case of the moon, Ptolemy's model became more complicated, and even more absurd than the two previously discussed models of the sun and the planets. In order to account for the observable motion of the moon, with its variations in the position of eclipses, the apparent motion of the moon on its epicycle without undergoing retrograde motion, and the variation in the size of the epicycle as it appears to the observer on Earth, he could not possibly account for all those variables with a relatively simple model as that of the sun or the planets. Instead he introduced the "sphere of the nodes" (figure 4.3) as an engulfing sphere for the moon and made it concentric with the Earth. He made this sphere responsible for the motion of another sphere inside it that he called the deferent, which was in turn eccentric with respect to the Earth. He made the sphere of the nodes move from east to west, while the engulfed deferent in the opposite direction. The moon was finally moved directly by an epicycle which was carried within the shell of the deferent but which moved in the direction opposite to that of the deferent.

But in order to create a variation in the size of the epicycle, especially when the moon was away from the mean sun by 90°, Ptolemy made the

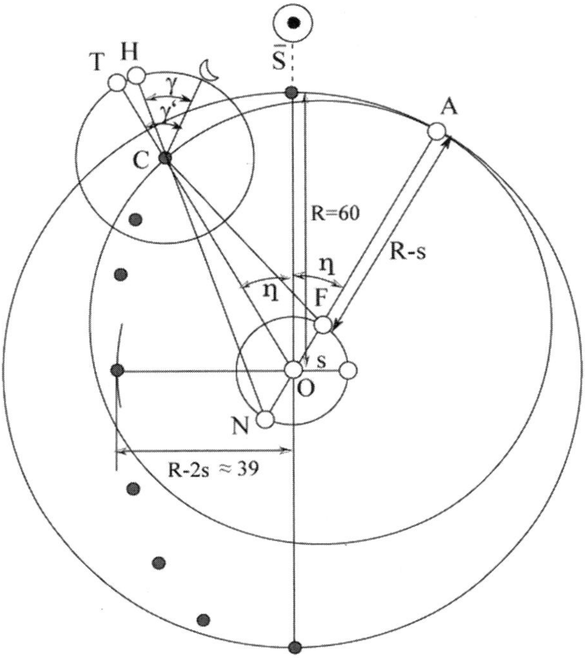

Figure 4.3
Ptolemy's model for the moon where the engulfing sphere that moves the nodes as well as everything contained in it is centered on the Earth at *O*. The deferent sphere, with center *F*, is moved by the engulfing sphere from east to west so that its apogee moves to point *A*. The deferent that carries the epicycle within its shell, here represented by the circle of the deferent itself, moves in the opposite direction to bring the center of the epicycle to point *C*. Note that the uniform motion of the deferent is measured around the center of the universe *O*, which means that it does not move uniformly around its own center *F*, thus producing an equant-like concept of its own. Now the epicycle moves by its own anomalistic motion in the same direction as the engulfing sphere. But its anomalistic motion is measured from the extension of the line that connects the center of the epicycle *C* to the ever-moving prosneusis point *N*. Furthermore, note that the distance of the epicyclic center from the Earth, when the epicycle is 90° away from the mean sun is almost half the distance it would have when the epicyclic center is in conjunction or opposition with the sun. That means that a quarter moon should look almost twice as big as a full moon, which is obviously untrue.

deferent move uniformly around the center of the Earth rather than around its own center, thus creating again an equant problem similar to that encountered with the upper planets. Furthermore, and in order to account for the second anomaly of the moon, he stipulated that the moon's own motion on its epicycle should be measured from a line that started at a point, *N* in the diagram, which was located diametrically opposite from the deferent center, *F* in the diagram, with respect to the Earth, and at the same distance from the Earth as the deferent center, but in the opposite direction. The point was called the *prosneusis* point (*nuqṭat al-muḥādhāt*) and the line joining it to the center of the epicycle, when extended to the circumference of the epicycle, constituted the starting point for the true anomaly of the moon. One should note here that this prosneusis point *N* was itself in constant motion as it was always defined by its diametrically opposite location from the moving center of the deferent *F*. That is, it was simply symmetrical to the deferent center with respect to the center of the universe. Thus the line that joined the prosneusis point to the center of the epicycle oscillated back and forth from the line that joined the center of the epicycle to the fixed center of the universe and never completed a full revolution. This oscillating motion was found objectionable as well, since there should not be any non-circular motions in the heavens according to Aristotle, and thus all revolutions should be completed.

For very instrumental reasons, this crank-like model, however, accommodated the observations rather well, as far as the longitude of the moon was concerned. But it failed miserably when it came to the apparent size of the moon. If taken seriously, and with its crank-like operation, the model required the moon to be pulled very close to the Earth when it was at 90° away from the mean sun. Accordingly, the moon's distance from the Earth at that point would become almost half the distance it had when it was at full moon or in conjunction with the sun. This would mean that for an observer, situated on the Earth, when the moon was quarter moon, it would then have to appear twice as big as when it was a full moon. This predictive aspect of the model was obviously untrue, and was fittingly described later on by Ibn al-Shāṭir as untenable since the moon was never seen as such (*lam yurā kadhālika*).

Furthermore, since the position of the *prosneusis* point itself was determined by the position of the symmetrically opposite position of the deferent center on the other side from the Earth, and since that center was itself moved around the Earth by the engulfing sphere of the nodes, that meant that both the deferent center as well as the prosneusis point, that depended on it, were in constant motion. This also meant that the line that joined the prosneusis point to the center of the epicycle's center was no longer fixed as well as we just said, and thus in ʿUrḍī's terms, was not fit to be considered the beginning line for the measurement of the anomalistic motion, since it would not constitute a fixed starting point.

With a deferent that moved around its own center, but measured its uniform motion around another center, now the center of the universe, thus repeating the same *equant* problem, and with the introduction of the moving *prosneusis* point that introduced an oscillating line that never completed a full revolution, and with the huge increase in the apparent size of the moon at quadrature, all *implied* by Ptolemy's model, one can see why this model attracted a large critical literature within Islamic civilization. Its own problem was often referred to as the *prosneusis* problem, in analogy to the *equant* problem that was used to describe the difficulties with the model for the upper planets. Several astronomers working in Islamic civilization tried to rectify the situation by creating their own models, some of which were more successful than others. In that tradition, Ibn al-Shāṭir's lunar model was by far the best, not only because it did away with the *equant* construction when it made all spheres move uniformly in place around axis that passed through their own centers, and reduced the variation in the apparent size of the moon, while keeping the increase of the size of the epicycle, but because it also turned out to be identical to the same model which was proposed by Copernicus (d. 1543) himself about 200 years later.[18]

In his usual fashion, Ptolemy said nothing about the difficulties of his model, and did not even draw attention to the fact that his model directly contradicted the real apparent size ever so blatantly. But things were moving from worse to worst, as the next model of Mercury and the models for the latitudinal motions of the planets were worst still.

Motion of the Planet Mercury

Because of the high speed of this planet, and because of its proximity to the sun, and thus the difficulty in observing it in a reliable fashion, Ptolemy's model for the motions of this planet reflected the faulty conditions of the observations.[19] As in the case of the moon, where Ptolemy's crank-like model predicted that the moon would come closest to the Earth twice during its monthly revolution (when the moon reached 90° or 270° from the mean sun, or when the moon was at the first or third quarter of its revolution), so was the case with Mercury, which was supposed to come closest to the Earth at two points during its revolution: when Mercury was 120° away from the apogee, on either side of the apsidal line. This meant that Ptolemy's model for Mercury would mimic some of the features of the lunar model.

In *Almagest* IX, Ptolemy proposed a model for the planet Mercury that had an engulfing eccentric sphere called the director (centered at *B* in figure 4.4), which in turn carried another eccentric sphere called the deferent (here centered at *G*). Needless to say, both eccentrics were in direct violation of the Aristotelian presuppositions. The director moved around its own center, in the direction opposite to the succession of the signs, i.e. from east to west, and carried with it the deferent in the same direction. The deferent, however, moved in place, inside the director, by its own motion but in the opposite direction, thus producing a crank-like mechanism that was similar to that which was employed in the lunar model. And like the lunar deferent, this one too did not move uniformly around its own center *G,* but around a center *E,* also called the *equant* as in the model of the upper planets, but placed, again without any proof, half way between the center of the Earth and the center of the director, instead of being on the far side as was the case with the upper planets. The epicycle, which carried the planet Mercury with its own anomalistic motion, moved in the same direction as that of the deferent, and was itself carried by the motion of the deferent in the direction of the succession of the signs.

Thus, in addition to two eccentrics (which one may have thought that Ptolemy could explain away in the same way he used the Apollonius theorem to explain the solar eccentric away) and one epicycle (unavoidable on account of the second anomaly), there was the same additional absurdity which had appeared twice before: the absurdity of having a sphere move uniformly, in place, around an axis that did not pass through its center. And as in the case of the model for the upper planets, there was the additional unproved statement of Ptolemy that the equant laid half way between the center of the world and the center of the director. One can see why such accumulated technical considerations would make Ptolemaic astronomy subject to the kind of severe criticism that was leveled against it once it came into Islamic civilization.

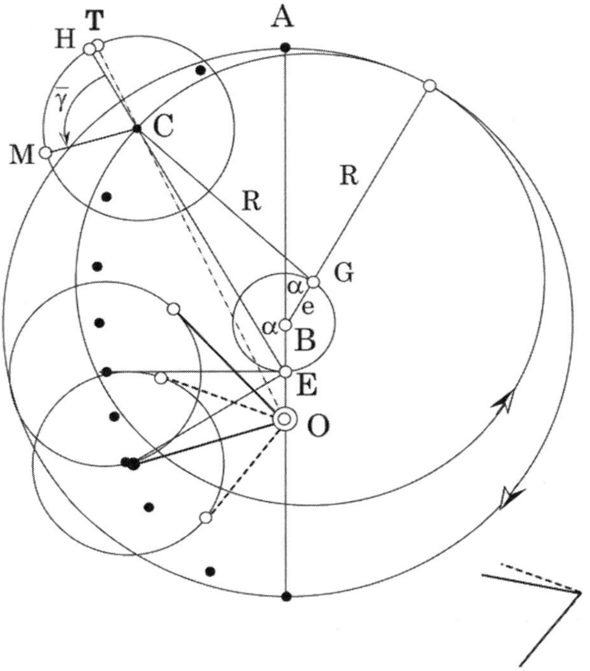

Figure 4.4
Ptolemy's model for Mercury. The observer is at point *O*, the center of the universe. The planet *M* is carried by the epicycle with center *C*, which is itself moved by the deferent with center *G*. The deferent, which is also moved by an engulfing sphere called the director and whose center is *B*, moves in the same direction as the epicycle, but measures its equal motion around the equant *E* rather than its own center *G*. The equant *E* is halfway between the center of the universe *O* and the center of the director *B*. For the observer at point *O*, Mercury's epicycle will *appear* at its largest when it is closest to Earth, at ±120° away from the apogee *A*, and not at quadrature, when it is only 90° away from the same apogee, as was thought by Copernicus. The two elongations are represented here by angles drawn with dotted and continuous lines.

Planetary Motion in Latitude

To make matters worse, the Ptolemaic models for the latitudinal motion of the planets further introduced some absurdities of their own. In this instance, and for purposes of computing the latitudinal component of the planetary positions, Ptolemy made a distinction between two groups of planets: He grouped Saturn, Jupiter, and Mars in one group and described

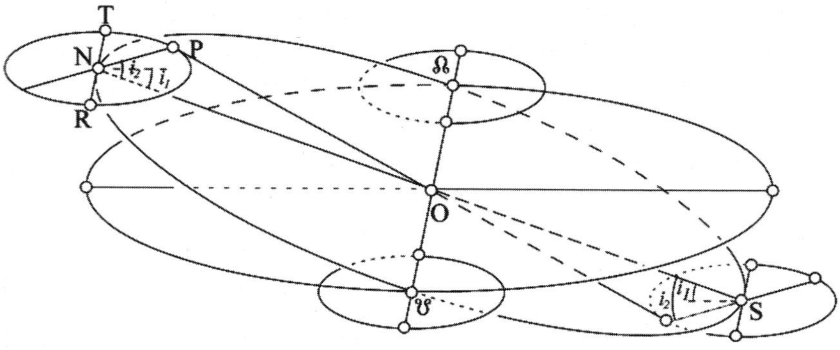

Figure 4.5
Ptolemy's model for the latitude of the upper planets. The observer is at point *O*, and the inclined deferent plane has a fixed inclination. The epicycle however had its own deviation from the deferent plane, whose value depends on the position of the epicycle along the deferent.

their latitudinal motion with one model, and grouped Venus and Mercury in another group that was the subject of a different, and quite offending, model. It should be stressed at this point, that in terms of longitudinal motions the models described by Ptolemy still yielded quite reasonable predictive results despite their physical absurdities. Those results were at least convincing enough to allow Ptolemy to make his pragmatic claim that he must have been following some correct conjecturing in configuring them out although he did not know rigorously enough how they worked.

For the upper planets (figure 4.5), Ptolemy proposed a model that included an observer at the center of the world *O*, that is the center of the ecliptic. He then proposed an inclined, eccentric deferent plane that intersected the plane of the ecliptic at a fixed angle. The line of the intersection between the two planes passed through the position of the observer and marked the two nodes. The epicycle that was carried by the inclined plane had its own *deviation* from that plane as well, but this deviation varied depending on the longitudinal position of the epicycle. At the northernmost end of the deferent the epicycle would have its maximum *deviation,* but as soon as the epicycle reached the position of the nodes it would lie flat in the plane of the ecliptic. At the southern end of the deferent it will have the same phenomenon of maximum *deviation,* but in the opposite direction. And although both *deviations* had the same value, the southern one simply looked bigger since it was closer to the observer.

In effect, then, the plane of the epicycle seemed to perform a seesawing motion of its own, as the position of the epicycle changed by the motion of the deferent. Since this kind of motion never completed a full circle, it was obviously deemed to be in the same category as the oscillating prosneusis of the moon. And thus it was as far as it could be from the uniform circular motion, which Aristotle would have required, since only full circular motions were the natural motions of the simple element ether of which all the celestial bodies were composed.

No word from Ptolemy in this regard. And although he had claimed that he was adhering to the Aristotelian cosmology, he still behaved in the same fashion as he did before when such violations were committed. That is, he still made no effort to explain them away as one would have expected. Instead, he invited the reader to imagine that the tips of the epicyclic diameters could be attached to pairs of "small circles." Those circles could be placed perpendicular to the plane of the deferent. And as the tips of the diameters moved along those "small circles" the resulting oscillating motions would produce the seesawing effects that were needed. He then had a problem synchronizing those motions of the tips of the diameters along their "small circles" and the motions of the epicycles themselves along their deferents, since the deferents themselves were eccentric, as we have already seen. To solve the problem, he resorted once more to the assumption that the diameter tips too had their own form of *equants* just like those of the other larger models of the planets, since they did not seem to be partaking of a uniform circular motion around their own centers.

Now, even if one could accept the motion of the diameter tips for the purposes of producing the seesawing effect, which was in turn "justifiably" required by the observations, one could easily see that any such "unnatural" seesawing would also create a wobbling effect that would interfere with the longitudinal component for which much pain was suffered when it was computed in the first place.

It was this specific feature of the latitudinal motion in the Ptolemaic model that led the thirteenth-century astronomer Naṣīr al-Dīn al-Ṭūsī (d. 1274) to proclaim, in his *Taḥrīr al-majisṭī* in 1247, that Ptolemy's speech was indeed intolerable, or, in his own polite words, beyond what was permitted in the craft (*khārij ʿan al-ṣnāʿa*).[20]

The latitudinal motion of the second group, the lower planets Mercury and Venus, did not fare any better in this regard. For them the inclination of the carrying plane itself varied as the epicycle moved along its circumference. One should always remember that those planes were supposed to be the equators of physical spheres, the only bodies capable of generating such motions in an Aristotelian universe. In the case of Venus, when the epicycle was at the northernmost end of the inclined plane that plane itself tilted northwards and the epicycle on its northern edge tilted away from the carrying plane along the eastern edge. But as the epicycle moved to the nodes one of its diameters coincided with the ecliptic, while the other still tilted along the eastern edge. When the epicycle reached the southernmost end, the whole inclined plane tilted in the opposite direction so that its southern end now pointed to the north, and the epicycle still inclined away from it along the eastern edge again. In effect then both the inclined plane as well as the plane of the epicycle itself would undergo the same kind of seesawing motion that was noticed in the model of the upper planets. And here again, the only solution Ptolemy had to offer was to propose the same kind of "small circles" to be attached to the seesawing diameters so that they could be forced to perform the latitude motion. And here again that arrangement would still force the whole plane to wobble and destroy the longitudinal component as before. We just saw what Ṭūsī would say of such an arrangement.

In sum, Ptolemy's models for the movements of the moon and the five planets introduced notions that were not only in violation of the Aristotelian presuppositions, but as we have seen, with the case of the motions in latitude of the planets, included arrangements that also destroyed the longitudinal components that worked rather well on their own. It looked like Ptolemy could not compute any component of the motion without destroying the other. In total exasperation, he ended up confessing that only gods were capable of such perfection, not the mere humans.[21] With this realization, the whole Ptolemaic configuration seems to fall apart, despite the fact that, on the computational level, it seems to have been able to predict the positions of the planets, and can still do, with a rather remarkable accuracy.

The reforms of this astronomy that were to take place in Islamic civilization after the thirteenth century went to great pain to retain that predictive value of Ptolemaic astronomy. But they definitely aimed to reform the conceptual arrangements of the spheres that were supposed to carry out these various motions. Toward the end of the twelfth century, when Averroes had to give his assessment of the Ptolemaic astronomy, an astronomy that was still the norm in his time, he had the following to say: "The science of astronomy of our time contains nothing existent (*lays minh*^{u} *shay*$^{\text{'}un}$ *maujūd*), rather the astronomy of our time conforms only to computation, and not to existence (*lā li-l-wujūd*)."[22]

Islamic Responses to Ptolemaic Astronomy: Creating an Alternative Astronomy

We have already seen several levels of responses to what was perceived as factual mistakes in the Ptolemaic tradition. Whether it was in simple mistakes in texts, or basic parameters, or even methods of observations, those were attended to and began to be fixed as early as the ninth century. New genres of writings addressing specifically the totality of those Ptolemaic problems, called *shukūk, istidrāk,* and the like were developed and sophisticated with time, so much so that they became subjects of discussion on their own by people who were not even astronomers by profession.

Serious attention to the philosophical and physical underpinnings of the Ptolemaic edifice, otherwise signaled as model building, and serious attempts to replace the inadequate Ptolemaic models did not begin until the eleventh century. But once it began, almost every serious astronomer felt that he had to take part in the enterprise. In the sequel I will only signal those who made fundamental shifts in the way astronomy was practiced to the neglect of others who kept the discipline alive by supplying the commentaries and the individual modifications that they saw the major shifts required, or simply made use of those shifts to overhaul the then current astronomy in order to incorporate those changes.[23] For example, when the trigonometric functions were introduced into the Islamic scientific tradition, and were perfected after being originally derived from the few functions already known in India, the tendency was to use those functions in any theoretical discussion of astronomy instead of the chord functions that were used in the *Almagest* and its translations.

It is these kinds of shifts that produced the astronomy that could then be called Arabic/Islamic astronomy, and whose example we hope the other disciplines had followed. In what follows, however, I will pay a special attention to the most subtle shifts that played, in my judgment, a catalytic role in producing other astronomical innovations, and became part of the universal legacy of astronomy. As was already said, I will neglect those who periodically took in those conceptual shifts and integrated them in their works without producing any shifts of their own.

The focus will then be on those astronomers who felt that they needed to invent new concepts, or more concretely new mathematical theorems, in order to solve the problems of Ptolemaic astronomy, and not that much on those who followed them by incorporating the latest theorems to build upon them in order to create new planetary models of their own. Of those who introduced such new theorems, the names of Mu'ayyad al-Dīn al-'Urḍī (d. 1266) and Naṣīr al-Dīn al-Ṭūsī (d. 1274) stand out on account of the fact that each of them supplied his own mathematical theorem while undertaking to overhaul the fundamental features of Ptolemaic astronomy.

The Work of 'Urḍī

'Urḍī thought that Ptolemy's models for the sun were adequate enough, and that Ptolemy's choice of the simple eccentric model was innocuous enough, that it did not deserve any special transformation. Neither 'Urḍī nor Ptolemy ventured to state explicitly what he really thought of the epicyclic model, posited by Ptolemy at least as an alternative to the "offensive" eccentric model. One suspects that the ubiquitous use of epicycles in all other planetary models, and the impossibility of their replacement, may have made their use a necessity that could not be avoided. But no one was willing to defend the use of those epicycles explicitly. Their final theoretical solution would not come until a century later with the works of Ibn al-Shāṭir (d. 1375) as we shall soon see.

As for the motions of the moon and Mercury, and the notorious *equants* in both models, in addition to the *prosneusis* point in the case of the moon, 'Urḍī felt that he could not let things be. Instead he decided to take advantage of the similarities between the two models, and tried to reconfigure them by adopting three new steps. First he decided to shift the directions of the motions of the various spheres. Then he adjusted the magnitudes of those motions. And finally he tried, in a global way, to avoid the Ptolemaic handicaps that plagued both models by making all spheres move uniformly on axis that passed through their centers. At this point he still restricted himself to the mathematics that was available to Ptolemy from Euclid's *Elements,* for example, without having to offer any new mathematical propositions of his own. There were times though when he would venture to say that he faulted Ptolemy for his inability to theorize (*ḥads*) properly, but would still express his full admiration for Ptolemy's observational and mathematical control of the data. Such shifts in theorizing, as long as they did not involve the introduction of new material like the trigonometric functions, were quietly introduced nevertheless without much fuss.

But when it came to the model of the upper planets, ʿUrḍī felt that the Ptolemaic model was no longer redeemable, and thus had to be reconfigured in a fundamental way. It was there that he proposed to introduce what has now become known in the literature as ʿUrḍī's Lemma in order to resolve the very thorny issue of the *equant* problem. At this point, ʿUrḍī's concern was no longer focused on the cosmological choices of eccentrics versus epicyclic models, but was focused on the more fundamental *equant* stipulation which forced the very sphere that was supposed to carry out the motion of the epicycle to loose its sphericity. This physical impossibility could not be tolerated, and still pretend to carry out astronomical theorizing, as these astronomers saw their functions to be. Instead, ʿUrḍī approached the problem of the model of the upper planets with the mathematically rigorous manner it deserved.

After demonstrating the physical failings of Ptolemy's model, he went on a tangent and said that in order to theorize better about the motions of those planets, he needed to introduce a new theorem, the statement of which could be rephrased thus: Given any two equal lines that form equal angles with a base line, either internally or externally, the line joining the extremities of those two lines would be parallel to the base line.[24]

Taken on its own, ʿUrḍī's Lemma looked like a generalization of Apollonius's theorem, in that the equal angles needed for the proof of the parallelism of the end line with the base line are no longer restricted to the exterior angles used in the construction of the epicyclic model. Instead ʿUrḍī could show that the internal equal angles would produce the same effect of parallelism and thus could be used to rectify the instance of the *equant* without losing its observational value that had forced Ptolemy to adopt it in the first place.

Instead of assuming (figure 4.6) that the epicycle is carried by a deferent that moved uniformly around an axis that did not pass through its center, as was done by Ptolemy, ʿUrḍī shifted the center of his new deferent to a point *K*, which was located halfway between the center of the old Ptolemaic deferent *T* and the equant point *D*. He then allowed this new deferent to

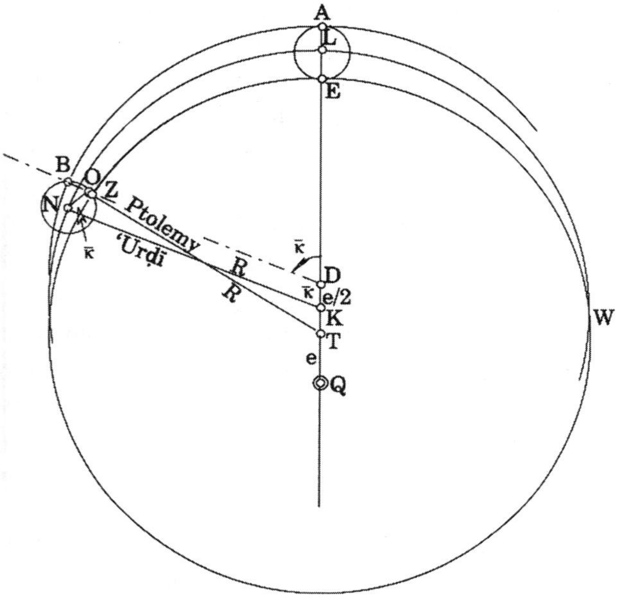

Figure 4.6

ʿUrḍī's model for the upper planets. By defining a new deferent with a center at *K*, halfway between the center of the Ptolemaic deferent *T* and the equant *D*, ʿUrḍī allowed that deferent to carry a small epicycle whose radius was equal to *TK* = *KD*. He made the small epicycle move at the same speed as the new deferent, and in the same direction. By applying his own lemma, ʿUrḍī could demonstrate that line *ZD*, which joined the tip of the radius of the small epicycle to the equant, would always be parallel to line *KN*, which joined the center of the new deferent to the center of the small epicycle. He could also show that point *Z*, the tip of the radius of the small epicycle, came so close to the point *O*, which was the center of the Ptolemaic epicycle, that the two points could not be distinguished. Then it was easy to see that the uniform motion of *O* that Ptolemy thought took place around point *D* was indeed a uniform motion around point *N* which in turn moved uniformly around *K*, thus making *Z* appear to be moving uniformly around *D* and satisfying the Ptolemaic observations.

carry a small epicycle whose radius was equal to half the Ptolemaic eccentricity, or equal to the same magnitude by which the center of the deferent was shifted in the first place. The small epicycle moved at the same speed as the old Ptolemaic deferent, and in the same direction, and in turn carried the Ptolemaic epicycle. The combination of the equal motions allowed the lines joining the extremities of the small epicycle's radius to points *K* and *D* respectively to be always parallel. This made the center of the Ptolemaic epicycle, now carried at the extremity of ʿUrḍī's small epicycle, look like it was moving uniformly around the Ptolemaic *equant*. In fact it moved around the center of its own small epicycle, and the center of that epicycle, in turn, moved around the center of the new deferent. With all spheres now moving uniformly, in place, around axis that passed through their centers, ʿUrḍī managed to avoid the absurdity of the Ptolemaic *equant* altogether, and, at the same time, still retain its observational value, as was required by the Ptolemaic observations.

ʿUrḍī's Lemma, introduced through the mechanism of the small circle in the model for the upper planets, proved to be a very useful tool for other astronomers and at other occasions as well. A whole host of astronomers ended up using it in order to construct their own alternative models to those of Ptolemy. Astronomers such as Quṭb al-Dīn al-Shīrāzī (d. 1311) used it in his lunar model. While Ibn al-Shāṭir of Damascus (d. 1375) ended up using it in more than one of his own models. ʿAlāʾ al-Dīn al-Qushjī (d. 1474) used it in his model for the planet Mercury, and Shams al-Dīn al-Khafrī (d. 1550) made a double use of it in his own model for the upper planets. Finally, Copernicus (d. 1543) used it for the same model of the upper planets. As it turned out, this mathematical tool became very fecund in the construction of all sorts of responses to Greek astronomy.

From that perspective, this relatively simple lemma proved to be an epoch maker, just like the Ṭūsī Couple, which will be mentioned later. And like the Ṭūsī Couple too, once it was discovered, it allowed several generations of astronomers to think differently about Ptolemaic astronomy, and about the possibilities with which this astronomy could be reformed.

As far as ʿUrḍī was concerned, it turned out that with one new theorem, and with small adjustments to the directions and magnitudes of motions, he could reconfigure the whole body of Ptolemaic astronomy, and still produce his own configuration that was free of the absurdities of the *equant* and the like. In that regard he ended up playing a pivotal role in the development of Arabic astronomy, a role comparable only to that of Naṣīr al-Dīn al-Ṭūsī (d. 1274) as we shall soon see. His work on the planetary latitudes, however, like the work of all the other medieval astronomers, up to and including Copernicus, it could not resolve the major issues with Ptolemaic astronomy as elegantly as it could solve the longitudinal component.

Naṣīr al-Dīn al-Ṭūsī

ʿUrḍī's former boss at the Marāgha observatory came to his own solution of the *equant* problem in a slightly different fashion. To him, the problem was not fundamentally a problem of an epicycle that moved uniformly around an equant point, thus creating the physical absurdity, but was more a problem of a uniform motion that was observed from varying distances thus appearing to be non-uniform. One way of thinking about it was to allow the center of the epicycle in the model for the upper planets to move uniformly while at the same time still allow it to draw close to the point of Ptolemy's *equant* when close to the apogee, and move away while at perigee. This motion would in effect duplicate the phenomenon that Ptolemy said was exhibited by the observations. Therefore one could solve the problem if he/she could devise a way in which a body moving in uniform circular motion could still be allowed to come close to a specific point and draw away from it while at the same time retain the uniform circular motion undisturbed. The net effect would be that the body would be perceived to move at varying speed in an oscillating motion with respect to that point when, in fact, it would in itself continue to move in uniform circular motion. The problem was to achieve an oscillating motion in the realm of spheres that were all supposed to move uniformly around their own centers.

The idea of an oscillating motion resulting from circular motion seems to have occurred to Ṭūsī when he was tackling the problem of the Ptolemaic latitude theory. This was apparently the same circumstance under which Copernicus reached the same connection between the two phenomena.[25] Later on in the *Commentariolus,* and while describing the motions of the planet Mercury, Copernicus goes further by clearly making reference to the second connection between the motions of Mercury and the motions in latitude. At that point he does in fact describe the same Ṭūsī Couple that he used for his own Mercury Model as being related to the motions that he had already described in the latitude theory.[26] This very connection between the genesis of the Ṭūsī Couple and the motion in the latitude theory first came about when Ṭūsī had already noted, some three centuries before Copernicus, in his *Taḥrīr* that the oscillating motions described by Ptolemy in the latitude theory could be accounted for by a combination of two circular motions. Once construed as such, the net effect of the motion of the Ṭūsī Couple could in addition account for the Ptolemaic statement regarding the inclined planes of the lower planets Mercury and Venus, which were supposed to seesaw in order to produce the latitudinal motion of these planets. The elegance and superiority of this solution of the oscillating motion, through a deployment of a Ṭūsī Couple, becomes very clear when we remember Ptolemy's alternative suggestion of having the tips of the diameter of the inclined plane be attached to two "small circles" so that he could achieve that seesawing motion—a motion that would, at the same time, destroy the longitudinal motion on account of the resulting wobbling necessitated by the "small circles." It was in that context that Ṭūsī felt that Ptolemy's speech was outside the craft of astronomy.

Instead Ṭūsī suggested that one could achieve a better seesawing effect, without having to accept the necessary result of wobbling. And in order to do that, Ṭūsī then produced a rudimentary construction of two small circles of his own, which were fitted in such a way that one of them rode on the circumference of the other, and had the tip of the diameter of the inclined plane attached to the circumference of the second circle as well. When the motions of the two circles were supposed to be in such a way that the one riding on the circumference moved at twice the speed as the other one and in the opposite direction, then the point at the very tip of the circumference of the riding circle, i.e. the tip of the diameter of the inclined plane, would end up oscillating along the joint diameter of the two circles as a result of their uniform circular motions. This produced at once an oscillating motion from two combined uniform circular motions, and allowed the tip of the circumference of the riding circle to oscillate along a straight line, thus keeping it from wobbling. The combined effect of Ṭūsī's two circles successfully produced a straight motion by combining two circular motions, a result that was to have tremendous effects on later astronomers.

About 13 years after he wrote the *Taḥrīr* (that is, around 1260 or 1261), Ṭūsī developed the idea further in his *al-Tadhkira fī al-Hay'a* (Memoir on Astronomy), and produced it in the form of a theorem, that is now called the Ṭūsī Couple (figure 4.7). He did reach the same conclusion a few years

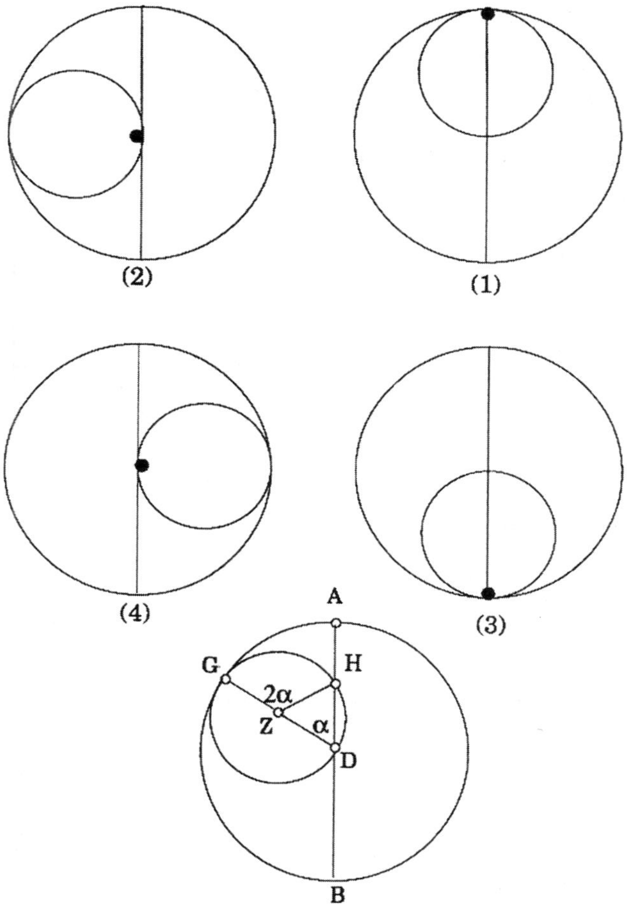

Figure 4.7

The Ṭūsī Couple. If two spheres such as *AGB* and *GHD* touched internally at point *A*, and if *AGB*'s diameter was twice as large as that of *GHD*, and if the larger sphere moved uniformly in the direction indicated, and the smaller sphere moved in the opposite direction, at twice the speed, then point *A* would oscillate up and down the diameter of the larger sphere *AB*.

earlier when he included the same theorem in his Persian text *Dhayl-i Muʿīnīya,* whose date of composition is still uncertain but has to be sometime between the publication of the *Taḥrīr* in 1247 and the *Tadhkira* in 1260/61, where the theorem is fully stated and proved.

The theorem itself spoke of two spheres, instead of circles, one of them twice the size of the other, and placed in such a way that the smaller sphere was internally tangent to the bigger sphere as in figure 4.7 (1). Then Ṭūsī went on to prove that when the larger sphere moved uniformly at any speed, while the smaller one also moved uniformly, but in the opposite direction, at twice the speed, then the common point of tangency would end up oscillating along the diameter of the larger sphere.

Once he had generalized that theorem, he knew he had found a fecund theorem that could be used whenever one needed to produce linear motion as a result of combined circular motions. He then went on to produce, in the same *Tadhkira,* a formal proof for the theorem, as in figure 4.7, and later applied it to construct two of his alternative models: the lunar model and the model for the upper planets. In this fashion he then managed to solve the *equant* problem in the two respective Ptolemaic models for those planets.

The success of this theorem had widespread repercussions. It ended up being used by almost every serious astronomer that followed Ṭūsī, including the Renaissance astronomers such as Copernicus and his contemporaries, as we have already hinted and shall see again in more detail later on. In contrast to Copernicus however, the only place where Ṭūsī failed to apply his Couple, was in the case of the planet Mercury, whose behavior was quite challenging for Ṭūsī as we have already seen. When discussing that planet's motions in particular, Ṭūsī unambiguously declared that although he succeeded in solving the *equant* problem of the models of the moon and the upper planets, he was hoping to complete his task later on by solving the *equant* problem of Mercury, to which he had no new things to add at the time.

Ṭūsī's student and colleague Quṭb al-Dīn al-Shīrāzī (d. 1311) made use of ʿUrḍī's Lemma twice, once in developing his lunar model, and the other time when he adopted the same model for the upper planets as that of ʿUrḍī. In the lunar model (figure 4.8), he avoided the use of the Ptolemaic equant by bisecting the eccentricity of Ptolemy's deferent for the moon, and adjusting for it by positing a small circle at the circumference that satisfied the

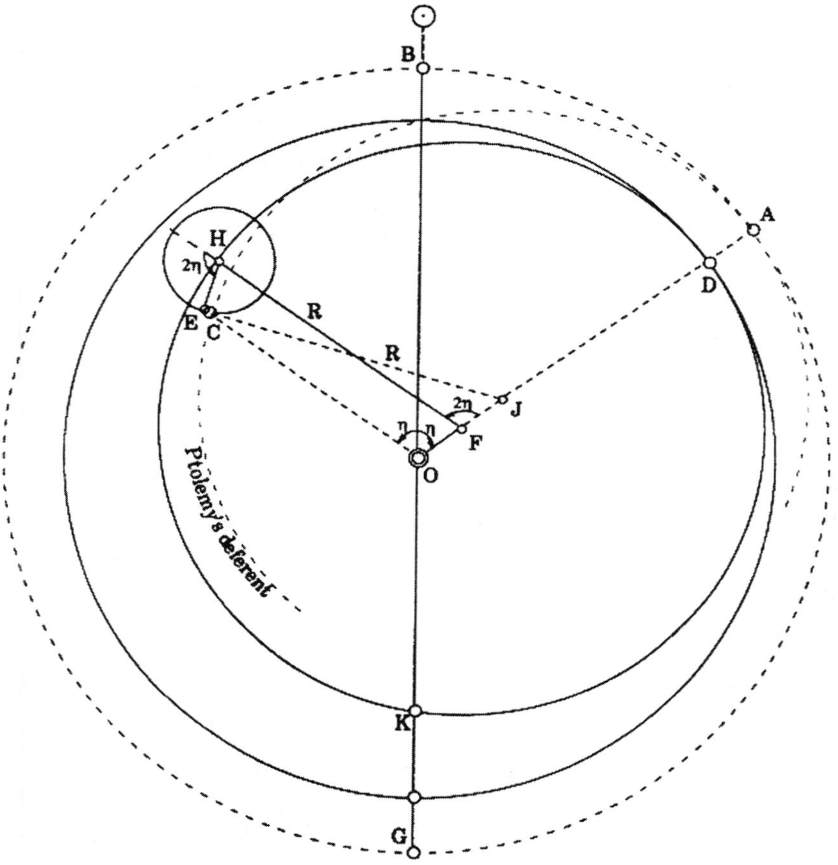

Figure 4.8

Shīrāzī's lunar model. By taking a new deferent whose center was halfway between the center of the world *O* and the center of the Ptolemaic deferent *J*, Shīrāzī compensated for that by introducing a new small epicycle, with center *H*, whose radius was equal to the same distance between the two centers of the respective deferents. By making the small epicycle move at the same speed as the new deferent, and in the same direction he managed to satisfy the conditions for 'Urḍī's Lemma, which could now be applied to lines *HE* and *OF*, thus making line *EO* always parallel to line HF, and making the epicycle center *C* appear to be moving around the center of the universe *O*.

same conditions ʿUrḍī's small circle satisfied in the case of the upper planets. That is, he allowed the small circle to move at the same speed as the deferent and in the same direction thus satisfying the condition of having two interior angles equal, and thus produced the parallel lines. The new arrangement, as suggested by Shīrāzī, made sure that the small circle, which moved uniformly around the center of its own deferent, also had the tip of its radius seem as if it was moving uniformly around the center of the world, which in turn was the observational condition Ptolemy's model satisfied. Thus by positing the Ptolemaic lunar epicycle at the tip of that radius the moon would then move around its own epicycle, but at the same time was seen to satisfy the same observational conditions it satisfied in Ptolemy's case.

In the case of the planet Mercury, Shīrāzī gave some nine models that could describe the motion of this planet. Those models are detailed in two of his most famous works, the *Nihāyat al-idrāk fī dirāyat al-aflāk* (the ultimate understanding regarding the comprehension of the spheres), and the *Tuḥfa Shāhīya* (the Royal Gift). In a still later work (*Faʿaltu fa-lā talum,* meaning "I had to do it thus don't blame"), he signaled that seven of those models were faulty. Furthermore, one of the last two was also wrong. But the determination of which one was left unsaid so that Shīrāzī could test the intelligence of his readers, as he boldly claimed. The chosen model, which he finally claimed was *the correct* one, involved the use of two sets of the Ṭūsī Couple, arranged in such a way that it could successfully avoid the use of the Ptolemaic *equant* but preserved its effect and the conditions it entailed. That is, the final center of Mercury's epicycle seemed as if it moved around the point designated as the *equant* by Ptolemy, without ever having to have that motion come about as the product of uniform circular motions of any sphere around an axis that did not pass through its center.

Shīrāzī did not offer any new theorems, but obviously he benefited from his two contemporary astronomers, and deployed their theorems to the best of his abilities. One wonders why, for example, he opted to use ʿUrḍī's model for the upper planets instead of the equally good model of Ṭūsī. But one has to also acknowledge that even if we cannot answer the question in the present circumstances, we can certainly affirm that Shīrāzī had at least two options and that he chose the one that deployed ʿUrḍī's Lemma for his solution of the lunar model as well as the model of the upper planets, and reserved the double use of the Ṭūsī Couple for his Mercury model. One has to also acknowledge that Shīrāzī's double use of the Ṭūsī Couple for the planet Mercury, was in itself a significant step, not only because he succeeded where Ṭūsī had already declared failure, but because he seems to have put into wider circulation a novel idea, such as the double use of the Ṭūsī Couple, which was itself a remarkable departure from the accepted Ptolemaic astronomy when it used once. This remarkable achievement of Shīrāzī does not only put him at the forefront of the astronomers of his time, but allows us to see how novel ideas began to take hold in the scientific culture. They apparently succeeded when they began to be accepted and deployed by their contemporaries.

Shīrāzī insisted that one could begin to think of solving the observational behaviors of the planets by applying different mathematical techniques and producing more than one mathematical model. At the same time though, Shīrāzī was still under the impression that some mathematical solutions are more "true" than others. This attitude will become considerably important later on when we consider yet another conceptual shift as the one taken in the works of Shams al-Dīn al-Khafrī (d. 1550). Properly speaking, Shīrāzī's work lies at the beginning of a tradition that began to seek alternative mathematical solutions to the same physical problem. But at that early stage, the tradition still sought *true* mathematical solutions that could properly describe the motions of the planets. This tradition would not mature until the time of Khafrī. But just by seeking alternative mathematical solutions, and thus new ways of thinking about the problems, allows us to group Shīrāzī with Urḍī and Ṭūsī, who also created new shifts in the articulation of responses to Ptolemaic astronomy.

But because Shīrāzī also tried to group together a series of solutions which were offered by his predecessors, a series that he called *uṣūl* (principles/hypotheses)[27], and which included such concepts as the eccentric versus the epicyclic models as two such principles, he can also be considered as the forerunner of the work of someone like Ibn al-Shāṭir (d. 1375), who came about half a century later, and who also used the solutions of his predecessors, and also globally called them *uṣūl,* as in his *taṣḥīḥ al-ūṣūl* (correction of principles). In the case of Ibn al-Shāṭir, he too went beyond his predecessors and managed to succeed where they failed, again as in the case of the Mercury model that was correctly solved by Ibn al-Shāṭir when Ṭūsī had failed to accomplish that. But Ibn al-Shāṭir did more still, and deliberately carved new directions for his astronomy that also proved to be very productive for the Renaissance scientists.

ʿAlāʾ al-Dīn Ibn al-Shāṭir of Damascus (d. 1375)

There are several features that distinguish the works of this remarkable astronomer who apparently spent his professional life working as a time-keeper at the central Umayyad mosque of Damascus. Although we do not know much about the details of his job description as a *muwaqqit* (time-keeper), his works, both the extant as well as the non-extant, lead us to assume that in his "spare time" he indeed managed to develop one of the most successful attempts at overhauling Greek astronomy. Not only did Ibn al-Shāṭir profit from the astronomers who preceded him with their own attempts to reform Greek astronomy, but he managed to produce some remarkable conceptual shifts of his own in the way astronomy was to be perceived and practiced.

To start with, Ibn al-Shāṭir went back to the very foundations of astronomy and insisted on resolving the very first problem of Greek astronomy: a choice between an eccentric and an epicyclic model. To him that choice was definitely limited, for one could not in any way justify the use of eccentrics. To his mind eccentrics represented a clear violation of the Aristotelian principle of the centrality of the Earth, which up till his time at least made perfect sense within the overall universal cosmology of Aristotle. On that count he insisted that all of his models would adhere to these principles, and would be strictly geocentric. Furthermore, all of his models also shunned the eccentrics completely.

That left him with the problem of epicycles, which, as we have seen, had to be used in all the other planets except the sun. On that round he had some very original remarks to make, and as far as I can tell he was the first, and probably the only one, to insist on making them. First he made the general observation that the sizes of some of the fixed stars were in fact much larger than the sizes of the largest planetary epicycles. Second, when it came to the nature of the epicycles themselves, he tossed the ball back to Aristotle's court and to the court of his followers. Aristotle and his followers had insisted that the epicycles were not permissible because they would introduce a center of heaviness around which the sphere of the epicycle would move, and thus constitute an element of composition in the celestial domain, which was supposed by Aristotle to have been fully made of the simple element ether. Here Ibn al-Shāṭir wondered how could that be, when everyone knew that the stars which were carried by spheres that were made of the same element ether emitted a light we can all see, while no such light was emitted by the spheres themselves? How could one part of the sphere, where the star was located, emit all that light, while the other part remain dark, or transparent, and still be called a simple sphere? How could the sphere and the star that it carried be both made of the same simple element ether and have such divergent appearances? Once it was admitted that such phenomena existed, and there was no way to deny something that everyone could verify for himself, it became obvious to Ibn al-Shāṭir that at least the lower celestial spheres of Aristotle, by which the stars and the planets were supposed to be moved, had to admit some kind of composition. Only the uppermost sphere, the one beyond the eighth sphere that was only responsible for the daily motion of the whole, but carried no stars of its own, that sphere could remain as simple as Aristotle would want it to be.

That reasoning allowed him to conclude that the composition introduced by the planetary epicycles should at least be as acceptable as the composition, which was already implied by the existence of the fixed stars and planets that everyone could obviously see in the skies.

Having "solved" the problem of the epicycles in this fashion, he then went ahead and systematically banished all eccentric circles from his models. Instead he substituted epicycles for the eccentricities in each case, in a manner very similar to the application of the Apollonius theorem where the epicyclic model could easily replace the eccentric one. As a result, he then managed to produce a set of models that were all unified by their strict geocentrism. And in order to achieve that throughout he used a combination of two well-known principles that had already been used: the Apollonius equation and ʿUrḍī's Lemma. The latter allowed him to adjust for the Ptolemaic equant by adding yet another epicycle, which was used by ʿUrḍī in the model for the upper planets.

And because all his models were geocentric and used the same two "principles" to solve the *equant* problem, they also managed to expunge from Ptolemaic astronomy the variety of approaches that were adopted by Ptolemy in his quadripartite model structures—different models for the Sun, the Moon, the upper planets, and Mercury. With the exception of the Mercury model, all the other models of Ibn al-Shāṭir had identical constructions but whose representation of the planetary motions were simply manipulated by the sizes and speeds of the various epicycles he had to deploy. In the case of Mercury, he only introduced an additional use of the

Ṭūsī Couple at the last step, but continued to use in it the two other principles just mentioned. This procedure was also followed some two centuries later by Copernicus, and for the same planet: Mercury.

One additional advantage resulted from this systematic use of geocentricity, which was to come in handy later on during the European Renaissance: the unification of all the Ptolemaic geocentric models under one structure that lent itself to the simple shift of the centrality of the universe from the Earth to the sun, thus producing heliocentrism, without having to make any changes in the rest of the models that accounted well for the Ptolemaic observations resulting from the *equant*. As we shall see later on, it may not have been entirely accidental that Copernicus ended up relying so heavily on the works of Ibn al-Shāṭir when he used, among other things, a lunar model that was identical to that of Ibn al-Shāṭir, and used the same Ṭūsī Couple, in the same fashion as was done by Ibn al-Shāṭir, in order to account for the motion of Mercury.

And despite common legends that claim that Copernicus was attempting to get rid of the *equant*,[28] by adopting Ibn al-Shāṭir's techniques, and just shifting the direction of the line that connected the sun to the Earth, he could in fact retain the observational value of the *equant*, without having to assume, as was done by Ptolemy, the existence of a sphere that could move uniformly around an axis that did not pass through its center.

In addition to the flexibility of Ibn al-Shāṭir's models, and his full control of the mathematics that allowed him to adjust his models so that they would fit the observations, Ibn al-Shāṭir also made another unprecedented step. He was the only one of the astronomers in the Islamic domain who seems to have devoted a whole book (*Taʿlīq al-arṣād*, meaning Accounting for Observations) to this particular relationship between observations and the construction of predictive models that could satisfy those observations. The book seems to be unfortunately lost, and thus we may never know the extent of his theorizing in this regard. But it is extremely significant that he did undertake the writing of such a book.

And even if we have to assume that *Taʿlīq al-arṣād* is lost to us, we still have some inkling about the methods and contents of that book from a few instances where such approaches have been followed in his surviving works, and in particular, in his *Nihāyat al-sūl fī taṣḥīḥ al-uṣūl* (The Final Quest Regarding the Rectification of [Astronomical] Principles). In this last book, we are explicitly told that Ibn al-Shāṭir had conducted his own observations in order to determine the apparent sizes of the two luminaries.[29] And we are also told that with those new results, which varied considerably from the values given by Ptolemy, he managed to construct a new model for the sun which was also at variance with the Ptolemaic model. In essence, this work demonstrates quite clearly Ibn al-Shāṭir's ability to construct theoretical models that were based on observational results, just as was done by Ptolemy, but without committing the inconsistencies of Ptolemy. It is in such instances that the centrality of Ibn al-Shāṭir's work can best be appreciated, and his relationship to Copernican astronomy can be better understood.

Shams al-Dīn al-Khafrī and the Role of Mathematics in Astronomical Theory

In all the previous Islamic responses to Greek astronomy one can detect a consistent tendency to solve the problems of that astronomy one problem at a time. From the prosneusis point, to the *equant,* and finally to the harmony between observations and predictive models, what the astronomers seemed to be doing was developing theorems and techniques that allowed them to reconstruct Ptolemaic astronomy along lines that would make that astronomy consistent with its own physical and cosmological presuppositions. What the predecessors of Khafrī seemed to be doing was trying to cleanse Ptolemy's astronomy from its faults, by using new mathematical techniques and theorems that were either unknown to Ptolemy or unnoticed by him. No one though seemed to think of the very act of mathematical theorizing itself and its relationship to the physical phenomena that were being described until Khafrī.

With Khafrī, Islamic astronomy moved to still newer territory. He was the first to begin thinking about the role of mathematical representation itself, the functionality of predictive models, and the relationship of all that to the actual physical phenomena. We had already noted the beginnings of this kind of thinking when we described Quṭb al-Dīn al-Shīrāzī's attempt at producing nine different models for the planet Mercury as a new step that marked the search for mathematical alternatives to the ones that were inherited from Ptolemy. But we also noted Shīrāzī's failure to pursue this line of thought when he seemed to have been still mired in the process of finding a unique solution, or say a unique representation, of the physical phenomenon that could be summed up in one *true* mathematical model. He spoke of such truths himself, for he was the one to alert his reader that of all the nine models he proposed for the planet Mercury in two of his books, seven were faulty by his own admission, while the eighth was left to the student to figure out its failings, and only the ninth was the *true* solution. So despite the fact that he was beginning to think that there were alternate mathematical techniques to describe the same physical phenomenon, he still thought that those techniques must climax in a unique solution that represented the truth of the matter.

With Qushjī (d. 1474) we can begin to see this trend pushed slightly forward. For although he must have known that Shīrāzī's model for Mercury solved the problems of the Ptolemaic model quite adequately, he still felt he could produce one more model that could do the same, which he did with his own new model. Was he thinking that his model, which relied entirely on ʿUrḍī's Lemma to solve Mercury's *equant* problem, was only an alternative to that of Shīrāzī, which used only the Ṭūsī Couple, in the sense that it was a deployment of an alternative theorem to solve the same problem, or was he in fact thinking that the problem itself admitted multiple solutions? At this stage, we do not yet know. But from the fact that Qusjī's treatise on the Mercury model is a very short treatise devoted to this model only, one can presume that he was thinking of it as an alternative to Shīrāzī's and thus as just another way of thinking of mathematics.

With Khafrī the issue becomes very clear. In one full swoop he produced four different models for Mercury's motions, which he called *wujūh* (approaches), all of them accounting for the observations in exactly the same fashion, and none of them similar to the others in terms of its mathematical constructions. It is as if Khafrī had finally realized that there was a difference between two ways of thinking about Apollonius's theorem. On the one hand it could be thought of as representing two different cosmological solutions for the conflict between Aristotelian presuppositions and observations, and it can be thought of as two different mathematical ways of speaking about the variation of the solar speed with respect to an observer on Earth. It is the latter understanding that was finally realized by Khafrī's *wujūh,* for all his four models were mathematically equivalent in the same way the eccentric representation was mathematically equivalent to the epicyclic one. And although it is not stated in quite those terms, one could almost hear Khafrī say that the mathematical models he was devising were only different linguistic phrases used to describe the same phenomenon.

Seen as a tool, mathematics in the hands of Khafrī would become just another language of science, a tool to describe physical phenomena, and nowhere required to embody *the truth* or the *correct* representation, as was apparently thought by Shīrāzī before.[30] Mathematics became just as simple as describing a phenomenon with poetic language, with prose, or with mathematical figures, and as such the language itself can then be isolated from the phenomenon itself.

Conclusion

By focusing on the major shifts in astronomical thought that characterized the Islamic responses to Greek astronomy, it is now easy to see in hindsight the major features of these shifts. We can see how important it was to explore the full technical details of the most sophisticated Greek astronomical texts (the works of Ptolemy), not only to correct their mistakes, observational and otherwise, but also to investigate their presuppositions and the manner in which they related the observed phenomena to the methods of representation that allowed for the prediction of those phenomena. This close look at the foundations of those texts gave rise, as we have seen, to a series of Arabic texts written specifically for the purpose of critiquing the shortcomings of this imported Greek tradition. From *istidrāk* to *shukūk,* to straight forward rejection, all this full exposure left the Greek astronomical tradition in desperate need of reform.

The most important transformation that took place during this time was the shift from Ptolemy's instrumental approach to astronomy (which satisfied itself with the pragmatic success of the predictive features of the mathematical models) to a more theoretical approach which required that predictive results be consistent not only with the observations but also with the cosmological presuppositions of the observations themselves. In other words, in Islamic astronomy, it was no longer sufficient to say that a specific predictive mathematical model, such as that of Ptolemy, gave good results about the positions of the planets for a specific time. The new requirement was that the model itself should also be a consistent representation of the cosmological presuppositions of the universe, in addition to its accounting for the observations. If the universe was composed of combinations of spheres, and if those spheres were, by their very nature, supposed to move in uniform circular motions, then it was no longer acceptable to represent those spheres with mathematical models that deprived those spheres of their essential properties of sphericities and satisfy one's self by saying that they yielded good predictive results.

What seems to have happened during the confrontation between the receiving Islamic civilization and the imported Greek tradition, which we know was very closely watched by various sectors of the society, was to subject this incoming tradition to all sorts of exacting criteria before it was allowed to survive the cultural critique to which it was subjected. In that context, astronomy was no longer a discipline that only supplied good answers about the positions of the planets, or good enough answers for an astrologer to cast a horoscope. But with the Islamic religious aversion toward astrology itself and toward the craft in general, astronomy had to define itself as a discipline that went beyond that simple predictive feature and had to pose itself as raising questions of much greater relevance to a wider and larger world view that required exacting measures at every turn. The astronomer had to attend to all larger intellectual questions that had any bearing on his craft. In that concern the astronomer could no longer afford to seem as if he was satisfied with a confused picture of the universe, as long as he could achieve reliable results for astrological prognostications. Astronomy had to prove its usefulness to the new social and cultural environment within which it had to struggle. It could only do so by engaging the theoretical criticism of the very foundation of Greek astronomy.

In that context, one can then understand why no one could continue to tolerate two different visions of the nature of the universe that were in direct conflict with one another. One could not isolate the results presented in the *Almagest,* as mere computational and mathematical tools that could predict the positions of the planets at specific times, and say that they were irrelevant to the physical universe presented in the *Planetary Hypotheses.* To be fair, Ptolemy never really claimed that. On the contrary; throughout the *Almagest* he repeatedly hinted to the necessity of keeping the universe of the *Planetary Hypotheses* in mind. But at the same time he still went ahead and violated almost every feature of that universe by representing it with mathematical concepts that were totally divorced from their very mathematical properties. The example of the *equant* spoke directly to this point, where the spheres of the *Planetary Hypotheses* lost their very properties of spheres, if one were to represent them only in the manner in which they were represented in the *Almagest.*

With those fundamental oppositions, the job of the astronomer in the receiving Islamic culture became focused on those very issues of consistency between the vision of the *Planetary Hypotheses* and the representations of that vision in the *Almagest.* In the first phase of the response to the Greek astronomical tradition, the problem was perceived as a problem of sophisticating the techniques of representation, that is, the deployment of the same mathematics that was used by Ptolemy in order to reconfigure the representations so that they would be more faithful to the objects that they were representing. Someone like Abū ʿUbayd al-Jūzjānī (d. ca. 1070), the famous student of Avicenna (d. ca. 1037) did just that in his failed attempt to reform the representation of what later on became the *equant* problem. It took about two centuries to realize that Ptolemy's mathematics itself was inadequate, and that new mathematics had to be invented for the purpose.

The works of ʿUrḍī and Ṭūsī ushered in the second, more important phase, when they spoke directly to that need of creating new mathematics. And each of them had a new theorem to add. Several astronomers, who used the newly enriched mathematics, and who also began to speculate about the various ways with which the physical phenomena could be mathematically represented, followed them. The nine attempts at representing the motions of the planet Mercury, which were devised by Ṭūsī's student and colleague Quṭb al-Dīn al-Shīrāzī, and the later attempt by ʿAlāʾ al-Dīn al-Qushjī to produce one more model for the motion of Mercury, fall in that category. This trend of re-defining mathematics as a language to describe the physical phenomena was to reach its climax with the works of Khafrī who finally gave concrete examples of four different mathematical models that described the motions of the planet Mercury, and yet were all exactly mathematically equivalent. In this fashion he could demonstrate, although never stated explicitly as such, that such physical phenomena did not yield unique mathematical solutions, but almost as many as the human imagination could conjure up, in exactly the same way a specific fact could be describe by an endless variations of the language.

With Ibn al-Shāṭir, the reorientation of astronomy took yet another turn, going back to the very cosmological foundations that were at the base of all phenomena as well as to the representations of those phenomena. Ibn al-Shāṭir ended up re-questioning the very use of the concept of eccentrics, and finding it cosmologically inconsistent with the cosmological foundations it was supposed to represent. When faced with the inevitable alternative of the epicycle, he insisted that such a tool be used despite the fact that the then current Aristotelian interpretation of his time thought of the epicycle as alien to the Aristotlian universe. Instead of giving up and pleading human imperfection, as was done by Ptolemy when the latter failed to find representations that were consistent with his own cosmological presuppositions, Ibn al-Shāṭir went back to the Aristotelian universe itself in order to criticize its inconsistencies and to point to the fact that such epicycles were, in a strict sense, consistent with an Aristotelian universe if the latter was properly understood.

Having banished all eccentrics in his representation of planetary motion, and having seen the essential similarities of all such motions in that they could all be represented by the same kind of model with a minor additional adjustment for the planet Mercury, Ibn al-Shāṭir went on to re-examine the relationship between the observed phenomenon and the mathematical models that were supposed to represent it. His readiness to adapt his mathematical models to match the observations speaks volumes about his priorities and about his ultimate re-definition of astronomy. To Ibn al-Shāṭir, astronomy was first and foremost a discipline that produced a systematic and accurate description of the behavior of the real universe around us. That description itself had to be a scientific mathematical representation that could only be a statement that described the reality of the observations.

Seeing these developments in Islamic astronomy in this fashion allows us to see how demanding the receiving culture was, and how its very demands required that its own scientific thought continued to be progressively defined and perfected according to the ever-changing criteria of precision and consistency that this culture imposed upon itself.

5 Science between Philosophy and Religion: The Case of Astronomy

The previous chapters focused on the social, political, and economic conditions that gave rise to and sustained science in the Islamic civilization. We had a chance to draw very broadly on the historical as well as the scientific sources themselves in order to illustrate with particular examples how these processes of motivation and encouragement as well as reward worked in order to enable certain scientific disciplines to be born, others to be abandoned, and still others to be maintained and reconstructed. We hinted several times already that we used the discipline of Astronomy only as a template simply because there was a methodological need to anchor the historiographic suggestions in a particular discipline in order to contextualize the much harder to document social forces at work.

At various occasions, reference was often made to societal forces that required new disciplines to be created, as in the case of *ʿilm al-mīqāt, ʿilm al-farāʾiḍ,* and *ʿilm al-hayʾa,* while at other occasions we hinted, again very briefly, to inner logical transformations within the disciplines themselves that gave rise to other disciplines as in the case of the development of trigonometric theory as a result of the need to satisfy solutions of spherical trigonometric nature. The natural consequence of the adoption of the new trigonometry was the demise of the old Greek chord functions, and the old Greek methods of solving spherical trigonometric problems.

At all those occasions this historiographic research was guided by the need to explain the historical and scientific facts as we know them from the extant sources. Again the emphasis was laid on the discipline of astronomy for illustrative purposes only; always hoping that colleagues, who work in other disciplines, would subject the general conclusions, which are reached in the context of the new methodological approach to the history of Islamic science that is adopted with the alternative narrative, to the test of the data they already know from their own particular disciplines. With this approach, it was possible to reconstruct the developments in the discipline of astronomy and to detect, almost at each juncture, the motivations behind most of the new breakthroughs that indeed took place during the long history of Islamic astronomy. Various stages of astronomical thought began to congeal and make much better sense when they were perceived from within this process of contextualization.

On several occasions, reference was made to general trends in the history of Islamic astronomy that were characterized as motivated by religious requirements. The identification of those trends and their twists and turns give hints of the necessarily complicated relationship between science and religion that I hoped to document within the context of the Islamic civilization. I needed to follow that path not only because we needed to know the extent to which certain religious ideas could motivate genuine scientific interest, or to know the role that was played by men of religion in the production of science, but I also needed to know if the prevalent model of antagonism between science and religion that seemed to work relatively well in the European context, as articulated by the ethos of the age of reason, would also work in the context of the Islamic civilization. And here again, there was constant recourse to the discipline of astronomy in order to illustrate the general developments with concrete examples at least from the scientific production of astronomical literature.

While still focusing on the discipline of astronomy, the previous chapter tried to explore the subtle shifts that took place in that discipline. It spoke of those shifts as having occurred, on the one hand, as a result of the mere historical circumstances, like the happenstance of observing the same astronomical phenomena, which were observed by Ptolemy during the second century, from the vantage point of ninth-century Baghdad, thus making use of the accidental passage of about 700 years that could definitely refine the earlier results. On the other hand, it spoke of some shifts that were necessitated by the developments within astronomical thought itself, thereby necessitating the deployment of new mathematical theorems, new mathematical techniques, and finally new perception of the role of mathematics in such disciplines as the astronomical disciplines. The latter realization of the role of mathematics as a descriptive language for natural phenomena could certainly be applied to other scienfitic disciplines that could either corroborate or negate the processes that seem to have taken place in the astronomical field.

In all instances, much emphasis was placed on the role of the dynamic social, economic and political forces in forging the new conceptions of astronomical processes that finally led to the development of a uniquely conceived Islamic astronomy that was not a mere regurgitation of the older Greek astronomy, nor was it a total break from it, and yet was in a position to lay the foundation for a revolutionary upset of that astronomical tradition. I was careful to note that all those developments, although they were conceived within the societal general context of struggling against the intrusion of the "foreign sciences" into the Islamic civilization, they were at the same time developments that were necessitated by the very shortcomings of the Greek astronomical tradition itself, whether on the practical observational level or on the more advanced theoretical one. But all those developments, twists and turns, were all symptoms of this double tension, which was mentioned earlier, as resulting from a discipline that was forced to negotiate its place within the general accepted epistemological frame of the society on the one hand, and within the general epistemological innovations of the discipline itself on the other.

This chapter will push this discussion a step further by focusing on the repercussions those developments generated, in terms of the new philosophical questions they raised, and will try to revisit the implications of those developments to the relationship between science and religion by using, once more, the illustrative and instructive role of astronomy.

The Philosophical Dimension[1]

All the theoretical astronomical works that we now know to have been produced in the Islamic civilization between the ninth and sixteenth century were conceived within the general determining parameters of Aristotelian cosmology. With the exception of those treatise that were generally titled *al-Hay'a al-sunnīya* (probably translatable as Orthodox astronomy)[2] and classified under religious astronomy, all other treatises, whether consciously or not, assumed a geocentric spherical universe, in which planets and stars moved in place, in circular motions, at uniform speeds, and so on. The contours of this universe were already defined by the Aristotelian cosmology that was embedded in, and in fact claimed to be the basis of, Ptolemaic astronomy itself.

In very general terms, one can characterize the whole tradition of Islamic theoretical astronomy, as a continuous attempt to save Ptolemy from his own folly, in the sense trying to make his work more harmonious with the same Aristotelian cosmological principles that he had accepted, and at the same time attempt to take issue with him, and with Aristotle behind him, for all the contradictions their cosmological visions brought forth. But ironically as well, the whole Islamic theoretical astronomical tradition was also an attempt to save Aristotle, whenever his ideas were not contradictory, and at the same time abandon Aristotle, whenever his thought was found absurd. So in a deep sense, one can say that Islamic theoretical astronomy was a continuous debate with Aristotle, but was guided by a real sense of commitment to the physical universe of which those astronomers attempted to make some sense.

One has to understand that this dialogue with Aristotle also took place in a culture that was first and foremost alien to the Greek culture of Aristotle, and had its own basic doctrinal premises that could not be violated. There was no way, for example, to disregard the organizing principle of religion itself, the existence of God, the revealed religion, etc., in any attempt to understand the universe itself. One did not need to speak directly to the issue of God's existence, while describing the motion of the planets and the creation of mathematical models that predicted their positions for any time and place. But one could not attempt to base these models and explanatory techniques on the assumption of God's absence from the universe. As long as Aristotelian cosmology did not come in direct conflict with such fundamental premises, the problem did not arise.

But when the Aristotelian vision conceived of the concept of change in the world around us through a process of generation and corruption, and that generation and corruption itself was in turn dependent on the motion of the celestial bodies, then it came in conflict with the fundamental religious principle which in turn perceived the Aristotelian cosmological vision as *the* founding vision of astrological theory. The implications of such a conflict can be very serious indeed. For to think of human activity as directly influenced by the action of the celestial spheres, as some astrologers would in fact interpret Aristotle to say, meant that one could relieve the individual of his religious obligations, or at best relieve him of the consequences of his actions. It is in that context that astrology became the Achilles heel of Greek thought in general, and had a detrimental influence on the discipline of astronomy to which it was very closely related in the Greek tradition.

In order to avoid charges of not heeding religious precepts, astronomers working in the environment of Islamic civilization had two choices to make:

Either disregard the religious authorities and continue to associate their discipline with astrology as was done in the Greek sources before, or redefine the subject of their discipline as seeking to know the positions of the planets without having to make any comments as to the astrological significance of those positions. For those who took the second option, their discipline became then willfully restricted to the empirical pursuit of planetary positions just because the problem of the determination of such positions was in itself a challenge that needed to be met. Of course, they also opted to support their work with the religious pronouncement that urged man to study the natural phenomena as signs of God's creation and indicators of the existence of God himself.[3]

Whichever justification they used for their discipline, the net result remained the same: they attempted to construct mathematical models, that were true to cosmological presuppositions, in this instance Aristotelian presuppositions, and yet were capable of predicting the true positions of the planets. At the same time, they willfully avoided the religious and astrological implications of that Aristotelian cosmology. In essence, they trimmed Aristotle down to their needs.

Defined in this fashion, astronomy no longer looked like its Greek counterpart, although it resembled it in many varied ways. As far as the computational part, and the mathematical calculations that connected the observed phenomena to the predictive models, the two astronomies were more or less the same. The only difference was that the later astronomers in the Islamic domain benefited from the passage of time to correct the flawed astronomical parameters that were embodied in the Greek tradition. But the most important difference lied in the purpose of the two astronomies: The Greek tradition needed to determine the position of the planets so that it could predict their influence on the world of change in the sublunar region, while Islamic astronomy restricted itself to the same description of the behavior of the planets, with the utmost accuracy they could muster, and yet refrain from asking about the planet's influence on the sublunar region in general or the human behavior in particular. It is in this environment that the new discipline *ʿilm al-Hayʾa* (science of astronomy) was born. And as such, of course it had no Greek equivalent. Its authors were fully aware of that, and for that reason restricted themselves to calling it by its newly coined name, which meant literally "the science of the configuration [of the world]."

Once the purpose of astronomy was conceptually redefined, then the pursuit of astronomical research was fully condoned within Islamic civilization. This did not mean that astrology was finally excluded from the social domain. In fact there are plenty of sources that speak to the contrary and some even attest to its flourish and its specific widespread acceptance within the political circles where it continued to guide the actions of potentates and their cohorts by the dictates of the planetary positions. But expurgating astronomy from astrological practice meant that astronomy itself could flourish among the religious elite who saw in it a complementary discipline to their own, and thus felt at ease with it, especially when this specific new astronomy began to direct its attention to the critique of Greek astronomy. This critical feature of *ʿilm al-hayʾa* marked the discipline from its very inception in the ninth century. In fact all the alternative planetary theories that we know from the Islamic domain were articulated in texts that identified themselves as *hayʾa* texts. And since *hayʾa* simply meant "configuration [of the world]," it meant that those texts were necessarily restricted to this descriptive aspect of astronomy, and never ventured as far as supplying actual tables that could be used for the actual determination of the positions of the planets as was done by the *Almagest,* for example. In that regard, the *hayʾa* texts looked more like Ptolemy's *Planetary Hypotheses* than either the *Almagest* or the *Handy Tables.*

And because of this newfound purpose of astronomy it could afford to keep its distance from the ancient Greek tradition, and took the full freedom to subject the latter to the strictest criticism whenever criticism was found necessary. It was after all Muḥammad b. Mūsā b. Shākir (c. 850), one of the most zealous sponsors of the translations of Greek scientific texts, who offered to make sense of the Ptolemaic attempt to account for two basic motions: (1) the daily rotation of the eighth sphere that produced the variations of day and night and (2) the motion of precession which was most observed by the sliding position of the vernal equinox. Working from the Greek philosophical precept that all celestial motions are produced by specific movers, in this instance individual spheres, those two motions then had to be accounted for by two separate spheres, since it was inconceivable that the same sphere could move in two separate motions at the same time, while still in place. In order to resolve the problem, Ptolemy assigned the daily motion to the eighth sphere of the fixed stars, and then added another concentric ninth sphere to account for the precession motion. One could reverse the order and assign the precession to the sphere of the fixed stars and then ascribe the daily motion to the ninth sphere. The order was not the problem.

Rather, for Muḥammad b. Mūsā b. Shākir,[4] the problem lay in the fact that the two last spheres, which were supposed to carry those motions, were concentric. And that arrangement in particular presented an important physical problem. For how could any sphere move another, if both spheres were concentric, and if both spheres were made of the same element ether that did not allow such properties as friction, dragging, and the like? By understanding ether in the strict Aristotelian sense, in that it was a simple element that did not have any of the features of the sublunar elements like heaviness, lightness, etc., it was then impossible for two spheres, made of this same element, to force each other's motions if they shared the same center. Muḥammad b. Mūsā b. Shākir had no difficulty accounting for one sphere forcing another eccentric sphere to move along with it, for that did not require physical friction and the like. But to him, "it was in no way possible" to have a ninth sphere whose motion would necessitate the motion of the eighth. As far as we can tell, Muḥammad b. Mūsā b. Shākir had no real solution to this predicament, but he definitely had a real objection to the Ptolemaic arrangement. And his objection was strictly philosophical in that it depended completely on the definition of the element ether.

Of course the apparent loss of the treatise in which Muḥammad b. Mūsā b. Shākir made his argument does not help us determine if he had a solution to the problem or not. The present study of that treatise is based on a fragmentary quotation from the work of an astronomer who lived centuries after Muḥammad b. Mūsā b. Shākir.

For the anonymous Andalusian (c. 1050) author of *Kitāb al-istidrāk,* whose extant work *Kitāb al-hay'a* is still preserved in Hyderabad, India,[5] the more global question was to firmly pinpoint the status of the new astronomy in whose writing he was now participating. In a critical passage on how this new astronomy ought to be pursued, he says:

> The one who works in this craft must obtain the [mean] motions, that are taken as principles, from the observations, and then consider through geometry how these motions could take place, and which configuration would fit them best. In his search he should not abandon the principles of this craft, which he should accept from natural philosophy. Accordingly, he should not depart from spheres and circular uniform motions and pass on to bodies that are not spherical or not circular. And if he were able to discover many configurations for the same planet, all of them yielding the same observable results of the particular motions, he should then chose that which is

simpler and easier, in a manner appropriate to celestial bodies, as was already done by Ptolemy who, in the case of the sun, opted for the eccentric model, that described only one motion, instead of the epicyclic one, which would have necessitated two.[6]

For this anonymous author, then, the astronomical universe within which all the planetary motions had to be understood and fitted, was a strictly Aristotelian universe that had its own premises that the astronomer was not allowed to violate. And while praising Ptolemy, the author used this language as an implicit critique of Ptolemy who did just that. According to all the authors of books on doubts (*shukūk*), Ptolemy definitely departed from the premises of that Aristotelian universe, and thus deserved to be criticized so severely by them.

Furthermore, the anonymous Andalusian author intended to stress that it was not only the principle of a spherical universe, with spheres moving uniformly, that had to be observed, but that the representation of the particular configurations of that universe also had to be consistent with the nature of that universe. In other words, he wished to advocate the main message of the *hay'a* writers, which could be summed up in the new requirement of consistency that all astronomical theories had to be subjected to. Simply stated, this consistency requirement demanded that the mathematics, used by the astronomer to describe the phenomena that one observed in the physical universe, must at no point depart from the mathematical characteristics of that universe. In these representations, for example, if one dared to accept the concept of a sphere that moved uniformly, in place, around an axis that did not pass through its center, then one might as well accept the absurdity of representing a sphere with the figure of a mathematical triangle.

In this context, one can define the main feature of the new astronomy of *hay'a* as an astronomy that was obsessed with this consistency between the premises of the field and all the ensuing constructions the field required.

The last section of the quotation underlines the importance of another aesthetic principle that was already known to the Greek authors, and which had nothing to do with observational astronomy proper, namely, that of the principle of simplicity and ease. Ptolemy himself already articulated that principle, in so many words, when he explained, in book III of the *Almagest,* why he opted for the eccentric model for the sun rather than the epicyclic one.

Other astronomers and philosophers working in the Islamic domain had other axes to grind with Aristotle himself, and sometimes with Ptolemy as a representative of that philosophy. After all, it was Ptolemy who had already started the debate by his unspoken options for the solar model. Both options violated Aristotelian cosmology. The first posited the existence of eccentrics whose centers by definition did not coincide with the center of heaviness around which everything moved, as was required by Aristotle. The second option assumed the existence of epicycles, out in the celestial realm, which had their own centers of motion, again contrary to what Aristotle recommended.

In the case of the sun, Ptolemy satisfied himself with the eccentric model and said nothing of the other option, except that it was an option. But in the case of the other planets, Ptolemy had no such simple options. He had to accept both models: the eccentrics as well as the epicycles. In this, every other astronomer working in the Islamic domain, with the exception of Ibn al-Shāṭir who rejected the eccentrics all together, followed him.

Under the circumstances, it becomes understandable why would someone like Averroes, who lived some two centuries before Ibn al-Shāṭir, object so vehemently to the astronomy of his days, when he said, "to propose an eccentric sphere or an epicyclic sphere is an extra-natural matter (*amr^un^ khārij^un^ ʿan al-ṭabʿ*)."[7] He then went on to say:

> The epicycle sphere is in principle impossible (*ghar^u^ mumkin^in^ aṣl^an^*), for the body that moves in a circular motion has to move around the center of the universe (*markaz al-kull*) and not outside it.[8]

He followed that with a more damning statement:

> The science of astronomy of our time contains nothing existent (*lays^a^ minh^u^ sha'^un^ maujūd^un^*), rather the astronomy of our time conforms only to computation, and not to existence (*hay'at^un^ muwāfiqat^un^ li-l-ḥusbān lā li-l-wujūd*).[9]

As has already been noted, it was Ibn al-Shāṭir who took these objections seriously, and who responded to the issue of the eccentrics by banishing them out of his system. But in the case of the epicycles he tossed the ball back to the Aristotelian yard to ask about the very nature of the ether as we have also said.

Then there was the issue of the Aristotelian spheres themselves, whether they would move by their own volition or be forced to move by something else.[10] The problem arose from the fact that the planets themselves do not have the same kind of motion, and seemed to exhibit individual motions of their own. But according to Aristotle, there were no such motions without movers that caused them in the first place. Thus every planet must have a sphere that caused its motion. And because of the complexity of those motions the spheres got multiplied, and so on.

These motions of the spheres led to a lively discussion that apparently started with ʿUrḍī (1266) in the thirteenth century and continued well into the sixteenth century with the works of Ghars al-Dīn b. Aḥmad b. Khalīl al-Ḥalabī (d. 1563). The essence of the debate is to point to the paradox in the Aristotelian thinking about those spheres. If those spheres moved of their own volition, as they seemed to do, then how could one anticipate their motions, and predict where the planets would be at a specific time? If the spheres, on the other hand were forced to move in predictable motions, then could they exhibit this variety of motions that we witness in the celestial realm? ʿUrḍī interjected:

If we were to admit that the mover of a planet could speed up and slow down, then we would have no need of constructing a configuration (*hayʾa*), and his own astronomy (*hayʾa*) (i.e. Ptolemy's) would be in vain. Any assumption that a planet would have more than one sphere would be an unnecessary excess, which is impossible.[11]

He continued:

If this were so then the motions of the deferents will have to be irregular by themselves, sometimes speeding up and at other times slowing down. And that is impossible according to the principles of this science (*uṣūl hādhā al-ʿilm*). . . . If one were to admit these kinds of impossibilities in this discipline (*ṣināʿa*), then it would all be baseless, and it would have been sufficient to say that each planet has one concentric sphere only, and any other eccentric or epicyclic sphere would be an unnecessary addition.[12]

The simple solution that was proffered by Ghars al-Dīn, for the voluntary motions of the spheres, and yet allowing for their predictability, simply stated:

Where would the need be for the particular spheres that you (meaning the Ptolemaic astronomers) have posited, which you have up till now failed to correct, with all the contrivances and circumvention implied by them? Let us then say that each planet has one sphere that moves by its own volition, sometimes speeding up, other times slowing down, becomes stationary, moves forward, and retrogrades, etc. What adds to its being natural is the fact that it follows a specific pattern.[13]

Incidentally, Ghars al-Dīn's solution of the problem of predictability and yet allowing for volition, by allowing the spheres to "follow a specific pattern," is reminiscent of the concept of custom (*ʿāda*) that was offered about

500 years earlier by Ghazālī who also had the same predicament of allowing miracles to take place and yet have the continuity of the world and the predictability that continuity entailed.[14] Dare we suggest here solutions derived from religious texts being applied to astronomical texts, as Ghars al-Dīn seems to be doing?

For the astronomer Naṣīr al-Dīn al-Ṭūsī (d. 1274), the necessity of developing a new mathematical theorem in order to resolve the Ptolemaic predicament of the latitudinal motion of the planets had other "unintended" philosophical consequences. When Ptolemy wished to allow the inclined plane (really the equator of the carrying sphere) of the lower planets of Venus and Mercury to oscillate north and south of the ecliptic as the planetary epicycles of those planets moved from the extreme north to the extreme south, he proposed to attach the tips of the diameter of that inclined plane to two small circles that were placed perpendicularly to the plane of the ecliptic. Ptolemy imagined that in this fashion he could have the diameter's tips move along those circles and as a result they would generate the required oscillating motion that will in turn explain the latitudinal motion.

At that point, Ṭūsī exclaimed that the kind of speech that Ptolemy was using was outside the craft of astronomy.[15] Not only because such attachments of the diameter's tips would produce a wobbling motion when it performed the required latitudinal motion, but because that same wobble would first destroy the longitudinal component of the motion that was painstakingly calculated and accounted for with the rest of the predictive mathematical model. Second, it would introduce into the celestial realm oscillating types of motions or motions that were not in complete circles. This last requirement would violate the very essence of the Aristotelian definition of the celestial spheres.

Ṭūsī proposed to resolve the problem by the introduction of his own theorem, now called the Ṭūsī Couple, which allowed for the solution of both of Ptolemy's problems: first it allowed for the oscillating motion as a result of complete circular motions, and second it avoided the necessary wobbling that was required by the Ptolemaic suggestion. With one theorem both problems were solved at once.

As an unintended consequence, this theorem confronted the Aristotelian dogmatic separation of the celestial world from the sublunar one. Aristotle had separated those two worlds on the basis of the nature of motion that pertained to either one of them. Linear motion was natural to the sublunar world, while the celestial world only moved in circular motion. Ṭūsī's theorem now presented the most glaring counter example. For here we have, with Ṭūsī's Couple, a universe in which, under the right conditions, linear motion could necessarily result from two circular motions. This would not only make the Aristotelian division of those two worlds completely artificial, that is unnatural in the Aristotelian sense, but it would also make the Aristotelian characterization of generation and corruption as a by-product of the contrary linear motions, particular to the sublunar world, also artificial and completely arbitrary. Furthermore, since the Ṭūsī Couple could also demonstrate that the oscillatory linear motion, which was produced by the two uniform circular motions, was necessarily continuous and uniform, then the requirement that there be a moment of rest between ascending and descending directions of oscillatory motion was also cast in doubt.[16]

Ṭūsī did not make any of those critiques of the Aristotelian universe at the time when he proposed his new theorem, for then he was more concerned with the damage Ptolemaic latitudinal motion was inflicting on the longitudinal motion. But his commentators, starting with his immediate student and collaborator, Quṭb al-Dīn al-Shīrāzī (d. 1311), noticed all the "unintended" philosophical implications the theorem managed to produce.[17] He articulated his observation concerning the moment of rest between two contrary motions in the following terms:

> This could be used as a proof for the absence of rest between two motions, one going up and one going down. This is obvious. And the one who asserts that there must be rest between the two motions cannot deny the possibility of such motions by the celestial bodies simply because he believes there must be rest and rest is not possible for the celestial objects. This is so because we shall use it whenever there is an ascending motion and a descending one as we shall see in the forthcoming discussion. We couldn't be blamed if we also used it to disprove that principle [i.e. the Aristotelian principle of rest between two opposing motions], as can be witnessed from observation. For if we drill a hole in the bottom of a bowl whose edge is circular, but of unequal height above its base, and if we pass a thread through the hole and attach a heavy object to it. Then if we move the other edge of the taut thread along the edge of the bowl, the heavy object will descend and ascend on account of the variation in the height of the bowl's edge, in spite of the fact that it does not come to rest because the mover does not come to rest by assumption.[18]

This example of producing oscillatory motion as a result of continuous circular motion is a variation on another example, dealing with the very same notion of rest between two contrary motions that was already offered by the twelfth-century philosopher Abū al-Barakāt al-Baghdādī (d. 1152). Al-Baghdādī stipulated that one could produce such oscillatory motion by drilling a hole in the middle of a ruler and passing a thread through that hole. If one were to attach at one end of the thread a plumb line, and hold the other end with his hand, then as one moved his hand continuously from one end of the ruler to the other, the plumb line would oscillate up and down without coming to rest in between the contrary motions since the cause of those motions did not come to rest.[19]

Commentators who came after Shīrāzī continued to draw attention to those consequences, but mostly focused on denying the moment of rest between the two contrary motions, rather than the production of linear motion as a result of circular motion. In the same vein, Galileo does the same thing and uses the very same Ṭūsī Couple, that he had learned from Copernicus's *De Revolutionibus, III,4,* to disprove the Aristotelian notion of a moment of rest between two opposite motions.[20]

Only Khafrī tried to raise the issue once more from a slightly different perspective. While he agreed with Shīrāzī and others that the circular motion of the Ṭūsī Couple did in fact produce linear motion, nevertheless that linear motion itself was not as uniform as the circular motion. His concern was that the point of tangency, which moved linearly along the diameter of the larger sphere of the Ṭūsī Couple, did not in fact move at constant speed as the circular motion that caused it did. In his usual mathematical acumen and insight, his analysis came very close to defining the concepts of limits and of acceleration, but did not do so in the strict sense. He simply said that the linear motion was not the same at all points, and thus wondered if this is the same motion that Aristotle was talking about so that now it was being refuted by the Ṭūsī Couple. In his opinion it was not the same. Thus all that one could say is that the circular motion did indeed produce linear motion, but cannot say that uniform circular motion would produce uniform linear motion.

With the status of the literature that we now have from medieval Islamic times, it is hard to determine if this last aspect of the theorem, which was mainly picked up by Ṭūsī's commentators, had itself initiated any discussions among the astronomers themselves, or whether this discussion crossed over to the philosophers. What seems to be certain is that examples given by one group, such as the example that was given by Abū al-Barakāt al-Baghdādī could easily cross over to the astronomers, with some variation of course.

The variation that was offered by Shīrāzī, however, is of some significance for it seems to connect both aspects of the theorem: its implication for the moment of rest between two contrary motions and its implication for circular motion producing linear motion. By introducing a semi-spherical bowl, rather than the ruler of Baghdādī, Shīrāzī introduced the circular rim, although at varying heights from the base of the inverted bowl. And by allowing the hand to move along the circular rim, it was that motion that produced the linear oscillation of the heavy object.

Although Abū al-Barakāt's example seems to have been the direct ancestor of this problem, all the later astronomers, that I know of, who cited this continuous oscillatory motion, would use a variation on Ṭūsī's Couple to illustrate it, i.e. always staying in the context of the theorem that secured the generation of linear motion from circular motion as the Couple stipulates. And as we just said, it is hard to trace the lines of intersection between the philosophers and the astronomers in this respect so that one can determine who owes what to whom. But what seems to be certain is that the very discussion itself, as it moved from philosophical circles to astronomical ones, and back, demonstrates very clearly the shared interest the two disciplines had in such philosophical issues.

At this point I return to the issue that was raised by Ibn al-Shāṭir again, in order to illustrate once more the direct relationship between astronomy and philosophy. I have already cited the words of Averroes who objected very vehemently to the concepts of epicycles and eccentrics. I also said above that Ibn al-Shāṭir was the only astronomer I knew of who rose to the challenge. By arguing for the permissibility of the epicycles, Ibn al-Shāṭir moved away from arguing about the nature of their motions and focused on the very nature of the Aristotelian celestial world that produced the problem in the first place.

In Ibn al-Shāṭir's view, to assume that the spheres that carried the stars and the planets were all, together with the stars that they carried, made of the same simple element ether, whose very nature exhibited circular motion only, presented a very serious problem when one considered that some of the fixed stars, which were huge indeed, and some were larger than the largest planetary epicycle, emitted light while the sphere that carried them did not. To put it simply, the visible fixed stars were obviously not the same as the invisible sphere that carried them, and thus could not be made of the same element. And if they were, then that element could not be simple. In Ibn al-Shāṭir's words, Aristotle must admit that there is "some composition" (*tarkīb^{un} mā*) in the celestial element. And if this composition is allowed in the celestial realm, as it seems to be by the existence of the fixed stars, then the existence of epicycles could be of the same nature and thus one is allowed to use them.

As for the eccentrics, we have already stated that Ibn al-Shāṭir had accepted that they indeed violated the Aristotelian principles and thus should be avoided at any cost. It is for that reason that all the models that were developed by Ibn al- Shāṭir for planetary motions were all conceived as strictly geocentric models, as we have stated repeatedly before. And thus Ibn al-Shāṭir gave himself the full freedom to use as many epicycles as was necessary to account for all the observable motions. And so did Copernicus after him, who apparently faced the same problem from a slightly different angle when he shifted the center of the universe to the sun.

Whether it was a problem of eccentrics, epicycles, or the nature of celestial motion itself, almost all of the astronomers who were engaged in addressing the conflicting demands of Aristotelian cosmology as against the Ptolemaic formulation of that cosmology were not only blaming Ptolemy for his failures but were also trying to explain the difficulty of understanding the Aristotelian universe. Each in his own way was beginning to make the case against the Aristotelian conception of the universe, and was exposing the inadequacies of that conception. To Ibn al- Shāṭir, for example, the very essence of the definition of the element ether was no longer adequate and had to be changed if one were to make sense of the natural phenomena around us.

Whether Ibn al- Shāṭir, or any of the other astronomers who were engaged in this enterprise had the "right" solution for those problems or not is immaterial at this point. The important fact is that they brought the discussion to the point of collision with the Aristotelian worldview and thus pressed for the need to change it. If modern science is to be understood as an expression of the final collapse of the Aristotelian worldview, then the roots of that collapse have to be sought in those elementary, yet daring steps that were exposing the inadequacies of the view.[21]

Astronomy and Religion

As for the intersection between religion and astronomy, and through it the intersection between science and religion, we have already seen that the new astronomy of *hay'a* was developed in tandem with the religious requirements of early Islam. In a sense this new astronomy could be defined as religiously guided away from astrology. With the pressure from the anti-astrological quarters, usually religious in nature or allied with religious forces, astronomy had to re-orient itself to become more of a discipline that aimed at a phenomenological description of the behavior of the physical world, and steer away from investigating the influences its spheres exert on the sublunar region as astrology would require. Most *hay'a* texts if not all, would systematically avoid any discussion of the obvious astrological doctrines. For that reason those texts continued to be accepted in the religious circles. One can even go as far as saying that the very discipline of *hay'a* was itself born within the critiques of the religious circles that frowned upon anyone who sought the guidance of the stars in the same way the astrologers did. Within that context it is not therefore difficult to see that most *hay'a* writers were also at the same time renowned religious scholars themselves, as we shall soon see.

But before recounting the examples of the *hay'a* writers who also served as religious scholars, it is important to remember that the religious critiques not only produced two other scientific disciplines (*ʿilm al-farāʾiḍ* and *ʿilm al-mīqāt*) but had a general impact on the other sciences. The simple requirement of having to face Mecca, every time one prays, definitely required the solution of one of the most sophisticated spherical trigonometric problems of the time, known as the *qibla* problem. The *qibla,* being literally the direction one must face while praying, and knowing that the globe is supposed to be spherical, meant that one had to solve for the angle his own local horizon makes with the great circle that passes through his own zenith and the zenith of Mecca. That calculation itself requires the deployment of such trigonometric functions as the sine, cosine, tangent, and cotangent. It also meant the development of the equivalent trigonometric laws that apply to the surface of the sphere.[22]

Such kinds of trigonometric functions were not known in the Greek tradition, and the ones that were known from the Indian tradition were insufficient to solve the problem completely. As a result a whole series of trigonometric laws, like the spherical sine and cosine laws, had to be developed anew. Once that was done, there was little left to discover in the field of trigonometry.[23] One can then say that such a religious commandment, despite its apparent simplicity of requiring the believers to simply face a specific direction, was one of the reasons that gave rise to a most sophisticated discipline of spherical trigonometry. This new discipline in turn became subservient to other religious requirements, as much of it was used in almost every branch of *mīqāt* literature.[24] It also served the mother discipline of astronomy just as much, and without it, much of astronomical research, till this very day would have remained cumbersome if not impossible to conduct.

In short, the discipline of trigonometry is the best example that demonstrates the intersecting interests between the practice of one's religion and the scientific thinking that had to be developed as a result of that practice. With this in mind, and seeing how scientists, and astronomers in particular, could pose as the experts for the practice of the religious prescriptions, it should not be surprising to find scientists at this epoch closely affiliated with the religious functioning of the society. At times they were even at the helm of religious offices themselves, as we shall soon see.

In another scientific field, quite different from astronomy, we also find a rapprochement between the religious precepts and the scientific practice. In a field such as medicine, where religious thought had laid great emphasis on the need to keep a healthy body,[25] and one could quote several sayings of the prophet himself attesting to that interest, it is very difficult to miss the relationship between medical and religious practice. As a result, it should not be surprising as well to find famous physicians practicing their religious functions at the same time, and at times lending as great an authority to their religious practices as they would do to the medical one. To confirm that close association no one is surprised to find the famous Ibn al-Nafīs (d. 1288), the author of the critical commentary on Avicenna's Canon in which he criticized Galen regarding the functioning of the heart, which in turn led to the discovery of the pulmonary movement of the blood, being at the same time a practicing Shāfiʿī lawyer, who even gave lectures on Shāfiʿī law at the Masrūrīya *madrasa*.[26] In light of what we know about the status of medicine in Islamic society, joining these two functions should not require any further explanation.

Returning to the astronomers, and in particular to the theoretical astronomers, whose works have been designated so far as *hayʾa* works, one should expect to find the same close association between their scientific functions and their religious ones, especially when they had already formulated their new astronomy of *hayʾa* specifically to cast astrology out of the domain of astronomy and to respond to the religious pressures of the society. In the new configuration, theoretical astronomy, which became the domain of *hayʾa* studies, became a close ally of religious thought. At one point, during the Iranian Safavid period and thereafter, it became another subject of religious instruction. In a separate publication I have argued for the interpretation of the phenomenon of the continuous use of the Arabic language in the production of *hayʾa* texts, even when the native language of the writers was Persian, as a phenomenon of integrating astronomy into the school curriculum. These school curricula had always weighed heavily in the direction of Arabic as the language of the primary religious texts.[27] From interviews with graduates of modern day Iranian seminaries, my understanding is that this incorporation of *hayʾa* texts in the religious school curricula still goes on till the present day.

With this alliance, it is not surprising to find one of the most productive astronomers, the same Naṣīr al-Dīn al-Ṭūsī (d. 1274)[28] who produced the famous Ṭūsī Couple in the context of his attack on Ptolemaic astronomy, being at the same time a great Ismāʿīlī scholar first and then an acknowledged authority on general Shiʿite thought. His own spiritual autobiography *Sayr wa-sulūk,*[29] as well as his doctrinal text *Rawḍat al-taslīm,*[30] speak directly to his authoritative status within the Ismāʿīlī religious thought. His *Awṣāf al-ashrāf*[31] and his *Tajrīd al-iʿtiqād*[32] also speak to his much more exalted status among the Sufi adepts and the twelver Shiʿites, respectively. To some (especially Shiʿite biographers not skilled in the astronomical sciences), he was primarily a religious figure who may have had a side interest in astronomy. His student and former colleague Quṭb al-Dīn al-Shīrāzī (d. 1311) also produced several voluminous works on theoretical astronomy, two of which were detailed commentaries on Ṭūsī's *Tadhkira,* with their own original contributions to the field. In addition, Shīrāzī also occupied the position of a practicing judge in the cities of Sīvas and Malaṭiya, in 1282, after his affiliation with the Marāgha observatory, and while he was still writing his first commentary on Ṭūsī's *Tadhkira.*[33] He also acted as an intermediary between the Ilkhānids and the Mamluks, once the Ilkhānids had converted to Islam.

His mission was obviously an exercise of his religious duty to bring peace between two warring Muslim potentates.

Shīrāzī's religious works are as impressive as his astronomical works. Since he had become a *ḥadīth* scholar in his own right, his book *jāmiʿ uṣūl al-ḥadīth* naturally became one of the main references for this type of religious literature at this later period. And so did his work on the prophetic tradition *sharḥ al-sunna*. But his elaborate commentary on the Qurʾān, *Fatḥ al-mannān fī tafsīr al-qurʾān,* definitely attests to his wide-ranging control of the many religious disciplines of his time.

It was also his astronomical as well as his religious teachings that triggered the interest of his own student Niẓām al-Dīn al-Nīsābūrī (d. 1328), known as al-Aʿraj (The Lame), to write on these two subjects as well. Nīsābūrī's two voluminous astronomical works, *Sharḥ al-tadhkira* (Commentary on the *Tadhkira*), and *Sharḥ al-majisṭī* (Commentary on the *Almagest*), both commented on the two works of Ṭūsī's that are mentioned in the titles. And both of Nīsābūrī's works continued to be taught in schools well after the death of the author. A remark made in a fifteenth-century text about the astronomical education in the school of the most famous potentate and astronomer Ulugh Beg (d. 1449) attests to the use of Nīsābūrī's astronomical texts in the instruction.[34] But Nīsābūrī's commentary on the Qurʾān, *Gharāʾib al-qurʾān wa-raghāʾib al-furqān* (the Unusual [expressions] in the Qurʾān and the appealing [features] of the Furqān [a synonym of the Qurʾān]) is by far the most elaborate of his works as it falls in several volumes in the printed version.[35]

Ibn al-Shāṭir (d. 1375) of Damascus, of the great astronomical fame, was only a *muwaqqit* (Timekeeper) at the Umayyad mosque in the same city.[36] As a functionary of the mosque he must have derived his livelihood from the religious endowment of the mosque, and just like a judge was also considered a religious functionary. He apparently conducted his theoretical research on planetary motions in perfect synchronism with his religious duties. Naturally, he also developed instruments, such as sundials and the like, to tell the appropriate times of prayers as part of his religious duties, but also must have enjoyed developing them for their mathematical projection interest. His astronomical work, however, has now become of great interest after it was demonstrated, in the late 1950's, that his lunar model was identical to that of Copernicus, and his technical treatment of the motion of the planet Mercury used the same Ṭūsī Couple that was used by Copernicus as well. His model for the upper planets, which was also adopted by Copernicus after shifting the center of the universe to the sun, also included the use of ʿUrḍī's Lemma, and continues to be at the center of the ongoing research that will one day determine the routes by which Copernicus knew of this astronomer's work and of the work of his colleagues from the Islamic domain.

Mullā Fatḥallāh al-Shirwānī (c. 1440) who also wrote a commentary on Ṭūsī's *Tadhkira,* also called *Sharḥ al-tadhkira,* from which we know a great deal about the astronomical activities at Ulugh Beg's school, was obviously first and foremost a religious functionary as his title Mullā implies. In addition he was apparently one of the brightest students of Qāḍīzādeh al-Rūmī in astronomy. Some of his religious works have also survived to attest to his engagement in the religious fields as well.[37]

Finally, the works of the most prolific astronomer of the sixteenth century, Shams al-Dīn al-Khafrī (d. 1550), which we have considered at more than one occasion before as examples of the latest sophistication in Islamic astronomy, were at the same time the best examples of the use of mathematics as a language of science. This brilliant astronomer was also a renowned religious scholar in his own right.[38] His biographers, who speak of him mainly as a religious scholar, only marginally mention his most famous astronomical work *al-Takmila fī sharḥ al-tadhkira* (Completing the Commentary on the *Tadhkira*). At one point in his career he apparently fulfilled the function of the official Shiʿite jurist in Safavid Iran. The same biographers also report his issuing juridical opinions (*fatwas*) on matters pertaining to the Shiʿī doctrines before the arrival of al-Muḥaqqiq ʿAlī b. al-Ḥusain al-ʿĀmilī (d. 1553) from Lebanon to that country in the early part of the sixteenth century.[39]

Conclusion

The intersections between theoretical astronomy with philosophy and with religion are too numerous to recount. It is certain, however, that both of those disciplines had a very fruitful interaction with Islamic theoretical astronomy, thus allowing the latter to cast doubt on much of the Aristotelian cosmology in the first instance, and to reconstruct itself as a religiously acceptable science in the eyes of religious authority in the second. This association with religion, contrary to what one would expect when using the

European paradigm of conflict between science and religion, was apparently very healthy, and continued to support astronomers, one after the other, even at times when the astronomers' only source of income was provided by the religious institutions in which they served.

With this image, it becomes very difficult to document a paradigm of conflict between religion and science in Islamic society. But this does not mean that all astronomical disciplines were treated in the same fashion. One can easily document a conflict between religion and astrology as we have said several times before, since astrology was perceived early on as the purpose of astronomical research in the first place in full conformity with the Greek tradition.[40] But this does not mean that astrology was completely banished from Islamic society.

One can also consider the tenuous relationship between the production of *zījes* that served as ephemerid tables for the astrologers, and the later production of *mīqāt* tables that served only religious purposes. Ibn al-Shāṭir's *al-Zīj al-jadīd,* for example, can function as a tool for astrologers, despite the author's original intention to use it mostly for his religious timekeeping activities. It is in such areas that the disciplinary borders begin to be blurred, and the difficulty arises when attempting to characterize a specific production one way or the other.

6 Islamic Science and Renaissance Europe: The Copernican Connection

The new mathematical tools that were developed by the astronomers of the Islamic world did not only prove to be very useful for the emergence of new ways of looking at theoretical astronomy, as we have already seen, but also allowed astronomers to manipulate mathematical models so that they could meet the observational requirements. We have also seen this trend culminate in the works of Khafrī who demonstrated a total mastery of mathematics so much so that he attained complete freedom to use whatever mathematical configuration he wished in order to represent the same physical observational phenomenon. Mathematics became a new language, and became an efficient tool of astronomy, at least as far as the works of Khafrī were concerned.

The works of Ibn al-Shāṭir, on the other hand, with their emphasis on the strict Aristotelian cosmological requirements of abolishing eccentrics, also liberated astronomical models from the often cumbersome multiplicity of shapes and forms and unified all the planetary models with one geocentric format that could be easily applied to one planet at a time by simply changing the parameters of the two epicyclic spheres deployed in each model. In a roundabout way, the unintended consequences of these unified models produced the "strange" development that allowed them to be transferred into heliocentric models, despite the fact that there was no shred of support for such heliocentrism in the then reigning Aristotelian cosmology. All that someone like Copernicus had to do was to take any of Ibn al-Shāṭir's models, hold the sun fixed and then allow the Earth's sphere, together with all the other planetary spheres that were centered on it, to revolve around the sun instead. As we shall soon see that was the very step that was taken by Copernicus when he seemed to have adopted the same geocentric models as those of Ibn al-Shāṭir and then translated them to heliocentric ones whenever the situation called for it.

All those shifts in astronomical thought that took place in Islamic civilization had some very serious consequences. Not only did they expose all the factual and observational errors of Greek astronomy, but also demonstrated, in the most convincing manner, the inconsistencies of that astronomy with its very own cosmological presuppositions. In the later centuries, when Islamic astronomy reached its theoretical maturity, starting with the continuous rise of analytical discussions of planetary theories after the thirteenth century, one could hardly find a serious astronomer who did not make an attempt at reformulating Greek astronomy. At that time, no one could hope to practice astronomy, and be taken seriously, if he did not make an effort to solve the thorny cosmological problems of Greek astronomy. One astronomer after another, tried their hand at devising new mathematical models that represented a much more consistent cosmological picture of the Greek astronomical tradition. At the same time, those models could account perfectly well for the same observations, which were used by Ptolemy, in the first place, to construct his own predictive mathematical models for planetary motions.

This constant search for more consistent representations of planetary motions came to characterize the whole field of Islamic astronomical research, especially in the later centuries following the thirteenth century. The movement of continuously reforming Greek astronomy became so important that it apparently attracted the attention of astronomers from outside the Islamic domain. We know, for example, that Byzantine astronomers, like Gregory Chioniades (fourteenth century) and others, would travel to the Islamic lands in order to learn of the latest developments in Islamic astronomy and to report their findings back to their compatriots in their own Greek language.[1] In fact, one can also document the dependence of the late Byzantine astronomy on Islamic astronomy by simply browsing through the technical terminology that was used by Byzantine astronomers at the time. This terminology demonstrates very clearly that it bore a much closer resemblance to the Arabic sources, from which it was derived, than to the classical Greek texts such as those of Ptolemy.[2]

With the fall of Constantinople in 1453 to the Ottoman Turks, and the ultimate demise of the Byzantine empire, a good number of Byzantine scholars escaped westward, at times together with their books. But by then the Byzantine civilization had been in direct contact with the Islamic civilization for centuries already. And as a result those books inevitably bore the marks of having been influenced by the intellectual production of the Islamic civilization, and thus contained some of the developments that had already taken place in that civilization. In a way, these Byzantine contacts with Europe were much more complex than the contacts that had already taken place during the Middle Ages between the Islamic world and the Latin West. In the first, we saw several Arabic works that were translated into Latin, sometimes undigested, and mostly restricted to the confines of linguistic contours of the texts. With the new Byzantine contacts one can now distinguish a new manner of transfer of texts. Arabic and Persian scientific texts were apparently already digested in the Byzantine Greek sources for a period of about two centuries or so, before those Byzantine texts were brought into Europe. This time, their contents were not apparently translated into Latin. Rather, because of the emphasis of the Renaissance intellectual environment on the Greek language, they were read in the original Greek. The best of their contents, which were originally Arabic and Persian could now be directly assimilated into the Latin texts, without having to translate the whole text into Latin. This method of transfer of knowledge constitutes by itself a new phenomenon that is rarely acknowledged by all those who study the transfer of knowledge across cultures. More importantly, this later transfer of knowledge from the world of Islam to Europe this time spoke directly to the contemporary science of the Renaissance, where its impact can be best detected, as we shall soon see.

About the same period that witnessed the various contacts between Byzantium and the world of Islam there were various other contacts as well. One should pay attention to the several European travelers who performed their pilgrimage to the Islamic world, either to visit the Holy Lands, or to simply seek knowledge from the lands of Islam. This contact too must have brought some of the findings of the Islamic world to the European countries. What did they bring in particular is a matter that is currently under investigation and promises to produce some very interesting results.

Our current state of knowledge, however, can already inform us about some of those contacts and the nature of the information that was exchanged. We already know, for example, that those contacts brought some very advanced theoretical findings from the lands of Islam to Renaissance Europe, findings that were apparently highly appreciated by the

European scientists who consumed this material and ended up incorporating it in their own works.[3] And yet our research in that particular area is still in its infancy and once it is completed, it promises to change much of our world view, about cultural transmissions, cultural contacts, the nature of the European Renaissance, and the earliest roots of modern astronomy.[4]

Connections between Renaissance Europe and the World of Islam

A sheer accident, in 1957, brought to the attention of Otto Neugebauer, who was then working on the mathematical astronomy of Copernicus, a text that contained the theoretical astronomy of the famous Damascene astronomer Ibn al-Shāṭir (1375). It did not take the genius of Neugebauer much, despite the fact that he did not read Arabic himself, to realize that Ibn al-Shāṭir's lunar model was indeed identical, in every respect, to that of Copernicus (1543) (figure 6.1). The former model, has survived in Ibn al-Shāṭir's text *Nihāyat al-sūl fī taṣḥīḥ al-uṣūl* (Final Quest Regarding the Corrections of the [Astronomical] Principles), and was brought to Neugebauer's attention by his close associate and friend Edward Kennedy. Kennedy was then a professor of mathematics at the American University of Beirut, and a distinguished historian of Islamic astronomy and mathematics in his own right. His own encounter with Ibn al-Shāṭir's work at the Bodleian Library was in itself a pure accident as well, and now belongs to the world of legends. But that discovery, together with its ensuing discussion with Neugebauer, gave rise to the publication of an article in *Isis* by Victor Roberts, a student of Kennedy, who called it "The Solar and Lunar Theory of Ibn al-Shāṭir: A pre-Copernican Copernican model."[5]

Naturally, such a finding jolted the scholarly community somewhat, for up till then the prevailing belief was that Renaissance science, unlike its medieval counterpart, was considered to be a European self-contained creation, almost *ex nihilo*. Or if one were to widen his horizons and look outside the particular confines of the European environment one was supposed to find Renaissance science taking its inspiration from the classical Greek sources, rather than any other source, least of all Islamic sources. Common opinion stipulated a European enmity with things Arabic and Islamic and thus no one would have expected a fruitful contact between the two.[6] For Neugebauer to find that there was a direct connection between the works of Copernicus and the Arabic planetary theories, which were produced in the Islamic world about 200–300 years before, was a discovery that was shock-

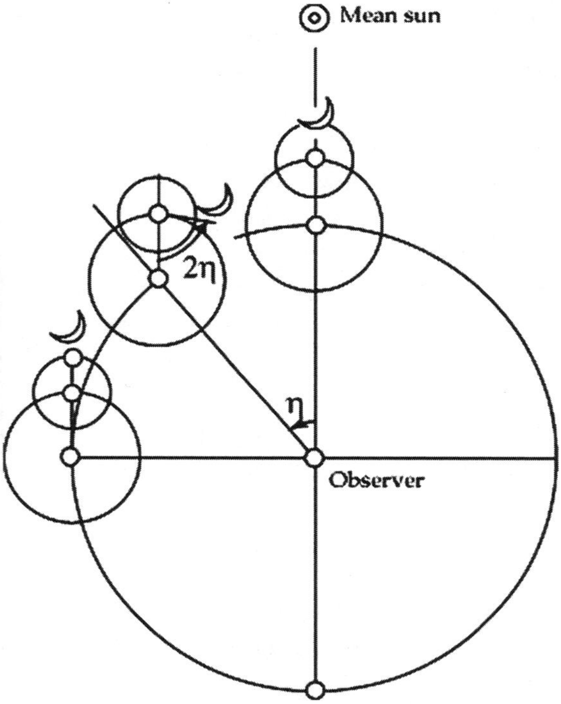

Figure 6.1
The lunar model of Ibn al-Shāṭir and Copernicus.

ing in its own time and has not been fully digested yet in the secondary sources dealing with the history of science in general. Only a handful of researchers seem to know about it still, and to appreciate its full significance.

But this finding opened the door to further investigation. It was also Neugebauer who widened the circle of research and began to look for other similarities of ideas between the works of Renaissance scientists and the scientists of the Islamic world. And it was in that context that he revisited a chapter from the *Tadhkira fī ʿilm al-hayʾa,* of Naṣīr al-Dīn al-Ṭūsī (d. 1274), which had already been translated into French by Bernard Carra de Vaux in 1893 and published under the title "Les spheres célestes selon Nasîr-Eddîn Attûsî."[7] In this chapter, which was originally written by Ṭūsī in 1260–61, Ṭūsī formalized as well as generalized and now supplied a rigorous mathematical proof to the famous theorem that has come to be known in the literature as the Ṭūsī Couple, which we have seen before.

We have also seen that the first articulation of this theorem, had already been proposed in 1247, in yet another work by Ṭūsī, the *Taḥrīr al-majisṭī* (Redaction of the *Almagest*), which is yet to be edited and published. And, as I have noted, this earlier first articulation was specifically proposed in order to respond to the failings of the Ptolemaic latitude theory of the planets. This background of the theorem was neither mentioned by Ṭūsī nor was it signaled by Carra de Vaux, on his own, nor was it apparently known to Neugebauer who did not work on the Arabic manuscripts directly.

Ṭūsī's Couple, as we have seen, offered a general solution to the problem of generating linear motion from a combination of circular motions. It was expressed in terms of the motion of two spheres, usually called in the Arabic astronomical literature that followed *al-kabīra wa-l-ṣaghīra* (the large and the small). As has been noted, one of those spheres was taken to be twice the size of the other, and in the initial setting the spheres were taken to be internally tangent at one point. With the motion of the larger sphere at any speed, and the motion of the smaller sphere at twice that speed, in the opposite direction, the point of tangency was then found to oscillate along the diameter of the larger sphere, thus producing the required linear motion. In 1260–61, after supplying the formal mathematical proof to this theorem, Ṭūsī went on to use it in the lunar model and then in the model for the upper planets, as we have already seen.

Carra de Vaux's translation of this chapter gave all the contents of the original Arabic in a rather faithful French, but was then concluded with an assessment by Carra de Vaux himself. In it, and on the basis of his encounter with this particular work of Ṭūsī, de Vaux summed up the general character of Arabic astronomy. In de Vaux's time, and because very little else was known then from the astronomical production of the Islamic civilization from these later periods, de Vaux was emboldened to say, that while Arabic astronomy did not hold Ptolemy's work with much regard [an understatement indeed about a chapter that was devoted specifically to critiquing the problems in Ptolemaic astronomy], it did not, on its own, have enough "génie" to transform astronomy altogether, and instead suffered from a general "faiblesse" and "mesquinerie" that did not allow it to develop further. From such a statement, one has to draw the conclusion that de Vaux could not fully appreciate the importance of the chapter that he was translating at the time. We shall have occasion to return to this issue when we speak about the so-called age of decline of Islamic science.

With a completely different attitude, and being immersed in the mathematical astronomy of Copernicus at the time, Neugebauer could immediately see the essence of Ṭūsī's problem, because he could also see that it was the same problem that was faced by Copernicus later on, in *De Revolutionibus* III, 4.[8] Both astronomers needed to utilize a mechanism that allowed them to generate linear motion from circular motion or combinations thereof, as I have said several times already. And both used the same Couple, except for one difference: Ṭūsī knew that he was introducing a new theorem in 1247[9] and again in 1260–61, which was nowhere to be found in any earlier Greek source, and said so, while Copernicus silently went ahead and described the same theorem and produced a very similar proof as we shall see, without mentioning that he had invented the theorem or the proof himself, nor that he had seen it in any other source. He only mentions the vague reference to a statement by Proclus,[10] referring to the latter's commentary on Euclid, in which Proclus says that linear motion could be gotten from circular motion. But for those who read Proclus closely will immediately realize that Proclus was talking about curved lines and straight lines being produced from one another and not oscillating motion resulting from complete circular motion as was required by Ṭūsī and Copernicus after him.

By 1973, Willy Hartner discovered a remarkable feature in Copernicus's proof of the same theorem.[11] By comparing Ṭūsī's proof, which was completed in 1260–61, with that of Copernicus, which was published in 1543, Hartner discovered that the two proofs (figure 6.2) carried the same alphabetic designators for the essential geometric points. That is, where Ṭūsī's proof designated a specific point with the Arabic letter "*alif,*" Copernicus's proof signaled that same point with the equivalent phonetic Latin letter "A," where Ṭūsī, had "*bā',*" Copernicus had "B," etc., except in one case where Ṭūsī had "*zain*" and Copernicus has "F." On the basis of the letter correspondences, letter to letter, Hartner ventured to say that Copernicus must have known about Ṭūsī's work while in Italy. The implication that was also spelled out by Hartner was that Copernicus must have had access to Ṭūsī's work in some indirect form, since as far as we know neither Copernicus could read Arabic, nor was Ṭūsī's text, in which the theorem appeared, was ever translated into Latin. To Hartner, it meant that Copernicus must have recruited someone who could explain to him the diagram, while he took notes and used those notes later when he came to write the *De Revolutionibus*.

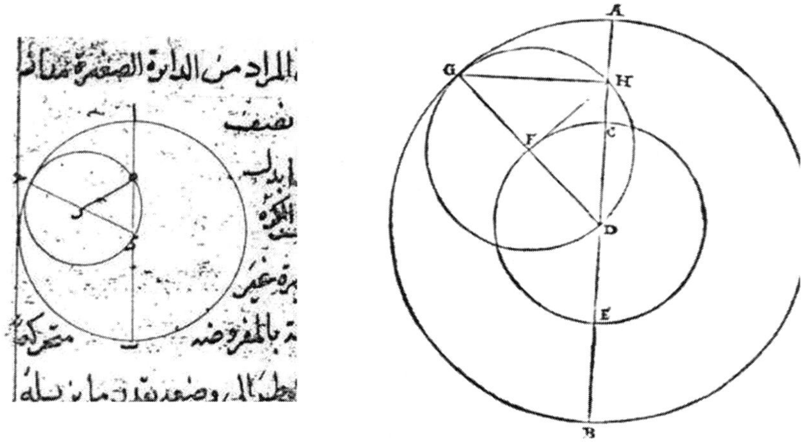

Figure 6.2
Proofs of the Ṭūsī Couple from the works of Ṭūsī (left) and Copernicus (right), showing the identity of the lettering of the diagrams. Wherever Ṭūsī had *alif* Copernicus had *A*, and wherever Ṭūsī had *bā* Copernicus had *B*, and so on, except that where Ṭūsī had *zain* for the center of the smaller sphere Copernicus had *F*. See figure 6.3.

In a more recent reassessment of Hartner's results, I added the Arabic manuscripts evidence to account for the variation between the "*zain*" and the "F" in the two proofs.[12] By comparing Arabic manuscripts from the medieval period, and noting that the two Arabic letters "*zain*" and "*fā*ʾ" that were usually used to designate geometric points, the similarities between those two letters were in fact so close that it would be quite easy for someone, who was not experienced enough with Arabic manuscript traditions, to misread the "*zain*" for a "*fā*ʾ" (figure 6.3). I ventured to say that either Copernicus himself or someone sitting next to him, looking at an Arabic text of the proof of Ṭūsī, simply misread the "*zain*" in the original Arabic manuscript for a "*fā*ʾ," thus leading Copernicus to introduce the sole variation in the lettering of the two proofs.

But misreadings and variations are on their own very useful for detecting textual transmissions. For as a result of these reading "mistakes" I became quite confident about the conclusion just drawn: that Copernicus was either himself working from an Arabic manuscript where he mistook the "*zain*" for a "*fā*ʾ," which is unlikely since we do not know that he knew any Arabic at all, or that he was reading, at least the diagram, with someone else's help

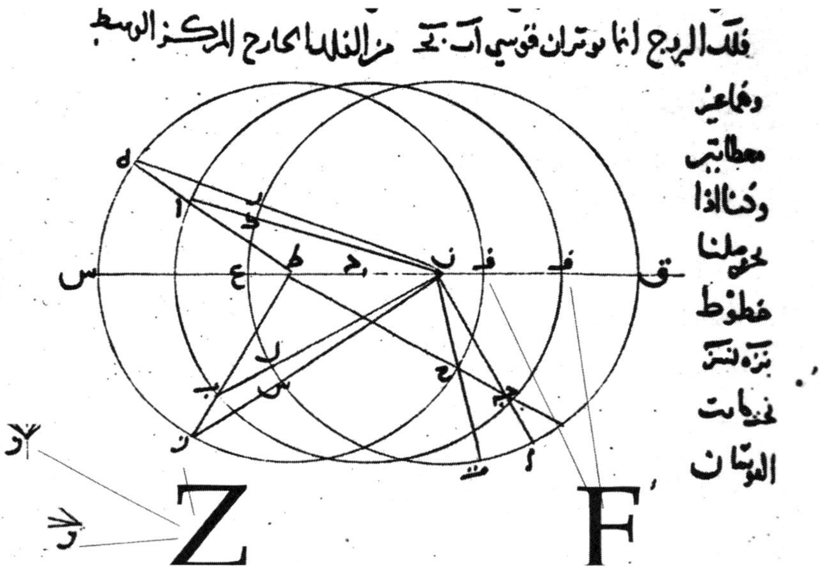

Figure 6.3
A medieval Arabic manuscript exhibiting the similarities between the letters *zain* = *Z* and *fāʾ* = *F*.

who made the same mistake. Furthermore, the complete conformity of the other geometric points between the two proofs now makes the issue of coincidence and independent discovery a most unlikely scenario.

Therefore, not only did Copernicus apparently seek to solve the same problem of the Greek astronomical tradition by adopting the same approach (that is, adding a mathematical theorem that allowed for the generation of linear motion from circular motion); he also used a theorem that had been invented by Ṭūsī about 300 years earlier, and supplied a proof that was very similar to the one supplied by Ṭūsī, with a slight modification in protocol, but still adhered to the same geometric points that were used by Ṭūsī in the original proof. All of this cannot be mere coincidence, as some people still like to think. And future research both in the world and works of Copernicus, as well as the world and works of the Arabic-writing astronomers who preceded him will, I am sure, eventually uncover the exact route by which Copernicus learned of the earlier astronomical findings of the Muslim astronomers.

Mu'ayyad al-Dīn al-ʿUrḍī (d. 1266) was a colleague of Ṭūsī, and a distinguished astronomer and engineer in his own right. His distinguished fame must have been the deciding factor for Ṭūsī when he hired him to build the observational instruments for the famous Marāgha Observatory.[13] The observatory itself was established in 1259 in the city of Marāgha in northwest modern-day Iran, under the patronage of the Ilkhānid monarchs.[14] Because of the concentration of the astronomers who worked at the observatory, and because of the recently-found connection between their works and the work of Copernicus, this observatory has now become very famous in the secondary literature. ʿUrḍī's fame, however, was obviously based on his most important work which was simply called *Kitāb al-hay'a* (Book on Astronomy).[15] In it he attempted to revamp the whole of Greek astronomy, having been obviously motivated by the same considerations, which had been discussed for generations before him within the intellectual circles of the Islamic Civilization. The most important problem for the time was still encapsulated in the discussion of the inadmissibility of the equant sphere, on account of the well-known absurdity this concept produced.

Trying his own hand at the resolution of this problem, ʿUrḍī proposed a new simple theorem which allowed him to reconstruct the Ptolemaic model[16] for the upper planets by adding new spheres and epicycles, but still accounted for the same observations that were reported by Ptolemy, without having any of the absurdities that were adopted by Ptolemy. In ʿUrḍī's model, all the spheres moved uniformly in place around axis that passed through their centers. One could therefore say that in this model ʿUrḍī managed to avoid the use of the Ptolemaic equant, but did not avoid accounting for its essential observational effects.

The theorem itself (figure 6.4), now known as ʿUrḍī's Lemma, is extremely simple. It stipulates that for any two lines (such as *AG* and *BD*) that are equal in length and that form equal angles with a base line *AB,* either internally or externally, the line *DG,* joining the other extremities of these two lines, is parallel to the base line *AB*.

In a sense, the way this lemma functioned in ʿUrḍī's model for the upper planets is very similar to the way Apollonius's theorem functioned in the solar model of Ptolemy. In the latter the Apollonius theorem allowed Ptolemy to equate the eccentric and the epicyclic models, and to replace one with the other. In the same theorem the radius of the epicycle was equal to the eccentricity of the solar model, and the epicycle itself moved at the same

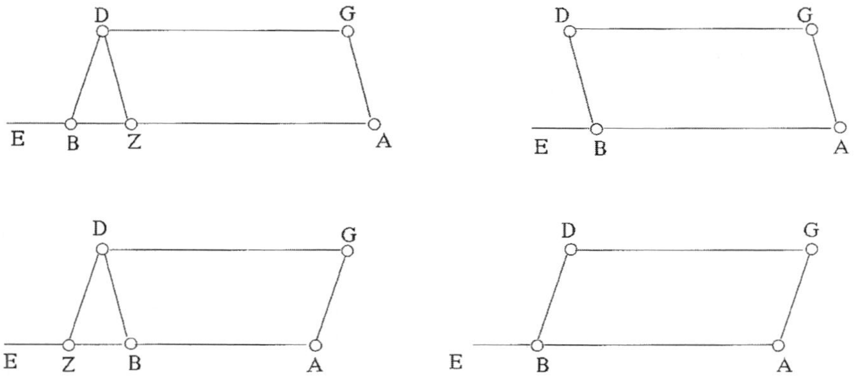

Figure 6.4
A general representation of the four cases of ʿUrḍī's Lemma as it appeared in the original manuscripts. This illustrates the four cases of possible equations between internal and external angles.

speed as the concentric sphere, but in the opposite direction, thus allowing the external angles to remain equal. In ʿUrḍī's Lemma, the additional epicyclet, which was added to the Ptolemaic model, had a radius that was equal to only one half of the eccentricity of the upper planets. And the motion of the epicyclet was in the same direction as that of the deferent and at the same speed, which required that the internal angles be equal for the parallel lines to be achieved. This, in turn, required that ʿUrḍī explicitly enunciate the lemma that produced these properties. Looking at it simply as a mathematical structure, one can easily see that, with its use of a small epicyclet, this lemma allowed for the transfer of half the eccentricity from the line of apsides to the circumference of the deferent, just as the Apollonius theorem allowed for the transfer of the whole eccentricity to the epicyclic radius at the circumference of the deferent. In both cases one can speak of epicycles compensating for eccentricities, and thus allowing the transfer of one mathematical model from one physical reality to another. And in this same sense one can speak of the Apollonius theorem as a special case of ʿUrḍī's Lemma.

Once enunciated and proved by ʿUrḍī, this lemma became immediately very popular. Someone like Quṭb al-Dīn al-Shīrāzī (d. 1311), who was primarily a student of Ṭūsī and a member of the Marāgha group, was probably the first to use it in a context other than the one for which it was intended.

He first used it in his lunar model, and then he incorporated it in the same model for the upper planets that he had borrowed from ʿUrḍī, theorem and all. In both of Shīrāzī's works, which were written few years apart in the 1280s, ʿUrḍī's model for the upper planets remained to be Shīrāzī's favorite model, despite the fact that both of the said works of Shīrāzī were themselves written as commentaries on the work of Ṭūsī, and not that of ʿUrḍī. By preferring ʿUrḍī's Lemma over the solutions that were offered by Shīrāzī's very own teacher Ṭūsī, which made use of the Ṭūsī Couple, Shīrāzī's choice can only be taken as a testament to the popularity of ʿUrḍī's Lemma.

Ibn al-Shāṭir (d. 1375), who lived a full century later, followed suit. After using the equivalent of the Apollonius theorem to shift the eccentricities to epicyclic attachments, as we have already seen, in order to return to strict geocentric cosmology, he added to what can be called the Apollonius epicycle another ʿUrḍī epicycle in order to account for the motion around the equant as was done by ʿUrḍī. In essence, Ibn al-Shāṭir's model for the upper planets is the same as that of ʿUrḍī, except for the transposition of the eccentricity that was used by ʿUrḍī, and which was one and a half times as large as that of Ptolemy, to an epicycle with the same radius. The rest of the model preserved the same properties. That is, it deployed the same dimensions for the ʿUrḍī epicycle, and the same motion conditions, exactly as was done by ʿUrḍī (figure 6.5).

As a new tool, ʿUrḍī's Lemma, was also used by other astronomers and in new areas of application as well, as we have also seen before, most notably by ʿAlāʾ al-Dīn al-Qushjī (d. 1474) and by Shams al-Dīn al-Khafrī (d. 1550) in their respective constructions of their models for the motion of the planet Mercury. Both of these astronomers could assume that they had this new tool in their repertoire, to use it whenever they pleased. The fact that it was used by the previous astronomers for some 200 years must have been taken as a proof that first, it withstood the test of time, and second, that it was a more general form of the Apollonius theorem. It clearly allowed for the transposition of eccentricities to deferent circumferences. But most importantly, it also allowed for the transposition of reference points for uniform motions, such as the motion of the equant, or any other center of motion that was required by the observations.

And since Copernicus had used the same model for the upper planets that was used by Ibn al-Shāṭir (figure 6.6), with the additional transposition of the center of the universe to the sun of course, in that sense Copernicus too ended up using ʿUrḍī's Lemma, as Ibn al-Shāṭir had done before him.

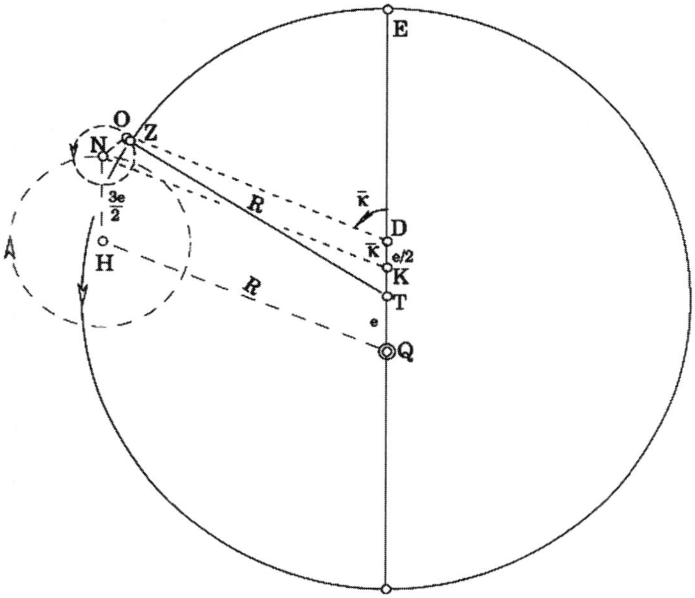

Figure 6.5
Ibn al-Shāṭir's model for the upper planets, clearly showing the incorporation of ʿUrḍī's Lemma.

Copernicus, however, did not apparently realize the full significance of the two components of Ibn al-Shāṭir's model (the Apollonius and the ʿUrḍī components), and simply used the model as a whole, by transposing it to heliocentrism as we just said. As a result he did not feel that he had to produce a formal proof for the ʿUrḍī component as he had done with the Ṭūsī Couple. It was Kepler (1630) who wrote to his teacher Maestlin (1631) to ask him specifically about this omission on the part of Copernican astronomy, as was already demonstrated by Anthony Grafton.[17] And it was Maestlin who supplied the proof of that specific case of the ʿUrḍī Lemma which applied to the model of the upper planets, without supplying the general proof as ʿUrḍī had done.

For our purposes, the almost unconscious use of ʿUrḍī's Lemma by Copernicus, in a construction that was identical to that of Ibn al-Shāṭir, minus heliocentrism of course, must raise doubts about Copernicus's awareness of the roots of all the mathematical techniques that were put at his disposal. Would he have proposed this very new theorem? And would he have offered a formal proof of it as was done by ʿUrḍī, and as he did for the

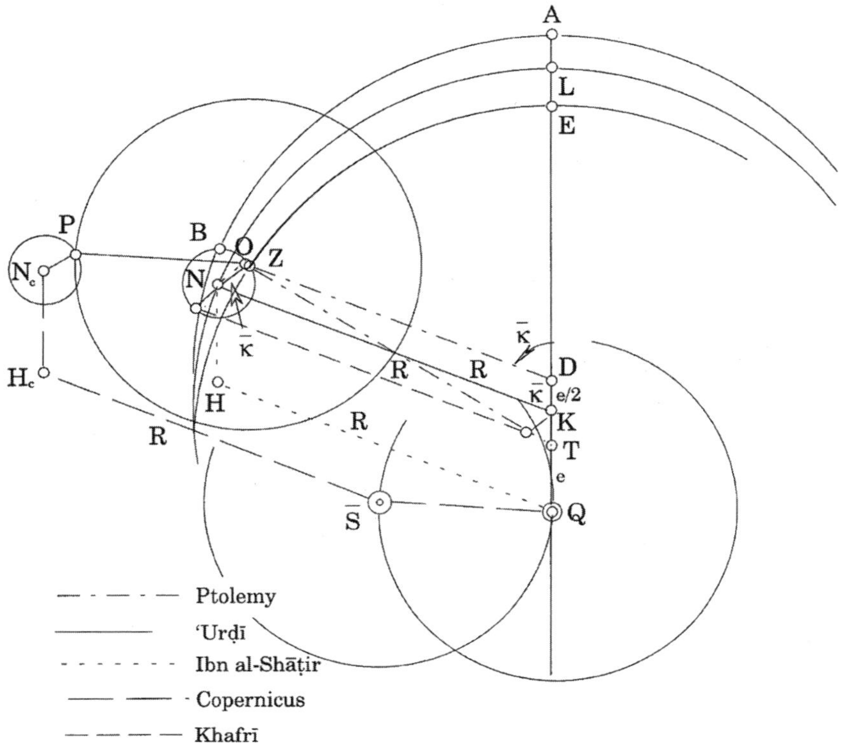

Figure 6.6
A schematic representation of the model for the upper planets as conceived by Ptolemy, ʿUrḍī, Ibn al-Shāṭir, Copernicus, and Khafrī. If one thinks of the radii of the spheres as vectors, all the models predict the same position for the planet *P*.

complementary Ṭūsī Couple that he also had to use, had this new theorem not been at his disposal from the Islamic sources? I doubt that very much.

But this example of the use of Ibn al-Shāṭir's model by Copernicus does not even begin to illustrate the extent of the technical interdependence between the two astronomers. For in addition to the identical construction of the lunar model, which we already discussed before, and now the identity of the model for the upper planets, Ibn al-Shāṭir and Copernicus also used identical techniques for resolving the last model of the classical planetary theory (the Mercury model).

If one were to compare Copernicus's model for the planet Mercury to that of Ibn al-Shāṭir, and if one were to allow for the simple mathematical transposition from geocentrism to heliocentrism and vice versa, one will be struck by the similarities between the works of the two astronomers. In this instance, both Ibn al-Shāṭir and Copernicus used a construction of a mathematical model that deployed in its last connection the use of a Ṭūsī Couple, in order to allow for the planet's epicycle to be brought close to the Earth, at the two perigees which were observed by Ptolemy, and to recede away at the apogee. The complete agreement on the technique of achieving this oscillatory motion, while moving the epicycle nearer and farther, raises the question of the possible influence of one astronomer over the other, especially when we already know of the other similarities that we have already witnessed in the other contexts. But the case of the Mercury model in particular brings some remarkable evidence for the case of the interdependence between the two astronomers; this evidence elevates the discussion of the similarities to a whole new level.

When Swerdlow studied the first version of Mercury's model in Copernicus's *Commentariolus,* which was itself written before 1514, he immediately realized that Copernicus was not aware of the full significance of the model he was describing. For example, Copernicus thought that the planet would have its largest orbit (i.e. the size of its epicycle would look the largest) at quadrature (i.e. when the center of the epicycle—or the Earth in Copernicus's language—was at 90° away from the apogee) while the model itself would predict two such largest appearances when the center of the epicycle, or the Earth, was at 120° on either side of the apogee, exactly as Ibn al-Shāṭir's and the Ptolemaic models would have predicted, and not at 90° as Copernicus now claimed. Having realized that, Swerdlow said:

> Copernicus's model for Mercury which, like his other planetary models, is identical to Ibn ash-Shāṭir's model except for the heliocentric representation of the second anomaly, is based on exactly this separation of the equation of center from the motion of the center of the eccentric in Ptolemy's model.[18]

While discussing the point, Swerdlow went on to explain why Copernicus did not seem to realize where his model would produce Mercury's closest position to the Earth (figure 6.7):

> There is something very curious about Copernicus's description. . . . Copernicus apparently does not realize that the model was designed, not to give Mercury a larger orbit (read *epicycle*) when the Earth (read *center of the epicycle*) is 90° from the apsidal line, but to produce the greatest elongations when the Earth (read *center of the epicycle*) is ±120° from the aphelion (apogee).[19]

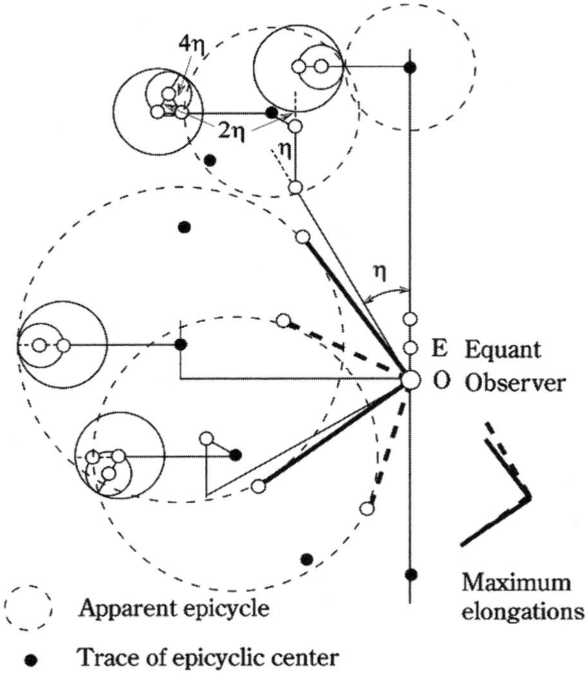

Figure 6.7
A model depicting the motion of the planet Mercury as described by Ibn al-Shāṭir. Copernicus adopted the same model without fully realizing the manner in which it functioned. Copernicus seems not to have realized that the *apparent* size of an object depended on the size of the object and on the object's distance from the observer. It seems that Copernicus confused the size of the planet's orbit, as marked by the dashed circles, with its appearance for an observer at point *O*. Although the planet's orbit indeed reaches its greatest size when the epicyclic center is 90° from the apogee, for an observer at point *O* the dashed epicycle does not *appear* the largest at that point, as Copernicus contends. Rather, it *appears* largest when the epicyclic center reaches ±120° from the apogee, as would be predicted by the observations of Ptolemy which were followed by Ibn al-Shāṭir, and as can be seen from the comparison between the maximum elongation angles at 90° (solid lines) and at 120° (dashed lines).

With all these problems laid bare, Swerdlow concluded:

This misunderstanding must mean that Copernicus did not know the relation of the model to Mercury's apparent motion. Thus it could hardly be his own invention for, if it were, he would certainly have described its fundamental purpose rather than write the absurd statement that Mercury "appears" to move in a larger orbit when the Earth is 90° from the apsidal line. The only alternative, therefore, is that he copied it without fully understanding what it was really about. Since it is Ibn ash-Shāṭir's model, this is further evidence, and perhaps the best evidence, that Copernicus was in fact copying without full understanding from some other source, and this source would be an as yet unknown transmission to the west of Ibn ash-Shāṭir's planetary theory.[20]

Later, while assessing Copernican astronomy in the context of Renaissance astronomy, Swerdlow returned to this very point of the connection between Copernicus and his predecessors, and particularly to the problems in the Mercury model:

The transmission of their [meaning the Māragha astronomers] inventions from Arabic in the East to Latin in the West is obscure. Yet Copernicus's lunar and planetary theory in longitude in the *Commentariolus,* right down to the additional complications for Mercury, is that of Ibn al-Shāṭir in nearly every detail, except for the heliocentric arrangement and the extraction of parameters from the *Alfonsine Tables,* and it is hard to believe in light of so many and such complex identities that Copernicus was entirely without knowledge of his predecessors.[21]

Taken together, all the previous evidence of the interdependence between the works of Copernicus and those of ʿUrḍī, Ṭūsī, and now Ibn al-Shāṭir, must at least strengthen the claim of a westward transmission of astronomical ideas from the world of Islam to Renaissance Europe. The works and newly invented theorems and mathematical techniques of ʿUrḍī, Ṭūsī and Ibn al-Shāṭir, were all organically and integrally connected to the preceding results of Islamic astronomy. This evidence clearly demonstrates as well how the totality of those earlier results had by the sixteenth century become the tools of the new astronomy that Copernicus was beginning to construct. When all that evidence is taken together, then it is in that sense that one must understand the statement of Swerdlow and Neugebauer, in their latest book on the mathematical astronomy of Copernicus, when they tried to see Copernicus as the last Marāgha astronomer rather than a completely disconnected figure who was forging a new astronomy completely based on new grounds of his own construction.[22]

Possible Routes of Contacts with Copernicus

All the evidence just cited, and the remarkable similarities between the works of Copernicus and the works of his predecessors from the Islamic world have not gone unnoticed as we have seen. In fact it continues to raise some very fundamental questions about the actual intellectual environment in which Copernicus conceived his path-breaking work. And like all good research results, this connection between Islamic and Copernican astronomies does not only raise new questions for the students of Copernican astronomy, as it must do, but also in turn it raises some very interesting problems for Islamic astronomy as well, as we shall soon see.

Even if one grants the existence of those connections on the intellectual level, still the problem of contacts between Copernicus and his predecessors, in the historical sense, remains further complicated by the fact that we have no evidence whatsoever that Copernicus himself knew any Arabic at all. We also have no evidence that any of the works of ʿUrḍī, Ṭūsī or Ibn al-Shāṭir, with which Copernicus seems to be in direct contact, had ever been translated into Latin in the same way that other Arabic sources were translated earlier into Latin. We cannot speak about the works of these astronomers in the same way we speak about the translations of the works of Avicenna or Averroes that went into Latin during the medieval period. We cannot even compare them to the translations that took place during the Renaissance, such as the fresh translations of the Avicennan works that were executed by Andreas Alpagus,[23] for there does not seem to be an equivalent Andreas for the astronomical works. And yet we know that the results that were produced in the Arabic astronomical works mentioned before, seem to have found their way to the technical repertoire of Copernicus so that he could use them so freely in his own construction of his own astronomy, and at times even use them without digesting them fully as we just saw in the case of Mercury.

Furthermore, we know that these same mathematical theorems and techniques, which must have seemed as novelties to Copernicus, were extensively used by Arabic-writing astronomers for centuries, as we just saw, well before Copernicus, as well as contemporaneously with him, and even after his time. They have a continuous tradition in the Islamic domain for which we find no parallel components in the Latin West. There is some mention of the use of the Ṭūsī Couple at the time of Copernicus in the Latin sources,[24] but that is almost the full extent of it. One does not find the multiplicity of similarities that were just discussed.

From a slightly different perspective, and for the purposes of the results that will be drawn later, we should note at this point that those same astronomical results that were established in the Islamic domain were expressly generated in the context of objecting to, reformulating, and casting doubt on the Greek astronomical tradition. In a sense, unlike the works of Avicenna and Averroes that could have been arguably translated into Latin in order to harvest the older Greek Aristotelian thought that they contained, those mathematical and astronomical results represented their own rebellion against those Greek sources. From them one could not recover Greek thought. On the contrary, one rather found the very critique of Greek thought. In themselves, those Arabic astronomical sources were creating alternatives to Greek astronomy instead of "preserving" it as we are often told. And most surprisingly, they all came from the period that has been dubbed, for centuries now, as the period of the deepest decline in Arabic thought.

So why a Renaissance scientist would be interested in recovering information from such sources, when these sources were representatives of a declining culture, if we were to believe the classical narrative of Arabic scientific historiography? Furthermore these sources were written expressly to counter Greek astronomical thought rather than preserve it. So why any Renaissance scientist would be interested in them, if the purpose of the Renaissance intellectual project was the recovery of the sources of classical Greco-Roman antiquity as we are also so often told?

On the other hand, when we remember the iconographic treatment to which Copernicus has been subjected, by those who name revolutions after him, and the portrayal of his revolutionary role as a path breaker, it will become difficult to imagine how and why would the same person seek results from sources that were steeped in saving Aristotelian cosmology, such as the Arabic astronomical sources seem to have been doing. If his purpose was to topple that cosmology altogether, as we are told by the literature about Copernicus, wouldn't he have looked somewhere else? Furthermore, if we are to believe that the works of Copernicus crystallized the spirit of modern Renaissance science, then we would have to contend that the basic technical foundations for this "modern" science were already laid in the Islamic world, centuries before, as we now realize that the only two theorems that were used by Copernicus to construct his own astronomy, that were not already found in Euclid or Ptolemy, were the theorems of ʿUrḍī and Ṭūsī. All these questions and reflections force us to review our standard historiography of Renaissance science first, and then, most importantly, the historiography of Islamic science itself.

Among the problems that the project of Islamic historiography must entail is that of explaining which Arabic work may have been available to Copernicus, if we were to continue to think that those astronomical results reached Copernicus directly from Arabic sources. The difficulty becomes critical when we realize that so far we can establish similarities between the works of Copernicus and the works of ʿUrḍī, Ṭūsī and Ibn al-Shāṭir, but in such a way that none of those Arabic sources could account for all those similarities. That is, if we were to assume that Copernicus knew of ʿUrḍī's work, we cannot explain from that work alone his knowledge of Ṭūsī's Couple. And if we assume he knew of Ṭūsī's work, then we cannot explain his acquaintance with ʿUrḍī's work through Ṭūsī's work. And if we assume that he knew of Ibn al-Shāṭir's work, who lived a century after ʿUrḍī and Ṭūsī, then we cannot explain Copernicus's insistence on proving the Ṭūsī Couple which is nowhere proved in the work of Ibn al-Shāṭir. And of course it becomes almost impossible to conclude that he knew of all these works individually and that he synthesized them himself when he did not know any Arabic, nor having had any of them available to him in Latin.

The best working hypothesis that can be proposed at this point is to think of Copernicus's acquaintance with some type of an Arabic astronomical work that contained commentaries on earlier works, such as the work of Quṭb al-Dīn al-Shīrāzī (d. 1311) where one could find proofs of Ṭūsī's theorem like the one reproduced by Copernicus, since the works of Shīrāzī were themselves commentaries on Ṭūsī's work. In addition the works of Shīrāzī also contained the adoption of ʿUrḍī's model for the upper planets, which was summarily adopted by Copernicus through Ibn al-Shāṭir's work and where he almost unconsciously adopted ʿUrḍī's Lemma. We just said that this model was the chosen model for Shīrāzī against the model of his own teacher Ṭūsī. But then Shīrāzī's work does not have the worldview of Ibn al-Shāṭir. And thus it cannot explain the identical lunar model of Ibn al-Shāṭir that was adopted by Copernicus, nor his deployment of the same Ṭūsī

Couple technique as was done by Ibn al-Shāṭir while describing the motions of the planet Mercury. If we follow the route of commentaries, then it will be for historians of Arabic astronomy to find a commentator, such as Shīrāzī, who must have lived after Ibn al-Shāṭir, and who must have probably written a lengthy commentary on Ibn al-Shāṭir and tried to place Ibn al-Shāṭir's works in the context of the earlier works of ʿUrḍī and Ṭūsī. But no one knows of such a commentary and the problem is yet to be solved.

The historiographic lesson, though, is the following: Had it not been for those similarities that have now surfaced between the works of Copernicus and the earlier astronomers of the Islamic domain, this problem would not have arisen in the first place, and we would not have even suspected that such commentaries may even exist.

There is yet another line of research that has to be pursued as well, this time taking a direction well rooted in the works of Ṭūsī. The later commentators on Ṭūsī's *Tadhkira* have already started this research but very few people have pursued it in the modern literature.[25] The research in question has to do with implications of the rupture of the Aristotelian universe by the Ṭūsī Couple. The rupture is in the following sense. As we have already said before, Aristotle had already divided he universe, on the basis of the natural motion of its elements, into two basic divisions: the celestial region (which moved by the natural circular motion of the element ether of which the celestial world is made) and the sublunar region (where linear motion predominated). With Ṭūsī's Couple, one can now demonstrate that circular motion could produce linear motion, and vice versa. Does that mean that the Aristotelian division has to collapse as a result, to what extent, and what can be saved of it if any? Only future research will uncover such repercussions.

Returning to the problem of contacts with Europe, the fundamental question for the intercultural science still remains: How could Copernicus know about those Arabic results, at such a late date, with all the known conditions of the Renaissance? The answer to this question has to presuppose other questions about Copernicus himself. Did he know Arabic at all? Was he in contact with Arabists? How well did the Arabists that he knew know about technical Arabic science? All these questions go to the very core of the intellectual environment during the Renaissance, and will have to be tackled from both sides of the Mediterranean.

The Byzantine Route

So far the assumption had been that Copernicus did not know any Arabic, and since the Arabic sources were not translated into Latin, he must have known about them through some other language to which they were "translated" and that Copernicus could read. Again it was Neugebauer who assumed correctly that Copernicus, like all other well educated Renaissance men, could read both Greek and Latin. Since Latin could not be considered as the language of translation for there was no evidence that the Arabic texts were translated into it, that left Greek as the only possibility, according to Neugebauer's implied reasoning. This train of thought led Neugebauer to examine the Byzantine Greek sources for clues to the solution of the problem of transmission. At the time when Neugegauer first came in contact with the Arabic material that signaled possible contacts with Copernicus, he had already been working on the Byzantine astronomical texts for his own *Studies in Byzantine Astronomical Terminology*. So at first glance, Byzantine Greek looked like a plausible route for such a transmission. Within few years Neugebauer's diligent search quickly yielded a very interesting fruit in the form of a Byzantine Greek manuscript now kept at the Vatican library as Gr. 211. The manuscript in question contained a Greek version of Ṭūsī's couple, and thus seemed like a good lead to pursue. But now that the same manuscript has been published,[26] we can clearly see that although it seems to have a qualitative description of the Ṭūsī Couple, it does not seem to have the proof of that Couple. And as we have seen before, there were remarkable similarities between the proofs of the Ṭūsī Couple in the Arabic work of Ṭūsī and the Latin work of Copernicus. And we have seen that both proofs depended greatly on the identical usage of the same letters of the alphabet to designate the same geometrical points. So the proofs are essential to explain this phenomenon and to clench the argument of the possible connections between the two. Nor does the Vatican Greek manuscript contain any of the material of ʿUrḍī or Ibn al-Shāṭir, which we have seen were very relevant to Copernicus and somehow made available to him.

All those issues and questions go beyond the accepted historiography of science, as it is now generally understood. Copernicus's connection to earlier Islamic material is such a new field of research that it has not yet had the chance to have an impact on the general history of science. But whatever information we now possess, inevitably leads us to the conclusion that there must have been an intimate connection, at least on the theoretical mathematical level, between the works of Copernicus and the works of his predecessors in the Islamic world.

A word of caution is in order so that these issues of contacts between Copernicus and his predecessors in the Islamic world will not be confused with Copernicus's genius idea of heliocentrism. None of the astronomers who worked in the Islamic world, and who have been mentioned so far, had any interest in such concepts as heliocentrism. In my estimate, they were so closely wed to the stubborn, but all comprehensive, Aristotelian cosmology, which dictated a geocentric universe, and which continued to reign supreme in the world of astronomy all the way till the time of Newton. This despite the hints and disparaging remarks one would hear about it from time to time by astronomers working on both sides of the Mediterranean.

In this context too, one is also forced to raise the question of the scientific legitimacy of heliocentrism itself in a pre-Newtonian universe, where no alternative cosmology was yet available. Copernican scholars, who have been busy trying to explain the origins of Copernican heliocentrism, have yet to explain how could Copernicus convince himself that he could shift the center of motion to the sun without having to propose some non-Aristotelian cosmological theory that could hold the world together in the same way the Artistotelian cosmology did.[27] That is, without the benefit of the Newtonian law of universal gravitation, how could he have hoped to maintain the system together?

What Copernicus's predecessors were doing was well within the limits of Aristotelian cosmology. And in that sense they were perfectly consistent in their attempt to replace the Ptolemaic models with alternative models that behaved much more consistently by avoiding the absurdities of the Ptolemaic models. But for Copernicus to complain, in the introduction of his *Commentariolus* about the *equants,* a complaint that made sense only in an Aristotelian universe, and then go ahead and abandon all that system, and retain the modified models that avoided the absurdity of the Ptolemaic equant, is very puzzling indeed. The models that were developed in the Islamic world to solve the problem of the Ptolemaic equant, were developed specifically so that the models would be consistent with Aristotelian cosmological considerations. So if one was willing to abandon the Aristotelian universe, then why retain those models? Such problems will have to be left for the Copernican scholars to handle.

All the attempts that were developed in the Islamic world to revamp Ptolemaic astronomy were, in essence, motivated by the simple and straightforward requirement of keeping astronomical theory consistent with its premises. That is, all those astronomers took Ptolemy at his word that he wished to develop astronomical models all based on a universe made of Aristotelian spheres, and all spheres moved around their own centers, in place, at uniform speed. When they found the Ptolemaic models wanting, they developed their own alternative models, for which, at times, they had to develop the right mathematical theorems in order to maintain the correspondence between those models and the observations upon which the models were based in the first place.

With that attitude, they managed to introduce the feature of consistency into astronomical theory, and to subsume mathematics as a tool of that theory. These astronomers who did not tolerate the Ptolemaic transgression into elegant mathematics at the expense of the physical nature of the presupposed spheres, would not in all likelihood have tolerated a whole new heliocentric system where the very foundations of the spheres, that were still retained by Copernicus, no longer made sense in the heliocentric world. Consistency between the presuppositions, the physical nature of the spheres, and the mathematics that represented the motions of those spheres, on the one hand, and models that served as predictive models for the behavior of the planets at any time and place, on the other, became the guiding principle of Islamic astronomy at this stage. Only at a later stage, i.e. toward the middle of the sixteenth century, would mathematics take its proper role as a tool of astronomical theory.

The fact that Copernicus too launched his own research, in the *Commentariolus,* with the same attitude of wishing to solve the problem of the *equants,* simply means that at the early date of the sixteenth century, if not well before, the flow of ideas across the Mediterranean was already in full gear. And now that we can document the similarities between the works of Copernicus and those of his predecessors in the Islamic world, they only confirm the fluidity of this traffic. Once that becomes clear, one can then readdress the question of "locality" versus "essence" of Islamic science by basing the discussion on concrete examples such as the ones that are being raised here.[28] If the solution of a problem that was developed in Damascus in the middle of the thirteenth century, and still made perfect sense to someone like Copernicus who was writing within the context of the Latin world of the Renaissance, then it becomes obvious that neither the passage of time nor the cultural borders could inhibit this motion of perfectly valid solutions. So what is "local" and what is "essence" about the solutions of such problems?

These findings do not only explain the background and motivation of the Copernican works; they also explain the continuity of thought from the Middle Ages into the Renaissance time, without having to make wild assumptions about ideas being born in abstract contexts. Their sheer number and complexity, as well as their technical nature also remove the possibility of coincidental discovery, and force us to agree with Swerdlow and Neugebauer that it is no longer the problem of "if" but "when, where and in what form" did Copernicus learn of those earlier works.[29] The answer to this question promises to change our common understanding of the history of science itself, as well as change our understanding of the nature of the relationship between Europe and the Islamic world at this crucial time in history.

The Renaissance Arabists

So far I have limited the discussion of the possible routes of contacts between Copernicus and the Islamic world to the language that Copernicus could read and into which the Arabic sources could have been translated: Byzantine Greek. But since the discovery of the Byzantine manuscript, Gr 211, of the Vatican, by Neugebauer, other hints of possible routes have come to light, mainly from the Arabic manuscripts themselves. One such manuscript, also kept at Vatican, Arabo 319, is another copy of the *Takhkira* of Naṣīr al-Dīn al-Ṭūsī, in which, of course, there is a chapter that included the proof of the Ṭūsī Couple. The manuscript itself was passed on to the Vatican Library as part of the legacy of a Frenchman by the name of Guillaume Postel (1510–1581) who was a younger contemporary of Copernicus.[30]

What makes this Vatican manuscript quite unusual is the fact that it is titled "*Epitome Almagesti.*" With Della Vida's help it was then determined that the identification was done by Postel himself. But more importantly, the manuscript also contains marginal annotations in Latin, also by Postel, that indicate his ability to read this highly technical astronomical text of Ṭūsī. He could clearly comment on it, although very briefly, which means that he understood what he was reading. The text is reasonably well preserved, especially around the chapter that included the statement and proof of the Ṭūsī Couple, and thus could have presented no material difficulty to someone who was capable of understanding its contents.

The existence of such a manuscript also indicates that there were Renaissance men who knew Arabic, and definitely knew of the contents of technical scientific texts.[31] The problem is to determine whether Copernicus himself ever came to know such men. For if he did, then it would be quite possible to assume with Willy Hartner that someone could have briefed him about the contents of such manuscripts, that is, bring him up to date on the latest in Arabic astronomy.

Such a scenario may inadvertently help solve the problem of Copernicus's indebtedness to more than one Arabic text and for which I had to speculate about the existence of a text that was written after the time of Ibn al-Shāṭir in the form of a commentary that included elements from the works of Ṭūsī, ʿUrḍī, and Ibn al-Shāṭir; texts that we know have come to the attention of Copernicus. Assuming the existence of such a colleague, to whom Copernicus could go for consultation, may solve that problem: it would make the colleague the gatherer of such information from various texts, and it could make him responsible for passing it on to Copernicus.

But once it was known that Postel owned at least one technical Arabic astronomical manuscript, it was reasonable to investigate other collections and see if they also included such manuscripts that were owned by him, in order to determine the extent of Postel's own commentaries. The hope was that this kind of research would shed light on the kinds of texts such contemporaries of Copernicus were reading. We would also know if the Vatican manuscript was an exception or a unique occurrence.

The search was then enlarged to include the astronomical texts that are still preserved in other European libraries. Luckily the first step in that research was immediately rewarded by the collection of the Bibliothèque Nationale of Paris. Among the Arabic manuscripts still kept in that collection there appeared another technical text, called *Muntahā al-idrāk fī taqāsīm al-aflāk* (The Ultimate Grasp of the Divisions of Spheres), this time written by Abū Muḥammad ʿAbd al-Jabbār al-Kharaqī (1138/9). The manuscript is definitely devoted to mathematical astronomy as is clearly indicated in the title. Furthermore, it is explicitly marked as having been owned by the same Gillaume Postel with the phrase "*ex libris guilielmi postelli*" clearly marked on the title page.[32] On the first flyleaf, the manuscript also states that it was bought in Constantinople in 1536, as it is clearly marked: "*G. postellus Contantinopoli 1536.*" The year 1536 also happens to be the year that culminated the mission of the delegation that had been sent to Constantinople by the French King François I (1515–47) to negotiate a treaty with the Ottoman Sultan, Suleiman the Magnificent (1520–66). The treaty was in fact signed in that year.[33] This Postel was apparently a member of the delegation. And we know that he was charged to buy Greek books by Budé, the Librarian of François I. But apparently Postel opted to buy Arabic scientific texts instead.

The earlier background of Postel, his childhood, education, and his acquisition of Hebrew and Arabic, as well as the other languages that he apparently knew, remain obscure. But the fact that he was selected to join the French delegation to Constantinople must mean that he had already acquired some fame as someone who knew what was then known as oriental languages, so that he could possibly act as an interpreter to the French delegation. It would be most interesting to find out the name of the Person who could have taught him Arabic in Paris in the early part of the sixteenth century. And was the Paris environment, in terms of exposure to such oriental languages as Arabic, much different from other cities in Eastern Europe and northern Italy where Copernicus spent his professional career, or was it the norm? Such a question bears directly on Copernicus's access to Arabic scientific material, which did not have to be translated into Latin.

Postel's trip to Constantinople was apparently quite successful, for in addition to the two Arabic manuscripts that he owned there were others that were signaled by Della Vida, which may have ended up in other European libraries.[34] And because of the signature of the treaty, which must have pleased the French king,[35] Postel was apparently rewarded with an appointment as professor of mathematics and oriental languages at the Collège Royal which later became the Collège de France. The philosophical Arabic manuscript, of the Leiden Library, clearly attests to this appointment since it is signed "Royal Professor of mathematics,"[36] which must refer to his official appointment to the Collège.

But Postel did not last long at the Collège, and for reasons that remain partially obscure he was dismissed of this post by 1543, the year when Copernicus died. From then on, his life took a dramatic turn as he began to pursue cultural and religious topics, but continued to make further trips to the Islamic world and to acquire other Arabic scientific texts, most notably between the years 1548 and 1551. Several of his trips took him through northern Italy, where he was finally involved in a spiritual conversion that may have cost him his demise and his eventual imprisonment by the Pope and his retreat to a convent near Paris where he finally passed away in 1581.

Other manuscripts at other libraries such as the Bodleian (Oxford) and the Laurentiana (Florence), some from Copernicus's lifetime while others from after his death, also contain similar marginal annotations, and sometimes even interlinear translations.[37] Such evidence attests to a widespread interest in most European cities in the Islamic sciences contained in those manuscripts.

The causes of this European interest in Islamic science at these latter ages remain very poorly studied. One can understand the reasons for it from the period of Copernicus's own lifetime, since the status of science in the European cities at that time was almost on equal footing with that which had been known in the Islamic lands. But the interest seems to continue well into the seventeenth and eighteenth centuries. And that becomes much more puzzling.[38]

Other questions remain of interest in this respect, and relate to the image of Arabic/Islamic science in those European cities in contrast to the image of the more ancient sciences. From the evidence that has survived so far, and this widespread interest in almost all fields of science, one may safely speculate that to a Renaissance person of the sixteenth and early seventeenth centuries Arabic science must have seemed quite advanced over and above the more classical Greek science, especially in the field of Astronomy. Such a person would have known from several sources, and particularly from the often quoted disparaging remarks that were made by Averroes himself in his own commentaries on the Aristotelian works, that Ptolemaic astronomy had been under attack in the Islamic world. Someone like Andreas Alpagus (d. 1522), who lived and studied in Damascus for about 15 years and who returned to Padua, probably near the turn of the sixteenth century, to assume the chair of medicine at Padua in 1505, thus possibly overlapping with Copernicus's sojourn in that general area where he acquired his last degree in canon law from nearby Ferrara, may have known of the attacks against Ptolemaic astronomy, or may even have heard about the remarkable reform of that astronomy that had been accomplished by Ibn al-Shāṭir (1375) of the same city of Damascus nearly 100 years earlier.

All these contacts with the Islamic world, of which we gave here the bare minimum by way of examples,[39] would have easily brought the news that the old Greek astronomy was already in great dispute in the Islamic lands, and that its results as well as its basic foundations were severely questioned and at times even overturned. A Renaissance person would then have every reason to seek information about these latest reforms that took place already in the Islamic world, and would in all likelihood only keep an antiquarian interest in the details of Greek astronomy. In such a setting, the image of Islamic science in Renaissance Europe would have attained a status similar to the one it attained in Byzantium in the early part of the fourteenth century, where astronomers would travel from Constantinople to Trebizond, in order to acquire the latest of Islamic astronomy, as was done by the author of the Byzantine Greek manuscript who brought the Ṭūsī Couple into Greek.

There is no doubt, then, that there were enough Arabists in various European cities who were not only writing Arabic grammars, as Postel did, but who were like Postel equally competent enough to read the technical contents of scientific manuscripts and to understand their import and thus pass them on either orally or even by request in a tutorial fashion. With Poland, where Copernicus was born, being so close to the borders of the Ottoman empire at the time, and with the free flow of books, trade, and scholars across the Mediterranean through the northern Italian cities, where Copernicus received his education, we must suspect that there were many people like Postel who could have advised or even tutored Copernicus on the contents of Arabic astronomical texts. Now that we have established the likelihood of another route, one hopes that future research will continue to explore it in order to explain the likelihood of such a scenario.

Contacts in the Field of Instruments

Lest we think that planetary theories were a special case of their own, and that contacts between the world of Islam and Renaissance Europe were restricted to connections with Copernican astronomy only, it is important to note that similar exchanges were taking place in a variety of other disciplines.[40] At this point, a few examples from cognate fields like the field of scientific instruments should be enough to make the point. Such supplementary evidence points to two curious instances that demonstrate a close connection between the instruments that were being produced in Renaissance Europe and those that were already produced in the Islamic world. Those instruments were produced centuries apart and their existence simply signals the range of contacts between the Islamic world and Renaissance Europe.

The first instance of contacts between the Islamic world and Renaissance Europe in the field of scientific instruments concerns Antonio de Sangallo the Younger (1484–1546), one of the most famous architects of Renaissance Italy. Among his papers, now still kept at the Uffizi in Florence,[41] there is one sheet that contains, on one face of it, a detailed drawing of an astrolabe that was made in Baghdad around the year 850, and on the back a drawing of the rete of the same astrolabe.[42] The reason we know such details about the drawing of this astrolabe is due to the meticulousness of de Sangallo, who not only copied the astrolabe on paper, face and back and rete, but, with great care, he also copied the name of the original maker of the astrolabe that was etched along the edge of the upper right hand quadrant on the back of the astrolabe. Unlike most other art objects that were produced in the Islamic domain, and did not usually carry the name of the artist, astrolabes were usually inscribed on the back with the name of the maker. So this astrolabe was not an exception.

The name of the original Baghdad maker was Khafīf. He apparently apprenticed with a more famous astrolabist who lived in Baghdad around the year 850, by the name of ʿAlī b. ʿĪsā.[43] Because of that relationship, Khafīf signed his name on the back of the astrolabe as "*ṣanaʿahu Khafīf ghulām ʿAlī b. ʿĪsā*," which means: "It was made by Khafīf the apprentice of ʿAlī b. ʿĪsā." De Sangallo dutifully copied this signature, which has no astronomical significance whatsoever. The question that this sole paper of the Uffizi poses is: Why was someone like de Sangallo interested, in the first place, in an astrolabe that was made some 800 years earlier? This, when we know that de Sangallo was in his own right a famous architect who was entrusted with the building of St. Peter's cathedral in Rome, a monument that continues to stand witness to his skill and mastery. My suspicion is that the scientifically oriented men of the Renaissance, especially during the sixteenth century, must have thought very highly of all scientific things coming to them from the Islamic world, even instruments that were made centuries earlier.

To complicate the puzzle somehow, and to point to directions already signaled in the case of the contacts with Copernicus, in astronomy, and with Michael Servetus and Realdo Colombo in medicine, here too, there is no evidence that de Sangallo knew any Arabic. My suspicion is that the drawing, which duplicates all the Arabic inscriptions from that astrolabe, down to the signature of the maker, only attests to his ability as a draftsman. And that in itself does not constitute enough evidence to conclude that he knew any Arabic, unless someone can demonstrate that de Sangallo ever learned Arabic, which would be very curious indeed.

The second instance concerns the Renaissance reception of this particular area of scientific instrument, and the extent to which this field was particularly interesting to Renaissance men.[44] The interest itself can be easily demonstrated by other contacts between the world of Islam and such famous astrolabe makers like the Arsenius family of astrolabists, who worked in northern Europe, mainly in the Flemish area, sometime toward the end of the sixteenth century. To illustrate the contact between this family of astrolabists and the Islamic world, consider the extant astrolabe (figure 6.8) that was originally made in Muslim Spain, and whose mater, back and plates were inscribed in Arabic by Muḥammad Ibn Fattūḥ al-Khamāʾirī in 619 A.H. = 1222 A.D. As is obvious from the picture, a member of the Arsenius family fitted the rete of this astrolabe with Latin inscriptions, and produced a plate that would work for the northern European clime.[45] The existence of this astrolabe, in this form, could only mean that some member of that family was in fact working with Arabic astrolabes, and must have been somehow competent in Arabic. Or say that at least he must have been bilingual enough in order to use the new rete properly with the mater that was made by Khamāʾirī. The reason is that the rete was inscribed with the Latin names of the star, while the rim, against which the altitudes of those stars had to be read, still carried the Arabic alphabetical numerals that were originally inscribed by Khamāʾirī. Therefore, we can only conclude that either Arsenius himself, the maker of the new rete and plate, or the user of the resulting hybrid astrolabe must have been able to read some Arabic at least, and that must illustrate some interest in the Islamic scientific instruments toward the end of the sixteenth century at such northern climes as the Netherlands.

Other such hybrid astrolabes are probably still waiting in private collections to be discovered. King's study of *Instruments of Mass Calculations*[46] has many examples of such influences and thus it is highly likely that such hybrids exist.

The same design of the retes that were commonly produced by the members of the Arsenius family (figure 6.9, right) may also demonstrate another connection between astrolabes that were made in the Islamic world and those that were made in Renaissance Europe and thereafter. In his most

Figure 6.8
A hybrid astrolabe that was once kept at the Time Museum. The mater was made by al-Khamāʾirī in 1222, as clearly signed in the picture on the right. The rete, which carries the standard design of the Arsenius family, was made by one of the members of that family toward the end of the sixteenth century.

recent publication, just cited, David King of Frankfurt raised the possibility that those designs may not at all represent tulips, as they are usually taken to do, but should rather be seen as skeletal representations of the Arabic calligraphic phrase *bism Allāh al-Raḥmān al-Raḥīm* (In the name of Allāh, the Compassionate and the Merciful), which is the opening phrase of most chapters of the Qurʾān.[47] As is obvious from figure 6.9 (left), the inscription of the phrase is beautifully interwoven among the leafy star pointers of the rete. The Arabic calligraphic design of this particular rete, on the left, comes from a slightly later astrolabe, which was made in Persia by Muḥammad Zamān in 1651–52. And the astrolabe itself is still preserved at the Metropolitan Museum of Art, in New York City. But despite the later date of the astrolabe, the rete design may have descended from an earlier astrolabe rete that utilized the same mirror image calligraphy of the phrase, or from a sim-

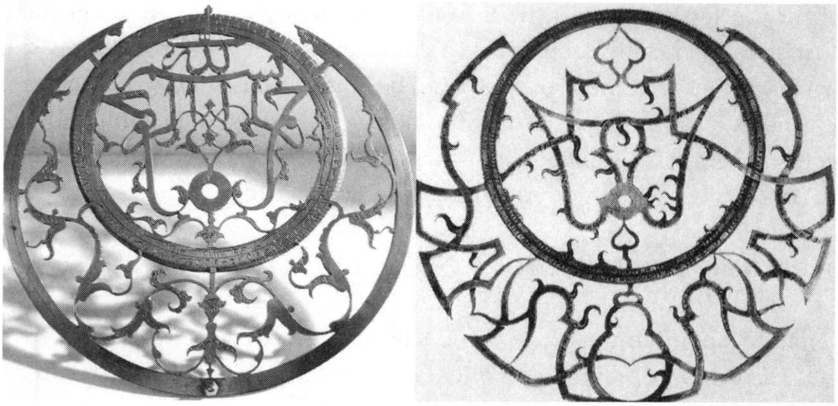

Figure 6.9
Right: A standard rete produced by a member of the Arsenius family. This rete is thought to represent the form of a tulip. Left: A rete produced by Muḥammad Zamān of Persia in 1651–52, which has the same design but for the Qur'anic verse *bism allāh al-raḥmān al-raḥīm.*

ilar design on other art objects that were produced in the Islamic world. The existence of calligraphic designs drawn in the shapes of animals or other objects are ubiquitously found among the artistic treasures of the Islamic world and may have influenced the production of such retes.[48]

The claim that I wish to make here is that the very similarity between the calligraphic design of the Arabic phrase and the shape of the tulip may have motivated the Arsenius astrolabists to produce such similar retes, thus at once paying a very clever homage to the Islamic tradition, which they knew rather well when they fitted retes for Arabic astrolabes, and to the tulip craze that hit the Netherlands during their time. The craze itself appears to have been occasioned by the importation of tulips from the sixteenth-century Ottoman domain.[49] The solution of this very intriguing problem has to wait for further work on Islamic metal works, astrolabes, and calligraphic designs in general, and on the routes that those works followed as they came into Europe. For now, the striking similarities between the two retes remain interesting as they demonstrate a certain relationship between the astrolabists of the Islamic domain and their European counterparts, even if that relationship may not be as well confirmed as the relationship of fitting a Latin rete on an Arabic astrolabe mater, as was done by one Arsenius astrolabist.

For those who work in the field of Instruments, very many other such instances will readily come to mind. And I am almost certain that they will agree with me that these examples can be multiplied manifold. But the two examples we have supplied so far should give us enough indication that the cognate field of instruments should also be investigated in the same context of contacts between the world of Islam and Renaissance Europe.

Traffic from "East" to "West"

Up to this point in the discussion, I have given few examples of the activities of European Arabists and orientalists in their pursuit of science from Islamic land, and tried to assess the reasons for such interests. I had not intended an exhaustive treatment of the subject, which is worthy of a whole monograph by itself.[50] I only needed to hint to the possible sites of interaction between Renaissance Europe and the world of Islam. But I have neglected to mention that we do have some evidence of men of science who crossed over from the Islamic lands into various European cities, and of course brought with them the sciences that they knew from their old countries.

The case of al-Ḥasan b. Muḥammad Ibn al-Wazzān, better known as Leo Africanus (d. ca. 1550), immediately comes to mind.[51] Although Leo came from the western part of the Islamic world, he nevertheless had traveled extensively over all of North Africa and parts of the east. What concerns us here is that he was a man of great intellect, and was apparently very well acquainted with the Islamic intellectual scene of his day. More importantly, Leo was a contemporary of Copernicus, and a man of great scientific knowledge, who also taught Arabic at Bologna.[52] He may have come across people, or even taught some, who knew Copernicus themselves. His teaching Arabic at Bologna is significant in itself as well. For Bologna fell along the famous corridor from Venice to Florence, along which many Renaissance intellectual activities took place. Leo's personal output is slightly better known than others on account of his geographical writings that included tidbits of his personal accounts. But his intellectual life and his impact on Renaissance scientists, as well as his role in introducing scientific ideas from Arabic into Latin, is still not fully investigated from the perspective of the Renaissance knowledge of Arabic Islamic science. A scientific biography of this distinguished pioneer scientist and belletrist is long overdue.

There were others too. For example, one could easily name members of the circle of the distinguished orientalist, Jean-Albert Widmanstadt (1506–c.1559), who was also a contemporary of Copernicus, and who may have also played a very important role in the transmission of Islamic scientific ideas to Europe; a role at least just as important as that of Guillaume Postel, whose input was already noted before.[53] A quick search for Widmenstadt's role revealed, to my pleasant surprise, that this Widmanstadt was himself a student of Leo Africanus,[54] and also knew much Arabic material as well as the scientific contents of Arabic astronomical texts. In our context his role should be seen as part of the influence of Leo Africanus on Renaissance thought, but should also be considered as part of the network of orientalists who were contemporaries of Copernicus and who may have known about the achievements of Islamic astronomy and were competent enough to bring it to the attention of Copernicus.

One can be certain that there were many more people who came in contact with Leo Africanus, and who may have either received information about scientific ideas directly from him or were guided by him to others who could supply the same. But until the field is fully explored with those questions in mind, we cannot be certain about the kind of information that was transmitted, nor about the people who played as conduits for this transmission. One thing we can be sure of is that there are much too many coincidences of ideas appearing first in Arabic texts, usually written between the twelfth and the fifteenth century, which reappear, without much explanation, in Latin sources of the sixteenth and seventeenth centuries. In most cases the original Arabic texts containing these ideas had never been "translated" into Latin in the strict sense of the word.

Others who followed similar routes as that of Leo Africanus, but at least under slightly different circumstances—if not of their own volition as far as we can tell—included people such as the Syriac Jacobite patriarch Niʿmatallāh, better known by his Latin name Nehemias (d. 1590).[55] This patriarch was involved in a series of conflicts in his native town, Diyār Bakr, of southeast modern Turkey, and in his own patriarchate of Antioch and All the East. At one point, his life became so endangered that he felt he had to flee to the Papal see via Venice. And in order to secure a generous Papal reception he used the excuse that he would help bring his followers back to the fold of the Roman church, and under the Papal flag. A note left at the margin of an elementary mathematical manuscript, still kept at the

Laurentiana Library in Florence, describes in some personal nostalgic terms the difficulties of his trip, saying that he was being tossed by the waves of the Adriatic Sea, during the year 1888 of the Greeks (1577 A.D.), on his way to Venice.[56]

Once in Venice, apparently without knowing a word of Latin or Italian, he was attached to an eastern "traveler" by the name of Paolo Orsini. Orsini, who then acted as Niʿmatallāh's interpreter, was originally a captured Turkish soldier, who, like Leo Africanus before him, accepted to convert to Christianity. The two went to Rome, of course via Florence, as most people were prone to do in those days. Along the way, or maybe more likely in Rome itself, he made the acquaintance of the Cardinal Ferdinand de Medici who later became the Duke of Tuscany. Like all the Medici's, Ferdinand could quickly recognize a commercial enterprise when he saw one. With the invention of printing nearly 100 years old, and with Arabic not yet being exploited for that purpose, the Arabic books that the Patriarch was trucking along, which were all in manuscript form were too tempting to Ferdinand. He saw in them the possibility of starting an Arabic press and using those books, as the bases for the printed versions.[57]

Of course, the excuse Ferdinand used, at least openly, was that he would use the press to produce reading material for the missionaries who could go out and convert the Muslims to Christianity. But the actual record of what was printed and sold at the Medici Oriental Press tells a different story.[58] While it may be quite understandable to produce 1,500 copies of the Arabic Bible for missionary activities, it would be much harder to justify the production of 3,000 copies of Euclid's *Elements* for the same purpose. And if one were to think that the publication of the *Elements* served a wider Renaissance purpose of recovering the scientific works of classical antiquity, one will be disappointed to learn that the *Elements* that were published by the Medici Oriental Press were not of the original Arabic translations of the Greek *Elements* (and two good translations are still extant), but rather a slightly modified version of the *Elements*. And what the Medici press published as Euclid's *Elements* was in turn a re-working of yet another re-working that was already produced toward the middle of the thirteenth century by the very same astronomer/mathematician Naṣīr al-Dīn al-Ṭūsī who was mentioned several times already.

Still, the disproportionate number of copies that were produced in the first place calls for a comment. Did the Medici Oriental Press prospector expect the missionaries to use more of the *Elements* than they would use the

Bible for the conversion activity? And if that was the purpose, the actual sales seem to support such a contention. The records show that the Arabic Bible sold 934 copies, while the re-worked *Elements* outdid that and sold 1,033 copies. Based on sheer numbers alone, could one draw the ironic conclusion that a re-working of Euclid's *Elements* served a better purpose in converting people to Christianity than the Bible itself?

Similarly, one has to wonder also as to why the first six Arabic books that were published by this press would include four that had something to do with linguistic or demonstrative sciences and not that much relevance to religious material. Such linguistic and scientific texts were abundantly available in manuscript form all over the lands of Islam, as any survey of extant library holdings can demonstrate. So what kind of profit a good Medici businessman could have expected to make by shipping those books to the lands of Islam?

When we consider the Renaissance environment, which apparently witnessed a great interest in Arabic scientific texts, one has to conclude that the real market for the Medici Oriental Press was in fact the European centers of learning who were calling for a return to the original Arabic rather than depending on translations. Didn't Andreas Alpagus (d. 1522) use the excuse of the unreliable medieval translations from Arabic in order to go to Damascus and learn Arabic so that he can produce new translations of Avicenna's works, a feat that he actually accomplished? And didn't Zacharias Rosenbach (c. 1614), when the press was still functioning, propose that the learning of Arabic be introduced in the Herborn Academy for the medical students so that they could read Avicenna's *Canon* in the original?[59] All these calls for Arabic texts must have sounded enticing for a good businessman seeking an investment, and Patriarch Niʿmatallāh's library came in handy as it supplied the raw material for such an enterprising publishing endeavor.

That most of Niʿmatallāh's books are still held in the Laurentiana Library speaks directly to this engagement between the Medicis and the Patriarch. But this was not the only contribution the Patriarch was to make to the intellectual life of the Renaissance. Sixteenth-century Europe had been obsessed with the problem of reforming the calendar as the celebration of Easter was continuing to slip backwards. And earlier councils, to which even Copernicus made a proposal for reforming the calendar, could not agree on the reform.[60] The job was finally left to the committee that was appointed by Pope Gregory XIII in order that it would specifically accomplish this task.

One of the distinguished members of that committee was the same Patriarch Niʾmatallāh. His role on that committee should be quite understandable as he was the one who had brought along with him astronomical books that contained values for the lunar month and the solar year that were much more refined than the values that were found in the old Greek sources, or the prevailing medieval European sources.[61] With his services on that committee, Niʿmatallāh became an actual participant in the making of the European Renaissance, just as much as his two predecessors Leo Africanus and Paolo Orsini did before.

What bearing does all this have on the works of Copernicus, and the problem of the transmission of Islamic scientific ideas to him, as most of these names and activities mentioned here date either to the late or to the post Copernican period? In fact, the more we can document the reliance on Arabic scientific sources from the period following Copernicus, when the whole world view was supposed to have been changed by him, and by others like him who created what is now called the Copernican Revolution, the more one is forced to ask why was there such a need for Arabic texts in the latter part of the sixteenth century and early seventeenth? And if one can document that interest, as the few examples we have given here seem to do together with many more that were left unmentioned, then shouldn't one expect even a greater eagerness on the part of Renaissance scientists to learn from those Arabic sources in the earlier period when the revolution had not yet taken place?

Conclusion

With all this evidence that was admittedly gathered here solely for the purpose of explaining the specific connections that seem to exist between the Copernican astronomical texts and the Arabic antecedents from the world of Islam, things begin to look like we unintentionally stumbled on a Pandora's box. And with very little effort in documenting the connections whole areas of research have come to life as a result. Churchmen like Postel and Widmanstadt, who seemed like they were involved in strict church activities, turn out to have been knowledgeable Arabists and men of science in their own right. We can even tell that they were following in the footsteps of other Arabists like Ambroseo Taseo (d. 1539), Andrea Alpagus (d. 1522), and before them Hieronimo Ramnusio (d. 1486 in Beirut) who were even much more glorious than them, and who may have laid the foundation for this intercultural exchange whose import we are now just beginning to appreciate.

But by looking at the works of these men, whether in the form of the fresh translations of Arabic scientific and philosophical texts by Andreas, or the still extant commentaries of Postel on more sophisticated astronomical texts, we cannot avoid but reach the conclusion that the Renaissance engagement with the Islamic world was of a completely different order than the engagement that took place during the Middle Ages. In the Middle Ages people relied more on the translations, and waited for them to be produced before they could use them. That was how the Latin translations of Averroes made their impact on Latin thinkers. But by the Renaissance time, men of science themselves apparently became Arabists and no longer needed the translations. They could go directly to the Arabic texts and exploit the ideas contained therein. Otherwise, how else can we explain the several occurrences we noted so far in astronomy and medicine as well as in the other sciences where we have original ideas that were developed in the Islamic world, expressly to object and reformulate the Greek classical scientific tradition, only to reappear a couple of centuries later in the works of Renaissance scientists without ever having those Arabic texts translated into Latin? Copernicus or his collaborator or instructors, Michael Servetus and Realdo Colombo in medicine, all seem to have followed that route.

This evidence can only lead us to look further into the works of these Renaissance men of science, not only to document those ideas, but in order to understand the nature of Renaissance science itself, and to understand the methods and techniques that were integral to the formation of that science. Most strikingly though, it looks like the Renaissance men of science were apparently looking to the world of Islam for the latest in scientific activities rather than looking to the Greek classical sources, especially for those sciences that were more of the empirical type like astronomy and medicine which needed to be constantly updated. In fact, one can hardly see an astronomical value adopted by a Renaissance scientist that was derived directly from the ancient Greek sources. For example, one no longer found a precession value that was as badly off as that of Ptolemy, or the inclination of the ecliptic as reported by Ptolemy, or the fixed solar apogee that was already proved wrong in ninth-century Baghdad. Even the kind of reasoning that was followed by Ptolemy, while constructing his mathematical predictive models, also became obsolete. Rather one found the latest results that were developed in the Arabic sources that could answer much better the same problems the Greek classical tradition had to answer.

In the final analysis, I do not think that we can understand the building blocks of Copernican astronomy, without paying close attention to the results that were already achieved in the Islamic world. Not only because those results preceded the works of Copernicus and hence it is legitimate to ask if there has been any transmission of ideas from east to west, but because in the Arabic tradition we understand better the accumulative process of scientific production, and can witness the slow growth of those ideas over the centuries, a feat that we cannot follow in the works of Copernicus, with the same rigor, if we assume that all those similarities were just coincidences. And when people think of the spirit of the Renaissance as characterized by that change in the scientific thought that shunned the ancient authority, now one can find the roots of that thought already documented in the works of generations of astronomers and scientists working in the Islamic world and writing their objections to Greek thought. And they did not only object, we now know that they were developing real alternatives to that thought. One can even go as far as to say that by the time Renaissance Europe came to know of Islamic science, especially as documented in the astronomical discipline, that science was by then a mature science on its own, confident of its ability to invest in the creation of new mathematical theorems to solve new astronomical problems, or even to deploy mathematics in more abstract ways in order to divest it of the physical truth it once laid claim to and return it to the realm of the descriptive language that could be applied to the physical phenomena.

By the time of the Renaissance, and if the words of Vesalius are any guide when he says "those Arabs who are now rightly as familiar to us as are the Greeks,"[62] we can conclude that Arabic science was by then a competitive science that stood at least on equal footing as the science of the Greeks, as far as Vesalius could see. But in matters of observational science, it looks like Arabic science was by then thought of as being definitely far superior to Greek science once all the mistakes of the latter had been laid bare.

7 Age of Decline: The Fecundity of Astronomical Thought

The previous chapter demonstrated very clearly the kind of results that were produced in the Islamic world and the impact those results had had on Renaissance Europe. In the chapters before that, where I talked about the encounter with the Greek scientific legacy and the innovations that encounter produced, I also noted that although the critiques of Greek thought began early on, the more mature criticism and the confidence with which the Greek scientific edifice began to be dismantled and replaced by more consistent alternatives, and far more sophisticated deployment of mathematics, did not really take place until the later centuries of the Islamic civilization, and mostly after the thirteenth century. Thus, based on what we have seen so far, one is justified in saying that those later centuries of Islamic civilization seem to have been centuries of great creativity, at least as far as the discipline of astronomy was concerned. In addition, one could also say that, that creativity was not apparently restricted to revamping all of the Greek astronomical theory but it seems to have had a seminal impact on Renaissance science as well.

But these are precisely the centuries that the classical narrative had earmarked as representing the total death of science, not to say the total death of rationality in Islam, which is more often used in connection with this period. Without paying any attention to the kind of evidence we have been reviewing, which was mainly produced during the latter centuries of Islamic civilization, or even indicating that such evidence existed, the classical narrative formulated its theory of decline by basing itself on two main assumptions. Those assumptions were held by two different groups of people. And although each group had its own analysis of the intellectual history of Islam, they converged, almost independently, on considering the the age of decline to have begun in the thirteenth century.

For those who looked, from the very beginning, at Islamic civilization as a continuous unfolding of religious thought only, and at the same time held the European paradigm of the conflict between religion and science, they attributed this death of rationality in the Islamic civilization, and in this later period, to an upsurge in religious thought, which they claim came about at the expense of scientific and philosophical thought. For those people, "progress" was defined by the very victory of science over the church, just as European progress was defined. Thus every civilization had to demonstrate that it had participated in this struggle before it could participate in this "universal" linear and constant search for "progress." Those civilizations had to have their science overcome their church, even if one had to redefine "church" in the particular terms of the said civilization. In the case of Islamic civilization, the struggle of the Muʿtazilites against the people of tradition (*ḥadıth*) exemplified, to a great extent, the conflict paradigm between "science" and "religion," without ever bothering to define the "science" of the Muʿtazilites, or the "church" of the people of *ḥadīth.* In that regard Ghazālī's (d. 1111) book *The Incoherence of the Philosophers* (*tahāfut al-falāsifah*) constituted a real milestone. Not only because this group of people saw in it the direct connection between philosophy and science in that period, and hence an attack on one is an attack on the other, but because they also rightly considered Ghazālī as the initiator of an Islamic Orthodoxy of sorts, and thus his book symbolized the triumph of religious thought. The conclusion that is usually drawn from the success of Ghazālī's religious thought is that this triumph must have caused the death of its counterpart, the rational scientific thought. Thus in a simple fashion, Ghazālī was single-handedly held responsible for the decline of rational, read scientific, thought in Islamic civilization in these later centuries.[1]

Thus, pinning the cause of the decline of Islamic science either on the conflict paradigm between religion and science, a paradigm that was first and foremost imported from the European example, or on the fatal blow that was single-handedly delivered by Ghazālī against the philosophers, has become so widespread[2] that those approaches continue to have their deleterious effects on the very reading of the scientific texts that were written both before and after the Ghazālī period.

Focusing on the conflict between science and religion before the Ghazālī period may have contributed to the lack of awareness that there were scientists working during that period and whose main concern was to combat the imported Greek scientific tradition, because of the errors and blemishes it harbored, and not because of the religious thought of their time. Muḥammad b. Mūsā's critique of Ptolemy, or Rāzī's *Shukūk* against Galen, or even Ibn al-Haitham's *Doubts* against Ptolemy, among many others discussed above, have gained some importance only recently as texts rebelling against the Greek scientific tradition, rather than texts rebelling against the religious authorities of their time. None of those texts made any significant impact on the group of people who saw Islamic history as an unfolding of religious thought, and in that sense those text were badly read if they were read at all. It is not accidental that both of Rāzī's book as well as that of Ibn al-Haitham's were edited in the latter part of the twentieth century, and not during the nineteenth century when most Islamic religious and juridical works were studied with great care by famous European orientalists.

But those same nineteenth-century orientalists summarily dismissed the scientific texts that were written after the Ghazālī period. And until very recently no one had ever bothered to investigate the kind of science they contained. In this sense those texts too were very poorly read if they were read at all. As an example of this misreading of texts, we have already seen the efforts of the two famous nineteenth-century orientalists who looked at two works from the post-Ghazālī period, and who read them very carefully, and still could not see the originality that was embedded in them, simply because those orientalists were not looking for any originality during this period.[3] And their self-fulfilling prophecies indeed materialized.[4]

The second group which saw Islamic history more in political terms, and thus portrayed it as a succession of dynasties and battles, with little attention paid to intellectual history, the bête noire that was made responsible for the decline of science in the Islamic civilization was after all Hulagu Khan.[5] Hulagu's devastating blow came at a time when he actually managed to destroy the city of Baghdad, in 1258, in his westward bid from Central Asia to conquer the rest of the world. Those who blamed Hulagu for the death of Islamic science took literally the anecdotes preserved in the historical sources, which were incidentally mainly written further west, in Mamluk areas that were not conquered by the invading Mongols. Those historical sources spoke of the water of the Tigris turning black from the dissolving ink of the manuscripts that were tossed into the river by that barbarian invader. They presented a scene of destruction that continues to stand in the collective memory of most Arabs, and Muslims in general, as the ultimate of disaster and the epitome of barbarity.[6]

In a sense, the dates of the death of Ghazālī (1111) and the devastation of Baghdad (1258) seem to allow for the conversion of the two historiographic traditions just mentioned, one of which saw Islamic intellectual history as an unfolding of religious thought and one of which saw it simply as a sequence of political events. No wonder then that most people could easily conclude that those two fateful centuries, the eleventh and the thirteenth, indeed ushered in the decline of Islamic civilization and with it the decline of science in general. This conclusion was especially true for people who also remarked that they no longer saw during those later centuries the emergence of religious legal schools that were anywhere similar to the four schools that had already emerged during the eighth and the ninth centuries, and was also true for people who no longer saw a continuity of the Islamic caliphate after the fall of Baghdad.

In that sense, the thirteenth century was in fact a fateful century, as it witnessed the final disappearance of a system of caliphate that had up till then functioned tolerably well. But as far as intellectual history is concerned, the extant scientific sources do suggest a different scenario. They suggest that the thirteenth century was an age where new creative scientific thought continued to prosper, and more importantly, they even support the claim that the disappearance of the caliphal system of government was almost a blessing in disguise. For the loss of that system did not seem to have brought the end of the scientific activity. On the contrary it seems to have opened up other centers of production in the lesser capitals, such as Diyār Bakr, Isfahan, Damascus, and Cairo, to name only a few, that continued to produce excellent scientific works.

In summary, and as has already been stated, none of those narratives of the age of decline can really explain the greater number of sources that seem to signal a real upsurge of scientific production both well after the death of Ghazālī and well after the Mongol devastation of Baghdad. And if one focuses on the discipline of astronomy in particular, as we have been doing up till now, the problem of pinning down the cause of the decline according to one of those two narratives becomes even much harder to solve.

In a separate book devoted to the study of one aspect of Arabic astronomy: the aspect of planetary theories, I went so far as to call that same age of decline the golden age of Islamic astronomy. In that book I traced the developments in Arabic planetary theories between the eleventh and the fifteenth centuries and demonstrated the fecundity of that discipline. That book, and the various articles that have appeared since then, dealing mainly with the work of the sixteenth-century astronomer Shams al-Dīn al-Khafrī, described an unparalleled originality during that period that would be difficult if not impossible to dismiss.

Critique of the Classical Narrative

If one takes either explanation of the age of decline, as offered by either group of the proponents of the classical narrative, one is then faced with problems that will not easily disappear. In the first case, and for those who hold Ghazālī responsible for the age of decline, they will have to explain the production of tens of scientists, almost in every discipline, who continued to produce scientific texts that were in many ways superior to the texts that were produced before the time of Ghazālī. In the case of astronomy, one cannot even compare the sophistication of the post- Ghazālī texts with the pre- Ghazālī ones, for the former were in fact far superior both in theoretical mathematical sophistication, as was demonstrated by Khafrī, as well as in blending observational astronomy with theoretical astronomy, as was exhibited by Ibn al-Shāṭir. Similar original production can be easily documented as well in mechanical engineering, in medicine, and in optics, to say nothing of the whole class of astronomers who were all working after the thirteenth century, and whose purpose was to push the frontiers of planetary theories into the realm of alternative astronomy or "New Astronomy" as was proposed by Ibn al-Shāṭir.

To take only few examples, compare the works of ʿIzz al-Dīn al-Jazarī (c. 1206),[7] who worked nearly 100 years after the death of Ghazālī, with those of Banū Mūsā in the ninth century.[8] Earlier in the ninth century period, Banū Mūsā focused on developing new devices and new techniques that were not known from the Greek tradition. For instance, we note the development of the conic valve in the works of Banū Mūsā, which is nowhere documented in the earlier Greek sources. We also note a shift from the Greek tradition that relied mainly on nature's abhorrence of void to animate the machines that they designed, to a more instrumental approach by Banū Mūsā where they used concepts of sources of power as running water, or flowing sand, to achieve similar animations. For Philo of Byzantium[9] or Hero of Alxandria,[10] for example, the siphon worked by water replacing the void, while for Banū Mūsā water flew and was interrupted by turning a conic valve on and off, by means of floats and other mechanisms that did not depend on the concept of void. It was in these later developments that they had to invent such tools as the conic valve and the like. This does not mean that Banū Mūsā did not understand the way void worked in nature and in the design of machines, but that they used it together with other techniques that they themselves developed.

Looking at the works of Banū Mūsā in comparison with the Greek tradition one cannot only detect a forward leap in the variation of techniques in their engineering designs, but can also detect their participation in the general cultural mood of early Islamic times that was critical of the Greek scientific tradition. However, when their works are compared to the work of Jazarī we note a remarkable maturity in the latter's work that is nowhere to be found in the works of Banū Mūsā. With Jazarī we begin to notice discussions regarding the real function of mechanical devices, and real appreciation of their significance as tools that did not only fulfill daily functions for the society, but that they were also tools that could demonstrate the way in which the natural physical principles worked.[11] His devices were examples of natural physical principles in action. And in the introduction to his book, he explicitly states that his devices were intended to actualize (*ikhrāj min al-quwwa ilā al-f'l*)[12]—the physical principles that were potentially there, waiting to be actualized. His full grasp of the Aristotelian approach to mechanical devices and their intrusion into the world of nature is far superior to Banū Mūsā's understanding of such principles as far as one can tell from their surviving writings.

Even the historical sources preserve for us anecdotes about the patrons of Banū Mūsā, especially al-Mutawakkil (ruled 847–861), and tell us that Banū Mūsā's devices had indeed enchanted him,[13] and that those skillful engineers produced for him such entertaining objects exactly to serve that very purpose. Contrast that with the patron of Jazarī who demanded of him, according to Jazarī's own introduction to his work, that Jazarī should compose the work in order to keep a record of the peerless "models" (*ashkāl*), things that he invented (*istanbaṭa*), and "illustrations" (*mithālāt*) that he brought forth. For anyone reading that introduction, the language reveals very clearly a fuller understanding of how mechanical devices operated and why they did. At times Jazarī would even explicitly state that he intended to illustrate the same principle with many devices, all in order to show the universal applications of those principles.

But as these texts have not been fully studied from those perspectives yet, one has to wait before passing any more detailed judgments about them regarding their relative merits. The impressions given here resulted from a first quick reading of the sources, and will, I am sure, finally withstand the test of analysis.

Or take the works of Ibn al-Nafīs in medicine. It was in his commentary on Avicenna's *Canon* that we find his remarkable remark that did not only depart from the teachings of Avicenna, whom he admired greatly, but went further to criticize Avicenna's original source, Galen himself. With his criticism, he ended up refuting the doctrines of Galen on the basis of his own observations, and thus laid the foundation for the eventual discovery of the pulmonary circulation of the blood.[14] It was in the post-Ghazālī period that such scientists seem to have gained a well-earned confidence in order to challenge their predecessors and through them attack the main Greek legacy that continued to be the site of contention, with such statements as "this is the common opinion, but according to us, it is false" (*hādha huwa al-ra'y al-mashhūr. wa-huwa ʿindana bāṭil*). These are echoes of Ibn al-Haitham, ʿUrḍī, Ṭūsī, and others who said, at one point or another, "This is the accepted opinion, but according to us it is false." In that regard, Ibn al-Nafīs reflects the same trend that was developing in astronomy, and had already had its roots in the works of al-Rāzī some four centuries before him. He also seemed to have been complementing the works of other scientists from other disciplines who were all engaged in a cultural revival rather than an age of decline.

Consider also the work of Kamāl al-Dīn al-Fārisī (d. 1320)[15] from the following century, which illustrates the same trend again, but from the field of optics. It was al-Fārisī's teacher, the great astronomer Quṭb al-Dīn al-Shīrāzī (d. 1311), who suggested to al-Fārisī that he study the work of the great scientist Ibn al-Haitham (d.c.1038) from the pre-Ghazālī period. Note that he was not advised to go as far back as the obsolete Greek optics for his study. Instead he was put to challenge the best and most recent production on the subject.

Neither the Greek tradition, nor Ibn al-Haitham had managed to explain the phenomenon of the rainbow properly, and thus up till al-Fārisī's time this problem had remained the site of competition and speculation. And it was al-Fārisī who finally put his mind to it, and developed the instrumentation in order to explain how the colors of the rainbow were in fact produced. And exactly like his predecessor, Ibn al-Nafīs, he too followed the same style, that is, he produced an elaborate commentary on the most advanced work of a pre-Ghazālī scientist. And in the context of that commentary he refuted the ideas of his predecessor and the ideas of the more ancient Greeks who had nevertheless been his models to some extent. In this regard, it should be remembered that although Ibn al-Haitham could be counted as a great scientist in his own right, he too did not hesitate to reject Greek ideas when they did not meet his exacting scientific standards. And yet it was al-Fārisī who had the final say in the matter of the rainbow.

In a sense,, this phenomenon is very similar to what took place in astronomy. Here too we find the formidable critique of Ptolemaic astronomy, which was leveled by Ibn al-Haitham in the middle of the eleventh century. We also find the same critique left as such, without anything more positive being done about it till the thirteenth century, when the astronomer Mu'ayyad al-Dīn al-ʿUrḍī (d. 1266), who lived more than a century after Ghazālī, complained against Ibn al-Haitham for not bringing anything new besides criticism. It was ʿUrḍī who undertook the development of a whole alternative astronomy that was destined to replace the Greek astronomy, and thereby produced his famous mathematical theorem that made its impact on almost all astronomers that followed him. Here again there is no comparison between the brilliance of ʿUrḍī and that of Ibn al-Haitham, startling and brilliant as Ibn al-Haitham was.

Still, in the discipline of astronomy this trend continued after ʿUrḍī, through Ṭūsī, Quṭb al-Dīn al-Shīrāzī (d. 1311), Niẓām al-Dīn al-Nīsābūrī (d. 1328), Ibn al-Shāṭir (1375) and his contemporary Ṣadr al-Sharīʿa al-Bukhārī (c. 1350), ʿAlā' al-Dīn al-Qushjī (d. 1474), Mulla Fatḥallāh al-Shirwānī (c. 1450), and finally to Shams al-Dīn al-Khafrī (d. 1550) to name only a few. Each and every one of those astronomers would take the works that were produced in the pre-Ghazālī period, and refer to them as the commonly known astronomy (*al-mashhūr*) only to attack them severely, and attack Ptolemy behind them as well. After having done that, they would then go on to build their own alternatives to that astronomy on grounds that were completely new, and on levels that were much more sophisticated than the levels found in the earlier period or the Greek sources themselves.

Like work in optics and medicine, the work of the astronomers was cast in the form of commentaries on each other's works, or at times on the Greek works themselves. They only used those commentaries as vehicles to produce their own alternative theories and to record their own scientific insight, much as was done by Jazarī, Ibn al-Nafīs, and al-Fārisī.

For people who had pre-judged this period as a period of decline in Islamic science, they saw in these commentaries a sign of decadence, without ever bothering to read them or appreciate the novel ideas that were contained therein. Even the most learned of the modern Arab intellectuals, once president of the Academy of the Arabic Language of Cairo, Professor Ibrahim Madkour, had this to say about this period, and the commentaries it produced: "Speculative thought was confined to increasingly narrow areas, scientific inquiry stagnated, and matters that had previously been studied and understood became obscure. Creative thinking and the spirit of discovery were replaced by sterile repetition and imitation, expressed in commentaries and studies of texts and stressing words rather than meaning."[16] How much more wrong could one be?

Professor Madkour, however, was not alone in that assessment. In fact, in the vast modern literature, one reads one such author after another all bemoaning the low state of Islamic science just because this period witnessed the production of commentaries instead of original works. Had these authors only read those commentaries, they would have realized that they could not have been much farther from the truth, for they would have also found one commentator after another saying: Ptolemy, or some other astronomer or Greek scientist, said this or that, but I say, and then insert their own novel ideas at the right context.

The problem with these kinds of judgments is that they clearly indicate that a true appreciation of the role of those commentaries has not been fully developed yet. In an earlier publication I had hinted to the fact that there is much to be found in those commentaries, and in that context gave only the example of the Ṭūsī Couple itself which I have demonstrated was first conceived in 1247, but in the context of a commentary on Ptolemy's *Almagest.*[17] Since then, and after reading many of the astronomical commentaries that followed Ṭūsī's, I have come to realize that those commentaries acted in a manner quite similar to our modern periodical literature. For when a modern author conceives of a bright novel idea, which was not conceived before, s/he would go ahead and compose an article that s/he would send to a specialized periodical announcing his/her own new idea. From then on, the new idea would enter the literature. And when enough of those ideas that deal with kindred subjects accumulate, they are then digested into a secondary book that popularizes them and finally allows them to enter the domain of public knowledge.

In regard to the popularization of new ideas, the medieval author has a greater advantage over the modern author, although his advantage has yet to be appreciated. For the medieval author, who had no access to specialized periodicals, as they did not exist at his/her time, the most efficient way of propagating his/her ideas would be to introduce them in the context of a commentary. For in such commentaries the new idea would in fact be properly contextualized and thus would gain a much greater significance than a lonely article in a journal that would need a lot of supplementary contextualizing information, not to say long years of waiting, before it could be fully appreciated, if at all. Needless to say, that on that ground alone, I now take the medium of commentaries to be a much healthier sign than even our own modern periodical scholarship.

Commentaries continued to be written throughout the post-Ghazālī period, and of course they continued to produce new ideas all the way till the sixteenth century, the last century, which has received a cursory study so far. This does not mean that they stopped then, for the later centuries have not even been investigated. Nor does it mean that there were no banal commentaries that were written during this period, for there were a lot, and many were indeed composed by mediocre minds that one finds in every age and place. But one can still document a series of commentaries, written by each and every one of the astronomers just mentioned, all building on each other's works, and all continuously taking up the challenge of perfecting the discipline of astronomy. By the sixteenth century, and with Shams al-Dīn al-Khafrī the commentaries reached such a sophistication that Khafrī could finally write two huge ones before he would eventually write an independent book, which he titled *Ḥall mā lā yanḥall* (Resolving that which could not be resolved). In the last, rather short text, he managed to produce a series of the most sophisticated solutions for all the problems that had plagued Greek astronomy and had become notoriously difficult to solve before his time.[18] In that instance, where most of the novel ideas were originally expressed in his commentaries before they were finally grouped together in his last independent original work, one can say that it was the commentaries that gave rise to originality rather than the other way around.

All this evidence points to one inescapable conclusion. Any one who takes the time to read the scientific production in the post-Ghazālī period would have to characterize this period as the most fecund, and in the field of astronomy in particular completely unparalleled. The disciplines that I have cited here, and the astronomers I have mentioned by name, all speak to a continuously ascending tradition all the way till the sixteenth century, being the last century that has been investigated although to a very modest degree.

As for those who still harbor the notion of the deadly struggle between science and religion, I only need to mention that with the exception of ʿUrḍī, whose religious credentials are yet to be determined, every one of the other astronomers mentioned, as well as Ibn al-Nafīs himself, were all religious men in the first place. Not in the sense that they were religiously practicing men only, but that they also held official religious positions such as judges, time keepers, and free jurists who delivered their own juridical opinions. Some of them wrote extensively on religious subjects as well, and were more famous for their religious writings than their scientific ones. This evidence leads me to conclude that the model of conflict between science and religion, which may have worked in Europe somehow, and I am not sure it did for it sounds too simplistic to contain the truth, this model does not seem to apply at least as far as the Islamic civilization is concerned. Nor does it particularly seem to apply in the post-Ghazālī period, when we witness more of the men of science being men of religion. Nor did it ever seem to be analytically useful as far as the discipline of astronomy is concerned for most astronomical works seem to have been produced by men of religion, and most of them were in fact employed in religious institutions.

As for those who think of history as a series of political events only, and a sequence of dynasties and wars, without paying much attention to intellectual history, they too can take little solace by relying so heavily on the Mongol invasion in order to justify their theory of decline. For although it was true that Baghdad was indeed destroyed at the hands of Hulegu Khan, it so happened that his vizier at the time was Naṣīr al-Dīn al-Ṭūsī, the astronomer he had captured in the conquest of the Islmāʿīlī fortress of Alamūt. It was this same Ṭūsī who had enough wisdom to save about 400,000 manuscripts before the sack of Baghdad. In addition, he even saved a young man by the name of Ibn al-Fuwaṭī, and took him along to what later became the Ilkhānid stronghold near Tabrīz. There, on a hill at the edge of the nearby city of Marāgha, Ṭūsī convinced the son of the same destroyer of Baghdad to grant him enough support in order to establish one of the most elaborate observatories the Islamic world had ever known.[19]

Of course, it helped that the same Ilkhānids soon converted to Islam, and granted Ṭūsī what he asked for. And there, in the city of Marāgha Ṭūsī managed to assemble the most distinguished company of astronomers ever assembled in one place. The mere gathering together of such astronomers, at such an active center, equipped with a new library of manuscripts that were rescued from Baghdad and other Iraqi and Syrian towns, together with Ibn al-Fuwaṭī as a librarian, they collectively and individually managed to produce the most sophisticated astronomical theories of Islamic times. Some of them had already made their contribution before they came to Marāgha. Just as Ṭūsī himself had done when he proposed his new mathematical theorem, the Ṭūsī Couple, while he was still at Alamūt, as we have already seen. And as ʿUrḍī too did when he had completed his most celebrated astronomical work, and his lemma, when he was still in Damascus. But their getting together at Marāgha produced the kind of astronomy that Shīrāzī was able to popularize by starting a tradition of dialogue with earlier astronomers, via the cumulative work of commentaries, and when he wrote two very long ones of his own within twenty years from the building of the Marāgha observatory.

More importantly, one should also remember that the foundation of the Marāgha observatory was commenced in the year 1259, that is, exactly one year after the destruction of Baghdad. The engineer who constructed the instruments, and in all likelihood who also built the structures himself, as most of them were masonry structures that doubled as astronomical observational instruments, was none other the famous astronomer/engineer ʿUrḍī himself. We are particularly fortunate to have a treatise written by this distinguished engineer, in which he detailed all the constructions that he accomplished at Marāgha, a treatise unequaled in its sophistication and utility, and the likes of which was unknown from the pre-Ghazālī period. Its importance can only be fully appreciated when we learn that it was also used as a guide for the construction of later observatories that were built in Samarqand during the time of Ulugh Beg (c. 1420) and in Jaypur, India, toward the end of the eighteenth century. The text of this treatise remains unfortunately un-edited in a modern scientific edition, and is only partially translated.[20]

Both of the decline narratives, therefore, that attribute the death of science either to the success of Ghazālī's religious thought, or to the destruction of Baghdad by the Mongols, do not seem to explain the brilliant scientific production we just mentioned. Furthermore, and in light of what we already know, these two causes do not seem to have even slowed down the production of science and did not seem to have set an age of decline. On the contrary, one may argue that the period that followed was marked by an increase in scientific production, and a remarkable upgrading of its quality, so much so, to make the production in the pre-Ghazālī period look much more modest in comparison. And as I have already said, in the field of astronomy alone, I have already argued that the golden age of that discipline, in terms of the production of planetary theories at least, should in fact be located in the post-Ghazālī period.

But if that were the case, and if the age of decline could be so easily reconstructed as an age of fecundity, at least as far as astronomical production was concerned, then when was this age of decline, and what is one to understand by the term decline in the first place? 'Decline' is a relative term implying a comparison between two levels, one perceived to be lower than the other. And as we have just seen, when we compared the scientific writings produced in the post-Ghazālī period, in several scientific disciplines, and found them to be more sophisticated than the ones that were written before, and at times even written in direct opposition to those earlier writings, we were at all times comparing one scientific production against another, and many factors were taken into consideration in this process. On the basis of those comparisons one would have dared to say that the post Ghazālī period witnessed a renaissance in comparison to the pre-Ghazālī period, and thus the latter could in turn be described as an age of decline.

Some of those factors that went into the comparison had to do with the quantity of production, for it is quite natural to expect that a few texts here and there would not make a trend, and thus would not constitute a shift in the type and quality of scientific production. It is for that reason that special attention was paid to the number of scientists in the post-Ghazālī period whose works were considered in the comparison process. And here again, the case of astronomy did not only prove that there were far more scientists from the later period who produced more creative material, but that there were more of them producing material in opposition to the earlier astronomers. One can detect a clear trend of new ways of doing astronomy, and thus one will have to admit that the post-Ghazālī period deserves a special consideration.

Then the number of fields in which this production was compared had to be considered, for it would also be natural not to think of excellence in one field as a sign of a trend that should characterize a historical period as more or less advanced. And there again, the circle was widened to include such disciplines as medicine, optics, and mechanical engineering. In all instances one could find results that were more or less in agreement with what seemed to be happening in astronomy.

And if one were to take a look at the field of scientific instruments, one could also document a similar flourishing activity in this later period, not only in the remains of large scale observatories such as the ones that were built in Marāgha and Samarqand, but also include those that were built by Jai Singh II (1686–1734) in India in imitation of those earlier ones.[21] One can also notice, from the sheer number of scientific instruments, still kept at Museums all over the world, whether astrolabes, quadrants, sextants, or what have you, that the number of more and more refined instruments kept increasing in the later period. To take only one example, the development of the universal astrolabe, of which we still have several samples, is not only a masterpiece of workmanship, but is also theoretically superior to the astrolabes that were constructed in the earlier period. And so was the case with many other instruments.[22] Thanks to the abundant results that are now in print, one can simply assert that the field of astronomical instruments also witnessed a "golden age" in the post-Ghazālī period, in complete synchrony with the field of planetary theories, although the two fields are only poor cousins.

I know the number of disciplines that I have attempted to list here is by no means exhaustive, and here I must admit that lack of competence in the history of other disciplines restrains me from passing judgments about them in the same fashion. But I certainly welcome colleagues who work in those disciplines to double check the results that have been so obviously achieved so far and decide for themselves whether this period can still be called a period of decline. My suspicion is that we have surveyed a good representative cross section of disciplines. And most probably the results we have already achieved, and the assessments we are now making, will withstand such additional tests from other disciplines.

While it is true that most of those results point to an increased volume of brilliant astronomical production, all coming from the post-Ghazālī period, nevertheless none of those results seem to come from the relatively later period, namely, the period beyond the sixteenth century, as far as I know. Again for the field of astronomy, I dare say that this may be due to lack of expertise on my part, since I have not thoroughly investigated the later works. I have focused over the last two decades or so on the works that were produced between the thirteenth and sixteenth centuries, and have not paid enough attention to the mostly inaccessible works of the later centuries. Disciplines other than astronomy may have suffered from similar handicaps, and thus may one day turn out new material that is still undiscovered.

But a quick survey of the readily accessible astronomical texts produced after the sixteenth century revealed an interesting phenomenon. Not only do we begin to see a slightly different astronomical production, one more concerned with purely religious astronomy such as *mīqāt,* or simplified astronomical texts, but also as early as the seventeenth century we begin to notice an incursion of European scientific ideas coming back into the world of Islam. We can even find echoes of things happening in Europe during the sixteenth century beginning to be acknowledged in the Islamic world, and sometimes incorporated. Here I am thinking of one of the later Syro-Egyptian astronomers: Taqī al-Dīn ibn Maʿrūf (d. ca. 1586), whose work included, in his own hand, an acknowledgment of his direct acquaintance with the multi-lingual dictionary of Ambrosio Calepino (1435–1510).[23]

Later works in geography from the seventeenth century reflect knowledge of the various astronomical systems of Copernicus and Tycho Brahe, and also demonstrate how such works began to mention the discovery of the new world. All these echoes come in the context of the translations of such works as the Atlas Maior and Minor, into Turkish.[24]

In sum, I am willing to accept the fact that a thorough investigation of this later period, say from the sixteenth to the twentieth century, will definitely demonstrate an increasing dependence on the scientific results that were produced in European centers of learning. Production that was by then making its way back into the Islamic world. This process apparently continued unchecked during these later centuries until the Islamic world was finally brought to rely completely on European science during the colonial era of the nineteenth and the twentieth centuries. That dependence has even intensified further in the present time.

The latter part of the twentieth century demonstrates this complete dependence extremely well. For after all, this particular century witnessed the "independence" of most Islamic countries after having already undergone a long period of colonization that was brought to an "end" during that part of the century. And now, most, if not all Muslim countries, depend for all their scientific education on the scientific output of European countries, their colonial centers of yesterday, or say western in order to include the United States of America. This is also the case at almost all universities in the developing world, with the Islamic world planted in its midst. They all depend, in their scientific curricula, on what is produced in the west.

With the end of the twentieth century we can see the pendulum moving all the way to the west. And if we look at the source of science, we would then be in a position to witness the extreme end of the spectrum. But we should also ask: When did this shift take place? That is, when did Europe stop being interested in the scientific production of the Islamic world, and when did it begin to export scientific production to it?

Determining the time of that shift itself may help us determine the onset of the age of decline. But let me be clear on the concept of decline itself. In this context, I wish to define such an age as an age in which a civilization begins to be a consumer of scientific ideas rather than a producer of them.

Going back to the sources, as we have been doing all along, those sources seem to indicate a break that took place sometime around the sixteenth century, and that century seems to contain the seeds of that age of decline, or at least seems to have been the time when such decline may have commenced. And if my reading of those sources is valid, then we must look for the events that surrounded this particular century, i.e. the sixteenth century, in order to determine, if we can ever determine at all, the causes of that decline.

For a better diagnostic look at the age of decline, one must constantly bear in mind the relative nature of that concept, and must appreciate the fact that social processes such as cultural decline or Renaissance and the like are rarely datable to a specific decade or even a century. They usually are indistinguishable at first, but with the passage of time trends begin to consolidate and remarkable differences begin to be noticed. As was already noted by others like Needham, who will be referred to again below, if one were to compare the scientific production in the world of Islam, China and what is now Europe, just about the beginning of the sixteenth century, one would have noticed that all three were almost on equal levels. Two centuries later, say by the beginning of the eighteenth, that comparison begins to weigh more heavily in the direction of Europe.

Those 200 years, say roughly between 1500 and 1700, witnessed the creation of one scientific revolution after another in Europe and marked the definite birth of modern science. And for that same reason, they gave rise to the multiplicity of questions that have been asked ever since, all attempting to explain why did modern science rise in Europe and not in the other two competitive cultures of the time. Many have sought answers in the social make up of the cultures concerned, others looked at the legal, religious, and political conditions. Some have even taken the conditions of modern Muslim societies and projected them, in very essentialist and nonhistorical terms, back onto the histories of those societies.[25]

And yet the question has persisted for a long time now. And because of its sheer emotional and ideological underpinnings, in a world constantly polarized, it has not become any easier to answer. But if we focus on the big picture, that is a picture drawn over a period of a few centuries where the trends become clearer to observe, and if we widen the scope to the multiplicity of factors that may have caused the lopsidedness of the scientific production that becomes clearer to observe as well, then we stand a better chance at understanding not only the nature of the decline of scientific production in the world of Islam, but may also gain some insight into the social economic context of science itself. By applying the same methodology, which attempted to explain the rise of science in early Islamic times in terms of socio economic conditions, this time the same methodology may yet help us understand why within a period of two centuries or so there arose a remarkable difference between the sciences that were produced in Europe and those that were produced in the rest of the world, and most notably the Islamic world. From that perspective, it would no longer be interesting whether Copernicus knew of the works of his predecessors from the Islamic world or not. Instead the focus would shift to the conditions that led to the works of Copernicus to be incorporated into later, more advanced works, and systems of thought that led to the demise of the old Aristotelian world order. Most importantly, those developments indeed changed the very nature of scientific production itself. Therefore, those two crucial centuries have a lot to teach us about the nature of modern science, its relationship to the cycle of capital investment, and its relationship to the fluctuating political and economic conditions. It is under those conditions that one has to seek the meaning of the age of decline in the Islamic world that has noticeably set in from the sixteenth century on, without necessarily pinning that decline, if one can help it, to a particular cause, a particular event, or a particular train of thought, be it religious or otherwise. So what happened during those two centuries?

Political history of that period may after all be useful in this regard, and it does reveal some very interesting features. By the middle of the sixteenth century we witness, for the first time, a large-scale split of political power within the Islamic world. That split produced three great Muslim empires. They all came into existence around the same time, and with the exception of the Ottoman they all passed away around the middle of the eighteenth century.

With the Ottomans (c. 1453–1920), finally conquering Constantinople in 1453, they swept through the eastern Mediterranean in 1516, as far down as Egypt and large parts of North Africa, in order to consolidate their stronghold over that part of the Muslim world. The Safavids (1502–1736), farther east in what is modern-day Iran, came into power by the beginning of the sixteenth century, and with time established a new empire with Shīʿism as its official religion, thus perpetuating a rivalry with the Sunni Ottoman empire to the west and a friendlier but yet competitive relation with the Mughal empire to the south-east that lasted for centuries. The Mughals (c. 1520–mid eighteenth century) themselves, originally a Central Asian dynasty, spread southward to establish one of the long lasting empires in the Indian subcontinent.

Besides the disrupting effect of this internecine competition and warfare, other factors came into play, but all leading to a weakening of the cultural cohesion of the Islamic world. The religious sectarian competitiveness played an important role on its own, as it still does today. But then there was also the very important event that took place toward the end of the fifteenth century, and which shook the whole world order to its very foundations. The event in question was the discovery of the New World, which not only disrupted almost all of the well-established Euro-Asian trade routes that used to siphon commercial wealth into the Islamic lands for centuries, but it also brought new raw material into European countries just as those same materials were almost completely depleted in the Islamic lands. It is not accidental that all three Islamic empires came into being around the same time, during the early part of the sixteenth century, and disappeared around the same time, end of the nineteenth beginning of the twentieth centuries, as we just said. In order to understand this phenomenon better, we must examine it in terms of the socio-economical and political global shifts that were taking place around that time.

By the beginning of the sixteenth century, the so-called "discovery" of the New World had just begun, and the westward orientation of European exploration, trade, and access to untapped natural resources as well as to human slave labor, both in the New World and (later) in Africa, created a major conflagration all around the world. Only to be followed by the "Age of Discovery" in the next century, which witnessed a dogged search for more lands to "discover," more resources to acquire, and more colonies and slave labor to entrap. All of these events of the sixteenth and early seventeenth centuries re-oriented wealth and trade around the Islamic world, or say circumvented the Islamic world, and mostly to its disadvantage. And while almost every European royal house and its dependencies, in one way or another, began to receive tons of gold and silver, as well as free slave labor and other natural resources from the colonies, the Islamic world found itself then blocked to the west by the rising powers of the European royal houses. Those royal and princely houses were now wealthy and well equipped with commercial and maritime navies.

Circumnavigating around Africa by the Portuguese helped them spread their trade in the south-eastern direction at first, and eventually to the east where Portuguese and later Dutch colonies began to sprout as far down as southern India, up north the Indian Ocean till the southern edges of the Arabian peninsula itself, and farther east by the Dutch till the eastern edge of the known world. Eventually, that colonial exploration that reached the South Asian and the Chinese theater in the Far East began to re-route even the trade of that eastern region *around* the Muslim world rather than *through* it.

Yes, there were some windfall profits that came to the Islamic world as a result of trade with the newly discovered wealth of Europe. But on the whole, Islamic lands lost the commercial initiative they once had, and became more and more dependent on whatever wealth the European merchants were willing to part with while trading with ports in the Islamic world. In essence the relationship began to shift from producing wealth to consuming wealth in return for whatever natural resources were still available. And these are the whole marks of an age of decline. Yes, there were Venetian merchants who brought some wealth to Damascus, for example, by commissioning household products and items of high culture to be produced there, but that meant that the Damascene worker began, even then, to enter into a relationship of dependency where he was working for a foreign master. The dependence and consumerism that was set then and later began to characterize the relationship between the Islamic world and Europe continues till this day.

I do not know of a good study that elaborates on the effects of the "discovery" of the New World on the intellectual life of European royal and princely houses. But it is not difficult to detect, in several European areas, and toward the beginning of the seventeenth century, the appearance of new institutions that had no medieval parallels per se, and their very creation may have had something to do with this new acquired wealth. During the first half of the seventeenth century, Europe witnessed the rise of scientific and royal academies, a phenomenon that was not known before as such, at least not to that extent where almost every royal or princely house had an academy of its own. The purpose of those academies seems to have been directed at assembling the most educated men of the time and to liberate those men from the financial worries and the like. In their very structure, the academies offered those intellectual elites an environment of scientific and intellectual competition. And as we have seen before, it is the healthy competition that is usually conducive to the production of science. But most importantly, this whole movement came about at almost no cost to the patronizing royal houses, for the capital and the slave labor associated with the investment usually came through many circuitous routes from the "discovered" colonies. In regard to those scientific institutions, we only note that the first of them was the Academia de Lincei, which was founded in Rome in 1603, only to be followed by the Royal Society of England in 1662, and the Academie des sciences of France in 1666.

The connection between those academies and the "discoveries" in the New World is not always readily apparent. But one should note that the oldest of them, the Academia de Lincei, soon enjoyed the membership of none other than Galileo ca. 1609, whose work for the Venetian commercial navy is well known.[26] And one of the earliest projects of the Academia de Lincei was the re-publication of the survey of the medical plants of the new colonies in Mexico, which was then called New Spain.[27] That survey was completed few years earlier by Dr. Francisco Hernandez (1515–1587), at the request of King Philip II of Spain (1527–1598). Instead of verifying the older herbs of Dioscorides that were well known in the "old world" and were obviously commercially fully exploited by then, the academy, and of course the earlier royal patrons before it, began to look to the New World for new sources of wealth, and the medical plants were apparently such well suited targets.

It is only natural that in such institutions as the academies, where men of science were financed to further their research, and to think out new ideas, in an environment of competition with other academies and royal houses, as well as competition among the scientists themselves, new scientific discoveries would eventually be produced. The situation was not too different from the conditions we described in early ninth-century Baghdad, minus the institution of the academies that quickly became the norm in Europe. If one scientist in a hundred produced something in those academies that had a commercial windfall, then the wealth accumulated from the new idea would be returned to fund other ideas, and, of course, allowing the patron to keep some of the profits aside.

In this manner, I believe that the major scientific developments in Europe during the sixteenth and seventeenth centuries were the product of this dynamic cycle of wealth, mostly initiated by the "discovery" of the New World. Wealth drove further production of science, and in turn science allowed the acquisition of more wealth, and so on. This pattern seems to have been set then. And for those who look at the close relationship between modern corporations and the production of modern science, can easily discern the main features of this same dynamic cycle that is still going on.

At this sped-up rate, the production of science in what is now Europe began to grow almost at a logarithmic rate, leaving the rest of the world to struggle with its own depleted resources and its old ways of doing science. The Islamic, as well as the Chinese worlds, had up till that time a similar scientific status as that of Europe, as was so aptly noted by Needham more than 50 years ago.[28] But with the onset of the new dynamic cycle we just mentioned, that was set in motion by the end of the sixteenth and beginning of the seventeenth centuries, European science began to surge on, and both of the Chinese and Islamic worlds were left behind.

Returning to the age of decline in the Islamic civilization, in my opinion, this age of decline was less caused by such factors as a book of Ghazālī or the invasion of the Mongols, than by the external world circumstances of the sixteenth century and thereafter. And since the term decline implied a comparative context, as was stated above, in my opinion too, what seems to have happened was the onset of a race between the European royal houses and the rest of the world, including the Islamic world. And in that race, the Islamic world lost. But no one should forget that the real race started in the sixteenth century as a result of the discovery of the New World, and that it was a race between Europe, on the one hand, and the rest of the world, on the other. This race continues to intensify till this very day. In relative terms then, when one culture begins to produce more and better science, for now it can afford to, the other culture will look like it declined, no matter what.

Of course, the translation of the European superiority, and now add the United States' as well, in commercial, scientific and technological terms, into further acquisitions of resources and manpower from the rest of the world, and actual subjugation of the rest of the world to military occupation between the eighteenth and twentieth century, the so-called age of colonialism, which is still going on in many places, did not help in leveling the field of competition. So naturally, all non-western cultures look like they are experiencing an age of decline in comparison. And their decline too started around the year 1600, nearly 100 years after the discovery of the New World and during which European royal houses learned how to translate the benefits of that discovery into political power.

As far as I can tell, neither Islamic science, nor Chinese science, for example, had managed to start a capital-driven cycle through their methods of production. In the Islamic world the institutions of science, such as observatories, hospitals, and even the various houses of science (individually called *dār al-ʿilm*) that were mainly patronized by wealthy individuals, and at times the ruling sultan himself, were never directed at acquisition of further wealth, and never attained a self sustaining economic status that would have guaranteed their survival and perpetuity. They could produce brilliant scientists, such as the ones whose works we have examined, ever so briefly, but they could not guarantee the continuous production of the scientists themselves through the security of their income and position. As a result, the scientific production of the Islamic world was mainly driven by individual genius but only when those geniuses could by accident encounter the right patron who would offer the support.

For modern times, the problem of catching up with western science is not only a problem for the Islamic world alone. Instead it has obviously become the problem of the whole so-called second and third worlds as well. And now they all seem to be locked in this competitive race for which the non-western world neither possesses the necessary capital, nor the infrastructure, nor the manpower to compete on fair grounds. Add to that the constant brain drain that continues to feed the first world at the expense of the second and third, a fact that makes this race even harder to win.

Notes and References

Chapter 1

1. See G. Saliba, "Science before Islam," in *The Different Aspects of Islamic Culture,* vol. 4: *Science and Technology in Islam,* ed. A. al-Hassan, M. Ahmad, and A. Iskandar, part 1, *The Exact and Natural Sciences* (UNESCO, 2001), pp. 27–49.

2. A brief formulation of this contact theory, as expressed by G. E. von Grunebaum, goes as follows: "The tendencies inherent in the origins of Islam were to mature under the influence of those, in a sense, accidental *contacts* which grew out of the historical setting of the period and, more specifically, the conquest by the Muslims of the *high-civilization* areas in Persia, Syria, and Egypt." (*Islam: Essays in the Nature and Growth of a Cultural Tradition,* Greenwood, 1981, p. 12f., emphasis added) For other formulations of these contacts, see Christopher Toll, "Arabische Wissenschaft und Hellenistisches Erbe," ed. A. Mercier (Herbert Lang, 1976), pp. 31–57. Even modern Arabic writings reflect this understanding. See, for example, ʿAbd al-Ghanī, Muṣṭafā Labīb, *Dirāsāt fī tārīkh al-ʿulūm ʿinda al-ʿArab,* vol. 1 (Dār al-Thaqāfa, 2000), p. 43.

3. In fact the cultural conditions in Byzantium during a century and a half (641–780) accompanying and immediately following the rise of Islam in the early part of the seventh century are usually designated by Byzantinists themselves in expressions ranging from centuries of "negligible traces" to centuries of "Dark Age." See, for example, Alexander Jones, "Later Greek and Byzantine Astronomy," in *Astronomy before the Telescope,*" ed. C. Walker (St. Martin's Press, 1996), p. 104: "The century and half between the reign of Heraclius and the beginning of the ninth century has left us negligible traces of astronomical writings." See also Warren Treadgold, "The Struggle for Survival (641–780)," in *The Oxford History of Byzantium,* ed. C. Mango (Oxford University Press, 2002). Some would even go back another century or two to include the sixth and seventh centuries as Dark Ages, as was done most recently by Timothy Gregory in *A History of Byzantium* (Blackwell, 2005). Later more "enlightened" centuries like the ninth century still formed the subject of symposia such as *Byzantium in the Ninth Century: Dead or Alive?* ed. L. Brubaker (Ashgate, 1998). See also Irfan Shahid, "Islam and Byzantium in the IXth century: The Baghdad, Constantinople Dialogue,"

in *Cultural Contacts in Building a Universal Civilization: Islamic Contributions*, ed. E. Ihsanoglu (Istanbul, 2005). As for Sasanian Iran not much is known of its intellectual production for the same period, and whatever sciences were cultivated there were only translated into Arabic during the latter part of the eighth and early part of the ninth centuries only to be discarded very quickly in favor of the classical Greek sciences that came to replace them. The few books that stand out from the pre-Islamic period, like the *zīj-i shāh-i*, or *zīj-i Shāhriyār*, of which we only have reports from later astronomers who worked in Islamic times, and the *Kalila wa Dimna* which was translated by Ibn al-Muqaffaʿ in the eighth century were indeed similar to the much more elementary astronomical and literary texts that one encounters during the dark ages of Byzantium, or the contemporary Syriac texts, which were very well known at the time.

4. Sure one could point to such works as those of Proclus (mid fifth century), especially his *Hypotyposis*, or the slightly later Ammonius (fifth century–mid sixth century), or Philoponus (early sixth century), or even Olympiodorus (mid sixth century), but those were either elementary works, or dealt with astrology rather than astronomy. Even Leo the Mathematician's work itself when looked at closely by Alexander Jones ("Later Greek and Byzantine Astronomy," in *Astronomy before the Telescope*," ed. C. Walker, St. Martin's Press, 1996, p. 104), was deemed "questionable [in terms of] Leo's ability to interpret highly technical works, for his own surviving writings are meager and unimpressive."

5. For the treatise of Ḥunain in which he recounts his searches for Galenic texts, see Gotthelf Bergstrasser, *Ḥunain b. Isḥāq, Über die Syrischen und Arabischen Galenübersetzungen* (Leipzig, 1925).

6. Examples illustrating this pocket theory abound. Any source that mentions the survival of Hellinism in such major centers as Alexandria, Antioch, Edessa, Ḥarran, Jundishāpūr, etc. would be a good candidate for that purpose. A very recent version of that theory is embedded in L. E. Goodman's essay "The Translation of the Greek materials into Arabic," in the *Cambridge History of Arabic Literature: Religion Learning and Science in the Abbasid Period* (Cambridge University Press, 1990). In modern Arabic writings those pockets were at times given more significance by referring to them as schools. See, for example, Rashīd al-Jumaylī, *Ḥarakat al-tarjama fī al-mashriq al-islāmī fī al-qarnain al-thālith wa-l-rābiʿ li-l-hijra,* al-kitāb (Tripoli, 1982), p. 178f. Despite the shortcomings of this theory one should, nevertheless, give it more credit but insist on adding the caveat that allows for a distinction between the survival of such cultural aspects as religion, art, and music, aspects that may have survived in those centers, and the more rigorous aspects of science and philosophy that require much debate and apprenticeship in an open society that would encourage such studies. The conditions of the Byzantine empire, even in its better (?) days during the ninth century, when it was already in contact with the more advanced Islamic world at the time, are best illustrated by the life of the famous Leon the Philosopher/Mathematician that is

excellently related by Paul Lemerle in *Le Premier Humanisme Byzantin* (Presses Universitaires de France, 1971), where we are told (p. 148f.) that Leon could only study grammar and poetics in the capital city of Constantinople. For more rudimentary instruction in the other sciences he had to travel to other less important places where he would obtain the basics of these sciences from one individual at a time. The story goes on to tell of the great efforts Leon had to exert in order to get instructed in the other sciences, only to be severely attacked for that endeavor by his own student Constantine the Sicilian who says of him that he "enseigné toute cette science profane dont les anciens se sont enorgueillis, et qui a perdu son âme dans cette mer d'impiété" (Lemerle, p. 173).

7. For a brilliant chapter on the first "Byzantine Humanism," which witnessed the revival of interest in Greek scientific manuscripts and their transfer to minuscule hand at an extensive scale during the ninth century for the translation market of Baghdad, see Dimitri Gutas, *Greek Thought, Arabic Culture* (Routledge, 1998), pp. 175–186.

8. In fact Alexander Jones ("Later Greek and Byzantine Astronomy," p. 105) says the following about those manuscripts: "But this paucity of corrections by scribe or owner also suggest that, for all their splendour, these manuscripts were more for display than for study. Original writings from the ninth and tenth centuries, whether in the margins of those extant contemporary codices or in later copies, are pitiful and scarce. One concludes that practical understanding of astronomy was sustained by few besides astrologers, whose working copies of the old texts were presumably more perishable than the bibliophile's treasures that have come down to us, but whose existence is revealed by the odd horoscope or anecdote."

9. See "Cosmology" in the *Oxford Dictionary of Byzantium*, Oxford University Press, 1991, p. 537, where the views of this Cosmas are claimed to have been characteristic of those of the "school" of Antioch, one of the major alleged pockets for the transfer of Hellenistic knowledge to Islam.

10. See G. Saliba, "Paulus Alexandrinus in Syriac and Arabic," *Byzantion*, 1995, 65: 440–454.

11. Still preserved in the work of the thirteenth-century bio-bibliographer Ibn Abī Uṣaybiʿa, *ʿUyūn al-Anbāʾ fī Ṭabaqāt al-Aṭibbāʾ*, ed. Richard Müller, Königsberg, 1884, vol. 2, p. 134ff.

12. Max Meyerhof, "Von Alexandrien nach Bagdad: Ein Beitrag zur Geschichte des philosophischen und medizinischen Unterrichts den Araben," *Sitzungsberichte der Berliner Akademie der Wissenschaften*, Philologisch-historische Klasse, 1930: 389–429.

13. Paul Lemerle, *Le Premier Humanisme Byzantin: Notes et remarques sur enseignement et culture à Byzance des origins au Xe siècle* (Presses Universitaires de France, 1971), p. 25.

14. See the detailed entry in the supplement of the *Encyclopedia Iranica* for Paul the Persian and his works, available at http://www.iranica.com. For much more extensive treatment and citation of relevant literature see also the article of Dimitri Gutas, "Paul the Persian on the Classifications of the Parts of Aristotle's Philosophy: A Milestone between Alexandria and Baghdad," *Der Islam* 60 (1983): 231–267, especially the note on p. 239 discussing the identity of this Paul, and the decision on the one Paul who apparently wrote "the Syriac introduction to Logic." The following note in the same article, and note 29 on page 244, suggest "that the Introduction to Logic also was initially composed in Pehlevī." If the latter remark is true then Syriac may have acted in this instance as an intermediary between Pehlevī and Arabic rather than Greek and Arabic, which may very well be the case. This does not affect, however, the elementary nature of the contents of the treatise when compared to the classical sources.

15. See Saliba, "Paulus Alexandrinus."

16. F. Nau, "Le traité de l'astrolabe plan de Sévère Sébokt, publié pour la première fois d'aprés un Ms. de Berlin," *Journal Asiatique*, 13 (1899), 56–101, 238–303.

17. For his astronomical works that relate more to ancient folk Babylonian science rather than to sophisticated Greek science, see V. Ryssel, "Die Astronomischen Briefe Georgs des Araberbischofs," *Zeitschrift für Assyriologie und verwandte Gebiete* 8 (1893): 1–55, and Otto Neugebauer, *A History of Ancient Mathematical Astronomy* [*HAMA*], Springer-Verlag, 1975, pp. 597, 707, 720.

18. See F. Nau, "La plus ancienne mention orientale des chiffres indiens," *Journal Asiatique* 16 (1910): 225–227.

19. Ibid.

20. See Goodman, "Translation," p. 478.

21. See the account of Muḥammad Diyāb al-Itlīdī, *Iʿlām al-nās bi-mā waqaʿa li-l-barāmika maʿa banī al-ʿabbās*, Beirut, 1990, p. 244f.

22. Gutas, *Greek Thought, Arabic Culture*, pp. 36–41.

23. See al-Nadīm, *al-Fihrist*, p. 393. In what follows the various references to this work will be to the edition of Yusuf ʿAlī Ṭawīl, Beirut, 1996, and will be simply designated as *Fihrist*, unless otherwise stated.

24. For a short analysis of this *zīj* and a bibliography of other assessments of its importance, see E. S. Kennedy, "A Survey of Islamic Astronomical Tables," *Transactions of the American Philosophical Society*, New Series, 46, no. 2 (1956), p. 129f. See also D. Pingree, "The Greek Influence on Early Islamic Mathematical Astronomy," *Journal of the American Oriental Society* 93 (1973): 32–43.

25. See C. Nallino, *ʿIlm al-Falk: Tārīkhuhu ʿind al-ʿArab fī al-qurūn al-wusṭā* (Rome, 1911), pp. 193–196.

26. Now available in the Arabic version with English translation and Greek fragments (D. Pingree, *Dorotheus Sidonius Carmen Astrologicum*, Teubner, 1976).

27. This story is elegantly parsed and deconstructed by Gutas (*Greek Thought, Arabic Culture*, pp. 75–105, esp. 97f.). I used the Arabic version of the story that was preserved in the *Fihrist*, p. 397f.

28. On both of these terms, the Muʿtazila and their *miḥna*, see *Encyclopedia of Islam*, new edition (hereafter cited as *EI*²), s.v. "Muʿtazila" and "miḥna."

29. *EI*², vol. I, p. 272f.

30. A note on the title page of one of the Arabic translations of the *Almagest*, now kept at the Library of Leiden University (Or. 680), reads "This book was translated by the order of the *imām* al-Maʾmūn ʿAbdallāh, the commander of the faithful."

31. See, for example, *Géographie d'Aboulféda*, ed. M. Reinaud, that is, Abū al-Fidāʾ, *Taqwīm al-Buldān*, Paris, 1840, p. 14.

32. For the status of the manuscripts in Constantinople at that time, and the level of science in that city, see Gutas, *Greek Thought*, pp. 175–186, and refer to the anecdote of Leo the Mathematician mentioned before.

33. See, for example, the reports about the patronage of the three sons of Mūsā during the caliphate of al-Mutawakkil as reported by Ḥunain b. Isḥāq, the main translator of the period, in the treatise edited by Bergstrasser, and their undertaking of the digging of canals which they apparently sub-contracted in one instance to their protégé Ibn Kathīr al-Farghānī (d. after 861). The account of this failed project and the rescue of Banū Mūsā by the mathematician and engineer Sanad Ibn ʿAlī (d. 864) is recounted in Ibn Abī Uṣaybiʿa's *Ṭabaqāt al-Atibbāʾ*, vol. I, p. 207f.

34. See, for example, the work of the eleventh-century polymath Abū al-Raiḥān al-Bīrūnī, *Chronology of Ancient Nations*, ed. E. Sachau, p. 263.

35. On the role played by those astrologers in appropriating the Indian sources under the patronage of the caliph al-Manṣūr, see Nallino, pp. 141–215.

36. On the surviving fragments from the works of those two astronomers, see David Pingree, "The Fragments of the Works of al-Fazārī," *Journal of Near Eastern Studies* 29 (1970), pp. 103–123; Pingree, "The Fragments of the Works of Yaʿqūb ibn Ṭāriq," *Journal of Near Eastern Studies* 26 (1968): 97–125.

37. See al-Nadīm, *Fihrist*, p. 437.

38. *Fihrist*, p. 400

39. For the report on the *Almagest* translation during the time of Khālid al-Barmakī, see *Fihrist*, p. 430. The less likely report about the role of al-Manṣūr in the transmission of the *Almagest* is preserved in several texts, among them the text of al-Masʿūdī, *Murūj al-Dhahab (Les Prairies d'Or)*, ed. C. Barbier de Meynard, Paris, 1874, vol. 8, p. 291.

40. See the edition and translation of this work by Frederic Rosen, *The Algebra of Mohammed ben Musa*, London, 1831, reprt. 1986, and for the originality of Khwārizmī see Roshdi Rashed, "l'Idée de l'Algèbre Selon Al-Khwārizmī," *Fundamenta Scientiae* 4 (1983): 87–100.

41. For a similar argument see Carl Boyer, *A History of Mathematics* (New York, 1968), reprt. 85, p. 252.

42. F. Sezgin, *Geschichte des Arabischen Schrifttums*, vol. VI (Leiden, 1978), p. 182.

43. Roshdi Rashed, *l'Art de l'Algèbre de Diaphonte*, Arabic text first edited and published in Cario, 1975, and then re-edited and translated into French by Rashed as well, and published in Paris 1984. See also Rashed, "Problems of the Transmission of Greek Scientific Thought into Arabic: Examples from Mathematics and Optics," *History of Science* 27 (1989), pp. 199–209, reprinted in Roshdi Rashed, *Optique et mathématique* (Variorum, 1992), esp. p. 203f.

44. For this specific projection of Ḥabash, see E. S. Kennedy, P. Kunitzsch, and R. P. Lorch, *The Melon-Shaped Astrolabe in Arabic Astronomy*, Stuttgart, 1999. For a survey of his trigonometric and astronomical works with further bibliographical references, see Marie-Thérèse Debarnot, "The Zīj of Ḥabash al-Ḥāsib: A Survey of Ms Istanbul Yeni Cami 784/2," in *From Deferent to Equant: Annals of the New York Academy of Sciences* 500 (1987): 35–69. For more precise developments in trigonometric functions, see Kennedy, *Survey of Islamic Astronomical Tables*, p. 151f. For a much longer study of the type of projections produced by Ḥabash and their impact on later generations, see David King, *World-Maps for Finding the Direction and Distance to Mecca: Innovation and Tradition in Islamic Science*, Brill, 1999.

45. A. S. Saidan, *The Arithmetic of al-Uqlīdisī*, Boston, 1978, p. 343.

46. To his credit a similar position is held by Roshdi Rashed in his "Problems of the Transmission," where he speaks of a translation period preceding Abbāsid times, but characterized as "individual initiatives," and goes on to claim that the translation movement had to wait for the second period of "incomparable importance . . . [when] translation has become a part of a much wider activity that may be designated by the evocative title "the institutionalization of science," p. 200. Unfortunately though, Rashed explains the translation movement of the early Abbasids as dependent on the desires of the caliphs and the abundant existence of scientists without explaining how such desires and scientists came about. A. I. Sabra also proposes a similar explanation in his thought-provoking article "The Appropriation," where to his credit too, Sabra speaks in this article of a process of "appropriation" rather than "contact" or encounter of "pockets" and one can read him to say that Islamic civilization sought the ancient classical Greek texts and did not satisfy itself with what was available in Byzantium at the time.

47. Ptolemy, *Almagest, I.15*. G. Toomer, *Ptolemy's Almagest*, New York, 1984, p. 72.

48. See *A Concise History of Science in India*, ed. D. Bose, S. Sen, and B. Subbarayappa, New Delhi, 1971, p. 107

49. See Kennedy, *Survey*, p. 145, for the value 23;33° in the *mumtaḥan zīj*, and pages 151, 153, and 154, for the value 23;35° in the works of Ḥabash, and al-Battānī from the next century.

50. For the values of 1/66 years or 1/70 years, see Kennedy, *Survey*, p. 146. For tabulations of the other values in the same and other *zījes*, see ibid., p. 150f.

51. For the theoretical critique of Muḥammad, see G. Saliba, "Early Arabic Critique of Ptolemaic Cosmology: A Ninth-Century Text on the Motion of the Celestial Spheres," *Journal for the History of Astronomy* 25 (1994): 115–141. For the critique of observational methodology by the three brothers or someone in their circle, see O. Neugebauer, Thābit Ben Qurra "On the Solar Year" and "On the Motion of the Eighth Sphere," translation and commentary, *Proceedings of the American Philosophical Society* 106 (1962): 264–299, and Régis Morelon, *Thābit Ibn Qurra: Oeuvres d'Astronomie*, Paris, 1987, pp. xlvi–lxxv, 26–67, 189–215.

52. Neugebauer, "Thābit"; Morelon, *Thābit*; Saliba, "Early Arabic."

53. Jazarī's major work *al-Jāmiʿ bain al-ʿilm wa-l-ʿamal al-nāfiʿ fī ṣināʿat al-ḥiyal (Combining Theory and Useful Practice in the Craft of Mechanical Arts)* was first translated into English by Donald Hill as *The Book of Knowledge of Ingenious Mechanical Devices*, Dordrecht 1974, and only later edited by Aḥmad Yousef al-Ḥassan, under the full Arabic title, Aleppo, 1979. See also my review of the field and the importance of Jazarī's Arabic title in G. Saliba, "The Function of Mechanical Devices in Medieval Islamic Society," *Annals of the New York Academy of Sciences* 441 (1985): 141–151. Furthermore, see Aḥmad al-Ḥassan and Donald Hill, *Islamic Technology: An Illustrated History*, Unesco and Cambridge University Press, 1986.

54. *EI²*, I, 98.

55. For the works of this astronomer see the critical edition of his major work and the various bibliographical references to his other works in G. Saliba, *The Astronomical Work of Muʾayyad al-Dīn al-ʿUrḍī (d. 1266): A Thirteenth Century Reform of Ptolemaic Astronomy, ʿUrḍī's Kitāb al-Hayʾa*, Beirut, 1990, 1995, third corrected edition 2001. See other references to him as well as to other astronomers of this later period in G. Saliba, *A History of Arabic Astronomy: Planetary Theories During the Golden Age of Islam*, New York, 1994.

56. Ṭūsī's major astronomical work *al-Tadhkira* is now edited with a translation and commentary by F. J. Ragep, *Naṣīr al-Dīn al-Ṭūsī's Memoir on Astronomy*, New York, 1993.

57. The prolific production of this astronomer is manifest, with his three major astronomical works, each exceeding 200 folios in the various copies of the extant manuscripts. That extensive production itself may have made the critical editions of

his works a prohibitive task. But enough of his novel ideas have been extracted and published in several publications. See, for example, E. S. Kennedy, "Late Medieval Planetary Theory," *Isis* 57 (1966): 365–378, esp. pp. 371–377, and G. Saliba, "The Original Source of Quṭb al-Dīn al-Shīrāzī's Planetary Model," *Journal for the History of Arabic Science* 3 (1979): 3–18. For more information on this astronomer, see *Dictionary of Scientific Biography*, vol. 11, New York, 1975, pp. 247–253.

58. Although much about the works of this astronomer is known, his major theoretical work on planetary astronomy, *kitāb nihāyat al-sūl fī taṣḥīḥ al-uṣūl* (The Ultimate Quest in the Rectification of [Astronomical] Principles), has just been edited by the present writer and still awaits publication. What has appeared in print are descriptions of this astronomer's works and had been gathered together by E. S. Kennedy and Imad Ghanim in *The Life and Work of Ibn al-Shāṭir*, Aleppo, 1976, to which should be added G. Saliba, "Theory and Observation in Islamic Astronomy: The Work of Ibn al-Shāṭir of Damascus (1375)," *Journal for the History of Astronomy* 18 (1987): 35–43, and *Dictionary of Scientific Biography*, volume 12, 1975, pp. 357–364.

59. One of the major theoretical astronomical works of this astronomer has been published by the present writer in G. Saliba, "Al-Qushjī's Reform of the Ptolemaic Model for Mercury," *Arabic Sciences and Philosophy* 3 (1993): 161–203.

60. The fecundity of this astronomer is similar to that of Shīrāzī, and like him many of his works remain unpublished. The present author has devoted a series of articles to him in an attempt to make some of his ideas at least known. The most important of those articles are the following: G. Saliba, "A Sixteenth-Century Arabic Critique of Ptolemaic Astronomy: The Work of Shams al-Dīn al-Khafrī," *Journal for the History of Astronomy* 25 (1994): 15–38; Saliba, "A Redeployment of Mathematics in a Sixteenth-Century Arabic Critique of Ptolemaic Astronomy," in *Perspectives arabes et médiévales sur la tradition scientifique philosophique grecque. Actes du Colloque de la S.I.H.S.P.A.I.* (Société internationale d'histoire des sciences et de la philosophie arabe et Islamique), Paris, 31 mars-3 avril 1993, ed. A. Hasnawi, A. Elamrani-Jamal, and M. Aouad, Peeters, Leuven-Paris, 1997, pp. 105–122; Saliba, "The Ultimate Challenge to Greek Astronomy: *Ḥall mā lā Yanḥall* of Shams al-Dīn al-Khafrī (d. 1550)," in *Sic Itur Ad Astra: Studien zur Geschichte der Mathematik und Naturwissenschaften, Festschrift für den Arabisten Paul Kunitzsch zum 70. Geburstag*, Harrassowitz Verlag, 2000, pp. 490–505.

61. *Dictionary of Scientific Biography*, vol. 7, 1973, pp. 212–219.

62. See *al-Jāmiʿ li-Mufradāt al-Adwiya wa-l-Aghdhiya*, Bulaq, 1874.

63. Although this physician's work is still relatively understudied, much can now be gathered from the article devoted to him in the *Dictionary of Scientific Biography*, vol. 9, 1974, pp. 602–606.

64. O. Neugebauer, "Studies in Byzantine Astronomical Terminology," *Transactions of the American Philosophical Society*, New Series, 50 (1960): 1–45.

65. David Pingree, *The Astronomical Works of Gregory Chioniades*, Amsterdam, 1985; Pingree, "Gregory Chioniades and Paleologan Astronomy," *Dumbarton Oaks Papers* 18 (1964): 133–160.

66. A. Tihon, "L'astronomie byzantine (du Ve au XVe siècle)," *Byzantion* 51 (1981): 603–624, and her other articles now gathered in A. Tihon, *Études d'astronomie byzantine*, London, 1994.

67. Maria Mavroudi, *A Byzantine Book on Dream Interpretation: The Oneirocriticon of Achmet and Its Arabic Sources*, Leiden, 2002.

68. Because of the prohibitive costs of publishing the inordinate amount of illustrations which I needed for that article I decided to publish it on the World Wide Web at http://www.columbia.edu, where I also provided links to some of the manuscripts that were annotated by Postel and others.

69. Paris, 1893.

70. Ibid., p. 338.

71. Ibid.

72. Paris, 1899.

73. Ibid., p. 32 tr, 36 syr.

74. Ibid., p. 42 tr. 47 syr.

75. Ed. A. Sabra and N. Shehaby, Cairo, 1971.

76. See *Dictionary of Scientific Biography*, s.v. Ibn al-Nafīs.

77. Ed. Mehdi Mohaghegh, Tehran, 1993.

Chapter 2

1. See *Al-Fihrist.*

2. Ibid., pp. 391–398.

3. The little we know about the author of *al-Nahmaṭān* can be found in Sezgin, *Geschichte des Arabischen Schrifttums*, vol. VII, p. 114, although there the book is called *al-Yahbuṭān.*

4. *Al-Fihrist*, pp. 391–393.

5. This mythological king of Persia seems to have been at the origin of all legends of Persian civilization, somehow playing a role similar to that of Hermes. His name was usually associated with first kingship, first writing, first building, etc. His father's name is spelled in a variety of ways: Earlier in the *Fihrist*, and with the same spelling, he was associated with the Persian language, *Fihrist*, p. 23. Ṭabari, *Tārīkh al-rusul wa-*

l-mulūk, Beirut, 1987, vol. 1, p. 109, and Ibn al-Athīr, *al-Kāmil fī al-Tārīkh*, Beirut, 1995, vol. 1, p. 52, spell the name as Uyunjihān (*wywnjhn*). Yāqūt, *Muʿjam al-buldān*, Beirut, 1979, vol. 3, p. 170, s.v. *sārūq*, spells it as Nūjihān. All these variations underline the legendary nature of this report as will be argued later.

6. Teukreus, Sezgin, *Geschichte des Arabischen Schrifttums*, vol. VII, p. 71f.

7. *Fihrist*, p. 393.

8. I am referring, for example, to the length of the Babylonian lunar month, 29;31,50,8,20 days, that was reported by Ptolemy in the *Almagest* IV, 2, as having been the value already adopted by Hipparchus before him. See also Asger Aaboe, "On the Babylonian Origin of Some Hipparchian Parameters," *Centaurus* 4 (1955–56): 122–125.

9. On Māshāʾallāh, see Pingree and Kennedy, *Astrological History of Māshāʾallāh*, Cambridge, 1971.

10. See supra, chap. 1, and Ibn Abī Uṣaybiʿa, *ʿUyūn al-Anbāʾ fī Ṭabaqāt al-Aṭibbāʾ*, ed. Richard Müller, Königsberg, 1884, vol. 2. p. 134ff.

11. Pingree and Kennedy, *Astrological History*.

12. *Fihrist*, pp. 393–395. On Abū Māʿshar himself, and his writings, see the extensive entry for him in the *Dictionary for Scientific Biography*, New York, 1970– , vol. 1, pp. 32–39.

13. *Nishwār al-muḥāḍara wa akhbār al-mudhākara*, ed. A. Shaljī, Beirut, 1971–1973, IV, p. 66, also quoted in G. Saliba, "The role of the astrologer in Medieval Islamic Society," *Bulletin d'Études Orientales* 44 (1992): 45–68, p. 53 n. 47, repr. in *Magic and Divination in Early Islam*, ed. Emilie Savage-Smith, Ashgate-Variorum, London, 2004, pp. 341–370.

14. For an echo of this story, see Bīrūnī, *Chronology*, p. 27f.

15. *Fihrist*, p. 395.

16. Ibid.

17. *Chronology*, p. 169.

18. For Abū Māʿshar's defense of astrology against the enemies of his day, see G. Saliba, " Islamic Astronomy in Context: Attacks on Astrology and the Rise of the *Hayʾa* Tradition," *Bulletin of the Royal Institute for Inter-Faith Studies*, 2002, 4: 25–46.

19. Ibn Khaldun, *The Muqaddimah*, Princeton, 1958, vol. 3, p. 263.

20. For an overview of such literature, see Kennedy, *Survey*, 1956, and more recently *EI*^{2}, s.v. *zīdj*, vol. XI, p. 496.

21. *Fihrist*, p. 395f.

22. For a more reliable account of this war and its aftermath see Peter Sarris, "The Eastern Roman Empire from Constantine to Heraclius (306–641)," in Cyril Mango, ed., *The Oxford History of Byzantium*, Oxford, 2002, pp. 19–59.

23. *Fihrist*, p. 396

24. See a similar report, in the work of another tenth-century author, al-Masʿūdī (d. 957), *Les Praries d'or*, Paris, 1914, vol. II, p. 320f.

25. This "first humanism" has to be nuanced with the same nuance that was used by its foremost student Paul Lemerle, in *Le premier Humanisme*, when he says: "Quel sens peut avoire l'humanisme, quand tout est tendu vers un dépassement de l'humain? . . . Les Grecs de Byzance . . . lisent peu, ils se contentent aisément de florilèges, de recueils de citations, de glossaries, de commentaries, de manuels; ils ne cherchent pas l'esprit, tout paraît se ramener à des procédés. Souvent leur erudition nous surpend: mais, à bien regarder, la literature antique est-elle pour eux autre chose qu'un vaste magasin d'accessoire, au service d'une "rhétorique" savant compliquée?" p. 306. See also the very pertinent discussion of the conditions in Byzantium at the time in Dimitri Gutas, *Greek Thought Arabic Culture*, pp. 175–178.

26. For the various missions to acquire books from Byzantium and the conditions in which they were kept there see al-Nadīm's account below, and al-Qifṭī, *Taʾrīkh al-ḥukamāʾ*, Leipzig, 1903, p. 29f; Youssef Eche, *Les Bibliotheques Arabes*, Damas, 1967, p. 28f. On Maʾmūn's mission to the king of Cyprus to acquire Greek books and the discussion regarding the effect of those books on the Christians and the desirability of sending them to al-Maʾmūn hoping to corrupt the Muslims with them, and the discussion regarding the nature of translation itself, see Ṣalāh al-Dīn b. Aybak al-Ṣafadī (d. 1362), *Al-Ghayth al-musajjam fī sharḥ lāmīyat al-ʿajam*, Beirut, 1997, p. 87f

27. Qifṭī, *Taʾrīkh*, p. 29.

28. For a detailed bibliography on this Syriac author, see Albert Abuna, *Adab al-Lugha al-Ārāmīya*, Beirut, 1970, pp. 231–233.

29. Ibid., pp. 363–365.

30. Ibid., pp. 375–377.

31. Ed. Mingana, Heffer, Cambridge, 1935.

32. F. Nau, "notes d'astronomie syrienne," *JA*, 2e ser. t. xvi, 1910, p. 225f.

33. *Fihrist*, p. 396f.

34. The text has *zādā nfrwkh* (Zādā Nifrūkh?), which most likely should be Zādān Farrūkh.

35. The text has *sabiy*, which must be *sababī* (my link = i.e. my cause of livelihood).

36. *Fihrist*, p. 397.

37. It is in that respect that Lemerle's term "appropriation," which was later used by Sabra, gains a special significance in its capacity to render the intent of this classical source of al-Nadīm.

38. See the brief account of this dream in the previous chapter, here quoted in greater detail on account of its importance to our present discussion.

39. *Fihrist*, p. 397.

40. Ibid., p. 397f.

41. Ibid., p. 398.

42. As Lemerle would say: "ils lisent peu, ils se contentent aisément de florilèges, de recueils de citations, de glossaries, de commentaries, de manuels;. . . ils ne cherchent pas." For them the classical texts of high science and philosophy had become "vaste magasin d'accessoire," as already noted again by Lemerle.

43. Most writers on the subject of the transmission of Greek science into Arabic refer to the stories about Khālid as legends. See for example, F. Rosenthal, *The Classical Heritage in Islam*, Routledge, London, 1965, p. 3, where he says: "Therefore, we should do well to relegate the precise story of Khalid's alchemical translation activity to the realm of legend." See also Manfred Ullman, *Die Medizin im Islam*, Leiden, Brill, 1970, where he says: "Dass der Umaiyadenprinz Hālid b. Yazīd (gest. 85/704) dafür gesorgt habe, dass alchemitische und medizinische Bücher ins Arabische übersetzt wurden, gehört wiederum der wissenschaftlichen Legende an," p. 22.

44. Abū Hilāl al-ʿAskarī, *Kitāb al-awāʾil*, Beirut, 1997, p. 185f.

45. Rosenthal, *Classical Heritage,* p. 4f, where he also sends the reader for further confirmation of this interpretation to the work of R. Paret, *Der Islam und das griechische Bildungsgut*, Tübingen, 1950.

46. Al-Jahshiyārī, Beirut, 1988, p. 29f.

47. Al-Jahshiyārī, *Kitāb al-Wuzarāʾ*, p. 29f.

48. Khwārizmī says that he composed his book on Algebra in order to answer to the need of men who "constantly require in cases of inheritance, legacies, partition, lawsuits, and trade, and in all their dealings with one another, or where the measuring of lands, the digging of canals, geometrical computations, and other objects of various sorts and kinds are concerned. . . ." *The Algebra of Muḥammad ben Mūsā*, tr. Frederic Rosen, London, 1831, p. 3.

49. See Rashed, "l'Idée de l'Algèbre."

50. Ibn Qutayba, *Adab al-kātib*, ed. Muḥammad al-Dālī, Beirut, second ed. 1996, p. 12f.

51. Ibn Qutayba, *Kitāb al-anwāʾ*, Hyderabad, 1956.

52. Abū al-Wafā' al-Būzjānī, *Mā yaḥtāj ilaih al-Ṣunnā' min 'ilm al-handasa*, Baghdad, 1979; Abū al-Wafā' al-Būzjānī, *Mā yaḥtāj ilaih al-kuttāb wa-l-'ummāl wa-ghairihim min 'ilm al-ḥisāb*, in A. S. Saidan, *Abū al-Wafā' al-Būzjānīī: 'ilm al-ḥisāb al-'arabī*, Amman, 1971.

53. Muḥammad b. Aḥmad b. Yūsuf al-Khwārizmī al-Kātib, *Mafātīḥ al-'ulūm*, ed. G. van Vloten, Leiden, 1895.

54. Ibn Mamātī, *Qawānīn al-Dawāwīn*, ed. Aziz Atiya, Cairo, 1943.

55. Muḥammad b. Muḥammad b. al-Ukhuwwa, *Ma'ālim* . . . , ed. Reubin Levey, Cambridge,1938.

56. Jahshiyārī, p. 30.

57. Ibid.

58. Ibid.

59. Ibn Qutayba, *'Uyūn al-akhbār,* Beirut, 1997, p. 48, Ibn Khaldūn, *Muqaddima,* vol. 2, pp. 102, 352.

60. In the case of Sergius, see G. Saliba, "Paulus Alexandrinus in Syriac and Arabic," *Byzantion* 65 (1995): 440–454, esp. p. 443 where Sergius says: "The exact position of the sun, however, is known from the second book of the *Tables*, and the computations of Claudius Ptolemaeus. But if one would like to know, in simple fashion, where the sun is to be found, he computes thus: "As for Severus Sebokht, and while arguing in a text now preserved at the Bibliothèque Nationale de France (Syr. 346) fol. 59v et seq.) that the phenomenon of eclipses was not due to some metaphysical being called Atalia, but that it was a natural phenomenon that could be calculated and predicted, he speaks thus: "Les calculs á l'aide desquels on trouve exactement ces noeuds (ascendant et descendant) avec leurs causes, sont dans le livre qui est nommé *Régle* (canon) *des calculs*, fait par l'astronome *Ptolémée* sur la course et le mouvement de tous les astres. Bien que de nombreux hommes l'aient précédé et l'aient suivi, *il a brillé à lui seul, dans l'art de l'astronmie, plus que tous les anciens et les modernes (ensemble).* C'est d'après sa pensée que nous venons de placer les causes exactes et véritables des éclipses, et car *nous avons pu puiser une petite goutte de la grande mer de sublime science qui est dans ses écrits*, pour adresser un rappel, c'est-à-dire un stimulant, aux amis du travail (φιλοπονοι) pour qu'ils s'appliquent encore et ne se relâchent pas de l'amour de la sagesse (φιλοσοφια) bien *que les adversaires ouvrent fortement la bouche et aiguisent la lange (contre eux).*" translated by F. Nau, in *Revue de l'Orient Chretien,* 3e serie 7, no. xxvii (1929–30 : 327–338, esp. p. 330.

61. G. Saliba, "Competition and the Transmission of the Foreign Sciences: Ḥunayn at the Abbāsid Court," *Bulletin of the Royal Institute for Inter-Faith Studies* 2 (2000): 85–101.

62. Ibn Abī Uṣaybi'a, p. 258.

63. *Fihrist*, p. 465.

64. Ibid., p. 465.

65. Bergstrasser.

66. Evidence of translations of non-scientific nature, as well as fresh compositions based on the same, can be also gleaned from such works as Mario Grignaschi, "Les "Rasā'il 'Aristāṭālīs 'ilā-l-Iskandar" de Sālim Abū-l-'Alā' et l'Activité Culturelle a l'Époque Omayyade," *Bulletin d'Études Orientales* 19 (1965–66): 7–83.

Chapter 3

1. See chapter 1 above, and allow for the early incorporation of Sanskrit and Persian medical and pharmacological material that may have taken place before the time of al-Manṣūr. On the astronomical technical level, much of the material discussed in this chapter had already appeared in a preliminary fashion in *EHAS*, pp. 59–83, and more recently in *Encyclopedia Italiana, Storia della Scienza*, ed. Sandro Petruccioli, Roma, 2001– , v. III, 2002, pp. 198–213.

2. See Gutas, pp. 30, 44 and passim.

3. Future research may change this consensus to allow for the evidence of earlier translations from the Hellenistic tradition as well. See, for example, Grignaschi, "Les "Rasā'il" on Hellenistic translations, and probably read as well the often used expression "old translation" (*al-naql al-qadīm*) of several Hellenistic scientific and philosophical sources, to refer to possible earlier translations again, that may have taken place during the 100 years stretching from the time of 'Abd al-Malik reforms ca. 705 A.D. and the early part of the ninth century. For the instances of *al-naql al-qadīm*, or *bi-naql qadīm* (in an old translation), see *Fihrist*, pp. 412, 431 and passim.

4. The last years of the Umayyads witnessed increased upheavals in the eastern provinces that were still governed from the army encampments of Iraq. Some of those upheavals took the form of open rebellion of major proportions. Although they were crowned by the successful takeover by the Abbāsids, their forerunners, such as the rebellions of al-Mukhtār and the Mawālī (*EI*[2], VI, p. 874f.), should not be underestimated. What they all had in common was that their leaders would rarely be non-Arabs as such, but usually discontented Arab chieftains who were successful in harnessing the widespread discontent of the mostly Persian soldiery in advancing their claims against the Umayyads. This discontent of the non-Arabs of the eastern provinces is not unconnected with the pinch that was beginning to be felt by those who then realized fully what it meant to be estranged from the *dīwān* revenues, as the *dīwān* jobs continued to move into the hands of those who laid their claims to power on their knowledge of Arabic which at times must have taken the form of racial designations, to yield later on to the *Shu'ūbīya* movement and sentiment.

5. Qifṭi, p. 35f. Here one should also remember the similar references to "old translations" (*bi-naql qadīm*) reported by al-Nadīm in his *Fihrist* 404–410, in connection with the Aristotelian corpus, which may have taken place just about the same time as Ibn al-Muqaffa'’s translations if not earlier.

6. *Fihrist*, p. 437.

7. Qifṭī, pp. 171–177.

8. Bergstrasser, p. 11 and passim.

9. *Fihrist*, pp. 401–409.

10. al-Jāḥiẓ, ʿAmr b. Baḥr (869), *Kitāb al-Bukhalāʾ*, Beirut, n.d., vol. 2, p. 4f.

11. This long-neglected field has luckily drawn the attention of my friend and colleague, David King, who has already devoted several studies that will definitely contribute to its revival, the latest of which is the monumental work: David King, *In Synchrony with the Heavens: Studies in Astronomical Timkeeping and Instrumentation in Medieval Islamic Civlization*, vol. I: *The Call of the Muezzin*, Leiden 2004; vol. II, *Instruments of Mass Calculations*, Leiden, 2005. *ʿIlm al-farāʾiḍ* still awaits a similar interest.

12. Al-Ḥajjāj's Arabic translation of the *Almagest* survives in several copies, among them the fragment in British Library manuscript, Add. 7474, fol. 75r (as marked although the folios of this manuscript are woefully out of order), where the "correct" lunar value is cited. This parameter is further discussed in a recent article by Bernard Goldstein, "Ancient and Medieval Values for the Mean Synodic Month," *Journal for the History of Astronomy* 34 (2003): 65–74.

13. Asger Aaboe, "On the Babylonian Origin of Some Hipparchian Parameters," *Centaurus* 4 (1955–56): 122–125.

14. The idea that the passage of time could help refine observational values was already known to Ptolemy, *Almagest* [I,1], Toomer, p. 37.

15. See Kennedy, *Survey*, p. 146 and passim.

16. al-Farghānī, Ibn Kathīr, *Jawāmiʿ ʿilm al-nujūm*, Amsterdam, 1669, Arabic text, p. 49–50, 53, 58, 60, 74.

17. A value that was widely used in Islamic sources, as we have already mentioned in chapter 1, as well as the value, 23;30°, and at times 23;35°. See Kennedy, *Survey*, p. 145 and passim.

18. Here too, Farghānī, *Jawāmiʿ*, p. 49, adopts the concept of the moving solar apogee, in agreement with the "moderns" and in opposition to Ptolemy.

19. This *zīj* has not yet been properly studied, and is only excerpted by Kennedy, in *Survey*, pp. 145–147, and the surviving manuscript reports a solar maximum equation

of 1;59, clearly at odds with the Ptolemaic value of 2;23, *al-Zīj al-Ma'mūnī al-Mumtaḥan*, Escorial, Arabic Ms., 927, fol. 13r.

20. Qifṭi, pp. 351–354.

21. *DSB*, vol. 7 (1973), pp. 352–354, s.v. Khujandī.

22. Debarnot, *EHAS*, 503–4.

23. Saliba, "The Determination of the Solar Eccentricity and Apogee According to Mu'ayyad al-Dīn al-ʿUrḍī (d. 1266 A.D.)," *Zeitschrift für Geschichte der Arabisch-Islamischen Wissenschaften*, 2 (1985): 47–67, reprinted in Saliba, *A History*, pp. 187–207.

24. *Almagest* [V, 14].

25. Ṭūsī, Naṣīr al-Dīn, *Taḥrīr al-majisṭī*, India Office, Loth, 741, fols. 27v–28r.

26. Saliba, *A History*, p. 233f.

27. Ibn al-Shāṭir, *Kitāb nihāyat al-sūl fī taṣḥīḥ al-uṣūl*, Bodleian, Ms. Marsh 139, fol. 3r.

28. Titles of such works are still preserved in the *Fihrist*, p. 141.

29. al-Ṣūfī, Abd al-Raḥmān (d. 986), *Ṣuwar al-Kawākib*, Hyderabad, 1953.

30. Laurel Brown of Columbia University is preparing a Ph.D. dissertation that will include a detailed analysis of this text and its implication for the Arabic critical tradition that was beginning to take shape during the tenth century.

31. See, for example, Ṣūfī, *Ṣuwar*, pp. 78, 218 and passim.

32. It is not surprising therefore to find that most Western major libraries include several copies of Ṣūfī's text among their holdings. See for example the various copies at the British Library, Or 5323, Or 1407, IOISL 621, IOISL 2389, and Add 7488, among others.

33. *Almagest* I,13, and *HAMA*, 26ff.

34. George Saliba, "The Role of the *Almagest* Commentaries in Medieval Arabic Astronomy: A Preliminary Survey of Ṭūsī's Redaction of Ptolemy's *Almagest*," *Archives Internationales d'Histoire des Sciences* 37 (1987): 3–20, reprinted in Saliba, *A History*, pp. 143–160.

35. Debarnot, "*The Zīj*," and Morelon, Régis, "Eastern Arabic Astronomy between the Eighth and the Eleventh Centuries," *EHAS*, esp. pp. 31–34.

36. We have already seen the similar process that took place in the field of mathematics, where we found the translator Qusṭā b. Lūqā himself deploying the algebraic technical terminology of his time within the translation of the *Arithmetica* of Diophantus which had no such expressions in Greek. See Rashed, *l'Art de l'Algèbre de Diaphont.*

37. Saliba, *A History*, p. 208f.

38. In Aristotle's words in *On the Heavens I & II*, ed. Stuart Leggatt, Aris & Phillips, Warminster, 1995, II, 3 [286a12–20]: "Why, then, is not the entire body of the world like this? Because some part of the body that moves in a circle must remain at rest—that part at the center—but no part of this body is able to remain at rest, either in general or at the center. For its natural movement would then in fact be towards the center, but it moves by nature in a circle; its movement would not be everlasting, since nothing counter-natural is everlasting. Now, the counter-natural is posterior to the natural, and the counter-natural is a displacement of the natural in the process of becoming. Hence, there must be Earth, since this rests at the center. For the moment, then, let this be assumed; later this will be proved of it." And on II, 14, [296b21–24] he says: "It is evident, therefore, that the Earth must be at the center and motionless, both for the reasons given, and because weights thrown straight upward by force return to the same point, even if the force flings them an unlimited distance."

39. See the most recent summation of those problems in George Saliba, "Greek Astronomy and the Medieval Arabic Tradition," *American Scientist*, July-August 2002: 360–367.

40. Saliba, "Early Arabic Critique,"

41. Saliba, *A History*, p. 20f; Saliba, "Critiques of Ptolemaic Astronomy in Islamic Spain," *al-Qanṭara* 22 (1999): 3–25.

42. Saliba, *A History*, p. 85f.

43. Quoted in ibid., p. 279.

44. Ibn al-Haitham (d. 1049), *al-Shukūk ʿalā Baṭlamyūs (Dubitationes in Ptolemaeum)*, ed. A. Sabra and N. Shehaby, Cairo, 1971.

45. See the literature reviewed in F. Jamil Ragep, "Duhem, the Arabs, and the history of Cosmology," *Synthese* 83 (1990): 210–214, and Ragep, *Naṣīr*, p. 47 nn. 5 and 6 and passim.

46. Ibn al-Haitham, *Shukūk*, p. 5.

47. Ibid., p. 5.

48. Ibid., p. 16.

49. Ibid., p. 16.

50. Toomer, *Ptolemy's Almagest*, p. 443.

51. Ibn al-Haitham, *Shukūk*, p. 26.

52. See Swerdlow, "Jābir Ibn Aflaḥ' s Interesting Method for Finding the Eccentricities and Direction of the Apsidal Line of a Superior Planet," in *From Deferent to Equant*, ed. D. King and G. Saliba, *Annals of the New York Academy of Sciences* 500 (1987): 501–512.

53. Ibn al-Haitham, *Shukūk*, p. 33f.

54. Ibid., p. 36.

55. Ibid., p. 38.

56. Ibid. In similar wordings this admission was in fact made by Ptolemy in *Almagest*, IX,2 (Toomer's translation, p. 422) where he said: "we may [be allowed to] accede [to this compulsion], since we know that this kind of inexact procedure will not affect the end desired, provided that it is not going to result in any noticeable error."

57. Ibn al-Haitham, *Shukūk*, p. 38f.

58. Ibid., p. 41f.

59. Ibid., p. 44.

60. Ibid., p. 46.

61. Ibid., p. 47.

62. Ibid., p. 59.

63. Ibid., p. 60.

64. Ibid., p. 57.

65. ʿUrḍī, *Hay'a*, p. 212.

66. Ibn al-Haitham, *Shukūk*, p. 54.

67. Ibid., p. 62.

68. ʿUrḍī, introduction p. 39, text, p. 218.

69. Ibn al-Haitham, *Shukūk*, p. 63f.

70. Quoted in Saliba, *A History*, p. 151.

71. Saliba, *A History*, p. 151.

72. *Almagest*, XIII, 2, Toomer, p. 600.

73. Quoted in Saliba, *A History*, p. 153.

74. Al-Akhawayn, fols. 4v-5r.

75. See Saliba, "Al-Qushjī's reform," pp. 161–203.

76. Signaled in Saliba, *A History* (p. 283f).

77. Quoted in Saliba, *A History* (p. 284f.).

78. Quoted in ibid. (p. 285).

79. Saliba, *A History*, p. 285.

80. Ibid., p. 286.

81. See Saliba, "A Sixteenth-Century Arabic Critique"; Saliba, "A Redeployment"; Saliba, "The Ultimate Challenge."

82. Quoted in Saliba, *A History* (p. 287f.).

83. G. Saliba, "Copernican Astronomy in the Arab East: Theories of the Earth's Motion in the Nineteenth Century," in *Transfer of Modern Science and Technology to the Muslim World*, ed. Ekmeleddin Ihsanoglu, Istanbul, 1992, pp. 145–155.

84. Saliba, *The Astronomical Work of Mu'ayyad al-Dīn al-'Urḍī*.

85. *'Urḍī*, p. 249f.

86. Ibid., p. 250f.

87. See the similar texts in the translation of Leggatt quoted before and add to it, Aristotle, *On the Heavens*, Loeb, 1939, repr. 1960, II, xiii, and seq.

88. See, for example, Bernard Goldstein, *Al-Biṭrūjī: On the Principles of Astronomy*, 2 vols., New Haven, 1971.

89. See Saliba, "Redeployment,"

90. The account of 'Abd al-Laṭīf al- Baghdādī is taken from his book *al-Ifāda wa-l-I'tibār*, ed. Aḥmad Ghassān Sabānū, Dār Ibn Zaydūn (Beirut) and Dār Qutayba (Damascus), 1984, p. 103f.

91. In his own words 'Abd al-Laṭīf puts it thus: "(In any case) observation (*al-ḥiss* literally feeling) is more valid than hearing (*sam'*, i.e. learning from books being recited). Observation is even more valid than Galen, despite his rank among the scientists in investigation and his meticulousness in all that he said or practiced, observation is still more true than him (*al-ḥiss aṣdaq minh*).

Chapter 4

1. See Saliba, "Islamic Astronomy in Context."

2. 'Urḍī, p. 214f. and passim.

3. Ibn al-Haitham, *Shukūk*, p. 33f.

4. See, for example, Louis Cheikho, "Risālat al-Khujandī fī mayl wa-'arḍ al-balad," *Mashriq* 11 (1908): 60–69; A. Jourdain, *Mémoire sur l'observatoire de Méragah et sur Quelques Instruments Employés pour Observer*, Paris, 1870; E. Wiedeman with T. Juynbol, "Avicennas Schrift über ein von ihm ersonnenes Beobachtunginstrument," *Acta Orientali* xi, 5 (1926): 81–167; Aydīn Sayīlī, *Ghiyâth al-Dîn al-Kâshî's letter on Ulugh Bey and the Scientific Activity in Samarqand*, Ankara, 1985.

5. For brief statements of such problems, see Saliba, "Greek Astronomy" and "Arabic Planetary Theories."

6. Abū al-Raiḥān, Muḥammad b. Aḥmad al-Bīrūnī (1048), *Kitāb al-tafhīm li-awā'il ṣinā 'at al-tanjīm* (The Book of Instruction in the Elements of the Art of Astrology), London, 1934. For a more complete text with a Persian translation, see Jalāl al-Dīn Homā'ī, *al-Tafhīm li-awā'il ṣinā'at al-tanjīm*, Teheran, 1362 = 1984.

7. Abū Ma'shar al-Balkhī (d. 886), *al-Madkhal ilā 'ilm aḥkām al-nujūm*, Jārullah (Carullah) Ms., 1058, published in facsimile, Frankfurt, 1985.

8. In fact there is much evidence that Aristotelian cosmology was not all that secure in the Islamic domain, as F. Jamil Ragep recently demonstrated in "Ṭūsī and Copernicus: The Earth's Motion in Context" (*Science in Context* 14, 200): 145–163) and "Freeing Astronomy from Philosophy: An Aspect of Islamic Influence on Science" (*Osiris* 16, 2001: 49–71).

9. For a more detailed description of the alternatives that were developed during medieval Islamic times, see G. Saliba, "Arabic Planetary Theories after the Eleventh Century, " in *EHAS*, pp. 58–127, and more recently in G. Saliba, "Alternative all'astronomia tolemaica," in *Storia della Scienza*, ed. Sandro Petruccioli, Roma, 10v, 2001– , v. III, 2002, pp. 214–236.

10. Toomer, *Ptolemy's Almagest*, p. 144, note 32, and Neugebauer, *HAMA*, p. 149g.

11. Obviously resulting from the motions of spheres in place. Emphasis added.

12. Toomer, *Ptolemy's Almagest*, p. 140.

13. Ibid., p. 141.

14. Ibid., p. 153.

15. Ibid., p. 145.

16. See the objections of Jābir Ibn Aflaḥ against this particular point in the Ptolemaic model, and his proposed solution for it, in Swerdlow, "Jābir Ibn Aflaḥ' s Interesting Method," supra.

17. In addition to the objections raised by Ibn al-Haitham and others in the Arabic astronomical tradition note the following statement of Copernicus in his earliest astronomical work, the *Commentariolus*, which speaks directly to the absurdity of such a proposal, called by Copernicus "not sufficiently in accordance with reason": "Nevertheless, the theories concerning these matters that have been put forth far and wide by Ptolemy and most others, although they correspond numerically [with the apparent motions], also seemed quite doubtful, for these theories were inadequate unless they also envisioned certain *equant* circles, on account of which it appeared that the planet never moves with uniform velocity either in its *deferent* sphere or with

respect to its proper center. Therefore a theory of this kind seemed neither perfect enough nor sufficiently in accordance with reason." Noel Swerdlow, "The Derivation and First Draft of Copernicus's Planetary Theory: A Translation of the Commentariolus with Commentary," *Proceedings of the American Philosophical Society* 117, no. 6 (1973), p. 434.

18. See the seminal paper of Victor Roberts, "The Solar and Lunar Theory of Ibn al-Shāṭir: A Pre-Copernican Copernican Model," *Isis* 48 (1957): 428 – 432, reprinted in E. S. Kennedy et al., *Studies in the Islamic Exact Sciences*, American University of Beirut, 1983, pp. 50–54.

19. The same chagrin was expressed centuries later by Copernicus in his Commentariolus: "But of all things in the heavens the most remarkable is the motion of Mercury which passes through nearly untraceable points so that it cannot easily be investigated." Swerdlow, *Commentariolus*, p. 499.

20. Quoted in Saliba, *A History*, p. 153.

21. In Ptolemy's own words (*Almagest*, XIII, 2): "Now let no one, considering the complicated nature of our devices, judge such hypotheses to be over-elaborated. For it is not appropriate to compare human [constructions] with divine. . . ." Toomer, *Ptolemy's Almagest*, p. 600.

22. Averroes, *Tafsīr mā baʿd al-ṭabīʿa*, ed. Maurice Bouyges, Beirut, 1948, p. 1664. This comes from Averroes commentary on Book Lambda of Aristotle's Metaphysics.

23. For a full survey of these attempts to create new planetary theories see Saliba, "Planetary Theories" and Saliba "Alternative," already referred to above.

24. For a full original statement of this theorem see G. Saliba, "The Original Source," reprinted in Saliba *A History*, pp. 119–134. Also see ʿUrḍī, *Kitāb al-Hayʾa* for the full context of ʿUrḍī's works.

25. In his discussion of the latitudinal motion of the planets, Copernicus makes the following remark. "If indeed this motion of libration takes place in a straight line, it is still possible that such a motion be composed from two spheres." Swerdlow, *Commentariolus*, p. 483. In his commentary on this passage, Swerdlow states: "In order to account for the libration of the orbital planes, Copernicus takes up one of the two devices for the generation of a rectilinear motion from two circular motions originally used, and indeed invented, by Naṣīr ad-Dīn aṭ-Ṭūsī, and used extensively by Ibn ash-Shāṭir and other of the Marāgha astronomers." *Commentariolus*, p. 488.

26. Describing the connection between the oscillating epicyclic center of Mercury and the oscillating motion of the latitude plane of the planets we described above, Copernicus says: "For by this composite motion, the center of the larger epicycle is carried on a straight line, just as we have explained concerning latitudes that are librated," Swerdlow, *Commentariolus*, p. 503.

27. Now see Robert Morrison for the latest edition and translation of Shīrāzī's chapter dealing with the *uṣūl*, in "Quṭb al-Dīn al-Shīrāzī's Use of Hypotheses,"*Journal for the History of Arabic Science* 13 (2005): 21–140.

28. This legend was elegantly stated and rebutted by Otto Neugebauer in 1968 ("On the Planetary Theory of Copernicus," *Vistas in Astronomy* 10: 89–103).

29. See G. Saliba, "Theory and Observation in Islamic Astronomy: The work of Ibn al-Shāṭir of Damascus," *Journal for the History of Astronomy* 18 (1987): 35–43, reprinted in Saliba, *A History*, pp. 233–241.

30. Saliba, "A Redeployment of Mathematics."

Chapter 5

1. For a more detailed alternative discussion of the philosophical dimension of Islamic astronomy, see G. Saliba, "Aristotelian Cosmology and Arabic Astronomy," in *De Zénon d'Élée a Poincaré*, ed. Régis Morelon and Ahmad Hasnawi, Louvain, 2004, pp. 251–268.

2. One such treatise was published by Anton Heinen, *Islamic Cosmology*, Beirut, 1982.

3. After all it was even Ptolemy himself who had stated in the introduction of the *Almagest* that "this science [meaning astronomy], alone above all things, could make men see clearly; from the constancy, order, symmetry and calm which are associated with the divine, it makes its followers lovers of this divine beauty, accustoming them and reforming their natures, as it were, to a similar spiritual state." Toomer, *Almagest*, p. 37. This sentiment is echoed by ʿUrḍī, around a millennium later, in his *Kitāb al-Hay'a*, where he says of astronomy: "Its subject-matter is the most amazing of God's achievements, the most magnificent of His creations, and the best executed of His deeds. As for its demonstrations, they are geometrical and arithmetical and therefore definitive. The benefit of this science is immense for the one who contemplates the celestial marvels and the heavenly motions. For through that the mind has an abundant domain and an indisputable proof of the existence of God the most exalted. It leads to theology and demonstrates the magnificence of the Creator, the wisdom of the Maker, and the immensity of His power. May God, the best of creators, be blessed." Saliba, ʿUrḍī, *Hay'a*, p. 27f.

4. The following account is taken from Saliba, "Early Arabic Critique."

5. See Arabic Ms. 520RH, Osmania University Library, Hyderabad.

6. Ibid.

7. Averroes, *Tafsīr*, p. 1661.

8. Ibid., p. 1661.

9. Ibid., p. 1664.

10. See Saliba, "Aristotelian Cosmology," p. 260f.

11. ʿUrḍī, p. 212.

12. Ibid., p. 218.

13. Ghars al-Dīn, Aḥmad b. Khalīl al-Ḥalabī (d. 1563), *Tanbīh al-nuqqād ʿalā mā fī al-hayʾa al-mashhūra min al-fasād* (Alerting the Critics to the Corruption of the Well-Known Astronomy), Istanbul, Yeni Jāmiʿ, Ms. 1181, fol. 148r.

14. Al-Ghazālī, *The Incoherence of the Philosophers*, tr. Michael Marmura, Provo, 1997, pp. 174–176 and passim.

15. See Saliba, "Role of the Almagest commentaries," reprinted in Saliba, *A History*, p. 153.

16. The repercussion of this theorem in terms of categories of motion is treated in greater detail in Saliba, "Aristotelian Cosmology," p. 263f.

17. See G. Saliba and E. S. Kennedy, "The Spherical Case of the Ṭūsī Couple," *Arabic Sciences and Philosophy* 1 (1991): 285–291, reprinted with minor mistakes in *Naṣīr al-Dīn al-Ṭūsī: Philosophe et savant du xiiie siècle*, ed. N. Pourjavadi and Z. Vesel, Institut Français de Recherche en Iran and Presses Universitaires d'Iran, Teheran, 2000, pp. 105–111.

18. Quṭb al-Dīīn al-Shīrāzī, *al-Tuḥfa al-Shāhīya*, Paris, BN Arabe, 2516, fol. 28r.

19. Abū al-Barakāt al-Baghdādī, *Kitāb al-Muʿtabar*, Hyderabad, 1938, volume 1, chapter 24, pp. 94–103.

20. See Galileo's text as cited in Edward Grant, ed., *A Source Book in Medieval Science*, 1974, p. 290.

21. On the continuity of this "nibbling" at the edges of the Aristotelian universe in the writings of the Islamic astronomers and philosophers, see Ragep, "Ṭūsī and Copernicus" and "Freeing Astronomy."

22. For a brief description of the development of the various solutions of the *qibla* problem, see David King, in *EI*², s.v. "Ḳibla: Sacred direction," reprinted in David King, *Astronomy in the Service of Islam*, Aldershot, 1993, section IX.

23. Reviewing the general developments of the exact sciences, E. S. Kennedy had this to say about the discipline of trigonometry: "This subject, the study of the plane and spherical triangle, was essentially a creation of Arabic-writing scientists, and it is the only branch of mathematics of which this statement can be said." E. S. Kennedy, "The Arabic Heritage in the Exact Sciences," *al-Abḥāth* 23 (1970): 327–344.

24. For a short survey of the discipline and the use of trigonometric functions, see D. King, *EI*^{2}, s.v. "Mīqāt: astronomical Timekeeping," reprinted in D. King, *Astronomy in the Service of Islam* and in King, *In Synchrony with the Heavens.*

25. A frequently repeated tradition from the prophet, as quoted by Shāfiʿī, says: "*al-ʿilm ʿilmān: ʿilm al-adyān wa-ʿilm al-abdān, yaʿnī al-fiqh wa-l-ṭib*" (science is of two kinds: a science of religion, and a science of bodies, meaning jurisprudence and medicine). See Ṣalāḥ al-Dīn Khalīl b. Aybak al-Ṣafadī (d. 1363), *Kitāb al-wāfī bi-l-wafayāt*, Wiesbaden, 1981, vol. 2, p. 174.

26. *EI*^{2}, III, 896.

27. See Saliba, *A History of Arabic Astronomy*, p. 45f. n. 51; G. Saliba, "Persian Scientists in the Islamic World: Astronomy from Maragha to Samarqand," in *The Persian Presence in the Islamic World*, ed. R. Hovannisian and G. Sabagh, Cambridge University Press, 1998, pp. 126–146.

28. For a short biography see *EI*^{2}, X, p. 746, s.v. "al-Ṭūsī."

29. English translation: *Contemplation and Action: The Spiritual Autobiography of a Muslim Scholar*, London, 1998.

30. al-Ṭūsī, Naṣīr al-Dīn, *The Rawḍatuʿt-Taslīm : Commonly called Taṣawwurāt*, Persian text, ed. W. Ivanow, Leiden, 1950.

31. al-Ṭūsī, Naṣīr al-Dīn, *Awṣāf al-ashrāf*, Beirut, 2001.

32. Ṭūsī, *Tajrīd al-ʾiʿtiqād*, Cairo, 1996.

33. *EI*^{2}, V, p. 547

34. See G. Saliba, "Reform of Ptolemaic Astronomy at the Court of Ulugh Beg," *Studies in the History of the Exact Sciences in Honor of David Pingree*, eds. Charles Burnett, Jan Hogendijk, Kim Plofker and Michio Yano, Boston, 2004, pp. 810–824.

35. The Egyptian edition (1962–1970) has 30 volumes in 10.

36. See David King, *Dictionary of Scientific Biography*, s.v. "Ibn al-Shāṭir," and Kennedy and Ghānim, *Ibn al-Shāṭir.*

37. See George Saliba, "Reform," and Tashküprüzadeh (d. 1561), *al-Shaqāʾq al-nuʿmānīya fī ʿulamāʾ al-dawla al-ʿuthmānīya*, Istanbul, 1985, p. 107f. For his works on *ʿibādāt*, see Saliba, *A History*, p. 47, n. 56.

38. See Saliba, "A Sixteenth-Century Critique."

39. Saliba, "A Sixteenth-Century Critique," p. 16f.

40. See Saliba, "The Role of the Astrologer."

Chapter 6

1. See David Pingree, "Gregory Chioniades" and *The Astronomical Works of Gregory Chioniades*. See also E. A. Paschos and P. Sotiroudis, *The Schemata of the Stars: Byzantine Astronomy from AD 1300*, World Scientific, 1998.

2. See O. Neugebauer, "Studies in Byzantine Astronomical Terminology"; G. Saliba, "Arabic Astronomy in Byzantium," *Journal for the History of Astronomy* 20 (1990): 211–215.

3. See G. Saliba, *Rethinking the Roots of Modern Science: Arabic Manuscripts in European Libraries*, Washington DC, 1999.

4. See Saliba, "Whose Science."

5. Victor Roberts, "The Solar and Lunar Theory of Ibn al-Shāṭir: A pre-Copernican Copernican Model," *Isis* 48 (1957): 428–432.

6. A glimpse of the status of Arabic in some European quarters, especially among the humanists, can be gained from the advise given by Hernan Nuñez, a professor at the University of Salamanca, a most friendly place for Arabic studies to Copernicus's contemporary, Nicolas Clenardus of Louvain (1495–1542), who had traveled, around 1530–1532, all the way from Louvain to Salamanca, in search of an Arabic professor, only to be told by Nuñez: "What concern have you with this barbarous language, Arabic? It is quite sufficient to know Latin and Greek. In my youth I was as foolish as you, and, not content with adding Hebrew to the other two languages, I also took up Arabic; but I have long given up these last two, and devoted myself entirely to Greek. Let me advise you to do the same." (quoted in Karl Dannenfeldt, "The Renaissance Humanists and the Knowledge of Arabic," *Studies in the Renaissance* 2 (1955): 96–117) The enmity towards things Arabic and Islamic is signaled in Dannenfeldt's conclusion of the same article where he says: "However, the religious uses of Arabic to elucidate the Hebrew words in Christian religious literature and documents and to facilitate the pacific crusade against Islam seem to have predominated in the views of most of those who studied this oriental language. In this last area, the Renaissance humanists continued an earlier medieval theme." Ibid. p. 117. See also Giovanna Cifoletti, "The Creation of the History of Algebra in the Sixteenth Century," in *Mathematical Europe*, ed. Catherine Goldstein et al., Paris, 1996, pp. 123–142, esp. 123.

7. See De Vaux, "Les spheres."

8. Nicolaus Copernicus, *De Revolutionibus: Faksimiles des Manuskriptes*, Hildesheim, 1974, p. 75r.

9. In response to Ragep's assessment in Ragep, *Naṣīr*, p. 429, where he claims that Ṭūsī, did not state explicitly that he invented the new theorem, one should remember that when Ṭūsī first stated the theorem in rudimentary form, in the *Taḥrīr*, he preceded it by his famous objection to Ptolemy's treatment of the subject starting

with *"aqūl"* (I say), obviously implying that all the section that followed, including the rudimentary form of the theorem, until the resumption of his treatment of the Ptolemaic text, were all the work of Ṭūsī.

10. For a treatment of this problem and the vague reference by Copernicus to "some people" before him who had used the theorem, and the connection to the statement by Proclus, see Ragep's longer discussion, together with several references, of this particular point (*Naṣīr al-Dīn al-Ṭūsī's Memoir on Astronomy*, pp. 430–432). Ragep's conclusion is further strengthened when read together with Swerdlow's "Copernicus's Four Models of Mercury," in *Studia Copernicana XIII*, ed. Owen Gingerich and Jerzy Dobrzycki, Warsaw, 1975.

11. Willy Hartner, "Copernicus, the Man, the Work, and Its History," *Proceedings of the American Philosophical Society* 117, no. 6 (1973): 413–422.

12. See G. Saliba, "Re-visiting the Astronomical Contacts between the World of Islam and Renaissance Europe: The Byzantine Connection" (forthcoming).

13. See Jourdain, *Mémoire.*

14. On the motivation of the Ilkhānids to construct such an institution, see G. Saliba, "Horoscopes and Planetary Theory: Ilkhanid Patronage of Astronomers," lecture delivered at a colloquium organized by the Los Angeles County Museum, June 2003, and to appear in the proceedings of the colloquium.

15. Saliba, *The Astronomical Work.*

16. See figure 4.6 in chapter 4.

17. See Anthony Grafton, "Michael Maestlin's Account of Copernican Planetary Theory," *Proceedings of the American Philosophical Society* 117, no. 6 (1973): 523–550.

18. Swerdlow, *Commentariolus*, p. 500.

19. Ibid., p. 504.

20. Ibid., p. 504.

21. Noel Swerdlow, "Astronomy in the Renaissance," in *Astronomy before the Telescope*, ed. Christopher Walker, St. Martin's Press, 1996, pp. 187–230, esp. 202.

22. Noel Swerdlow, and Otto Neugebauer, *Mathematical Astronomy in Copernicus's De Revolutionibus*, New York, 1984, p. 295.

23. See Marie-Thérèse d'Alverny, *Avicenne en Occident*, Paris, 1993, esp. sections XII–XIV.

24. For the use of the Ṭūsī Couple by Giovanni Batista Amico in 1536, see Noel Swerdlow, "Aristotelian Planetary Theory in the Renaissance: Giovanni Batista Amico's Homocentric Spheres," *Journal for the History of Astronomy* 3 (1972): 36–48.

25. See the tantalizing hints by Willy Hartner in "Naṣīr al-Dīn al-Ṭūsī's Lunar Theory," *Physis* 11 (1969): 289–304, and more recently, Ragep, *Naṣīr*, p. 432f., and G. Saliba, "Aristotelian Cosmology and Arabic Astronomy," in *De Zénon d'Élée à Poincaré*, ed. Régis Morelon and Ahmad Hasnawi, Peeter, Louvain, 2004.

26. Paschos et al., *The Schemata*.

27. For the latest and most convincing attempt to explain the roots of Copernicus's heliocentrism, see Bernard Goldstein, "Copernicus and the Origin of His Heliocentric Universe," *Journal for the History of Astronomy* 33 (2002): 219–235, and the very relevant section in Noel Swerdlow, *Commentariolus*, pp. 474–478, and for the seriousness of the problems remaining, Swerdlow, "Astronomy in the Renaissance," pp. 200–202.

28. Such issues of "locality" versus "essence," as discussed by Abd al-Hamid Sabra, "Situating Arabic Science: Locality versus Essence," *Isis* 87 (1996): 654–679, do not seem to have benefited from the implications of the evidence discussed here.

29. Swerdlow and Neugebauer, *Mathematical Astronomy*, p. 47.

30. See Giorgio Levi Della Vida, *Ricerche sulla formazione del più antico fondo deu manoscritti orientali della biblioteca Vaticana*, Studi e Testi, Biblioteca Apostolica Vaticana, Citta del Vaticano, 1939, p. 307 and passim. (This reference was brought to my attention by my friend and colleague Giorgio Vercellin of Venice. His help is gratefully acknowledged. On Postel himself there are few biographies: Georges Weill and François Secret, *Vie et caractère de Guillaume Postel*, Milan, 1987, and Marion Kuntz, *Guillaume Postel: Prophet of the Resittution of All Things, His Life and Thought*, Hague, 1981. Much can also be gained from the proceedings of the 400 years commemoration of Postel simply published as: *Guillaume Postel 1581–1981*, Paris, 1985.

31. For a very detailed account of such men, see Dannenfeldt, "The Renaissance Humanists."

32. See the title page of Kharaqī's manuscript (Bibliothèque Nationale de France, Arabe 2499).

33. For a brief description of the conditions that led to that treaty, and the privileges it granted to the French, both commercial and military, see V. H. H. Green, *Renaissance and Reformation: A Survey of European History between 1450 and 1660*, London 1954, repr. 1975, p. 363.

34. See for example the other Arabic manuscript at Leiden University Library, Or 2073, which was also signed by Postel as having been among his possessions. I owe this reference to my friend Dr. Maroun Aouad of the CNRS, Paris.

35. *Encyclopaedia Brittanica*, 2003, Francis I.

36. That is, "Mathematatum Professoris Regii" as quoted by Maroun Aouad, from Leiden Ms. Or. 2073, in *Averroès (Ibn Rušd): Commentaire Moyen à la Rhétorique*

d'Aristote, Édition critique du texte arabe et traduction française, 3 vols., Paris, 2002. See also Kuntz, *Guillaume Postel*, p. 29.

37. See, for example, Laurentiana Ms. Or 218, which contains interlinear translations of a commentary on the *Conics*, and dated 1581, mentioned in G. Saliba, *Rethinking the Roots of Modern Science: Arabic Manuscripts in European Libraries*, Washington DC, 1999, p. 21. The Bodleian Ms. Selden A. 11, which contains a book by ʿAlī b. Sulaimān al-Hāshimī (ninth c.) called *Kitāb fī ʿilal al-zījāt*, contains several marginal Latin annotations as well. See E. S. Kennedy, Fuad I. Haddad, and David Pingree, *The Book of the Reasons behind Astronomical Tables*, New York, 1981, pp. 41, 43, 48 and passim.

38. See, for example, the translation of the elementary treatise by Ibn al-Haitham "On the Elevation of the Pole," which was translated by Jacob Golios in 1643, still preserved at the British Museum Ms. Add. 3034, dated 1646, and the publication of Rāzī's treatise on the Smallpox, which was published in London in 1760, with Latin and Arabic on facing pages, See Rhazes *de variolis et morbillis*, London, 1760.

39. Other earlier contacts involving Regiomontanus (1476) have been tentatively put forward by F. Jamil Ragep in "ʿAlī Qushjī and Regiomontanus: Eccentric Transformations and Copernican Revolutions," *Journal for the History of Astronomy* 36 (2005): 359–371. For other contacts that were contemporary with Copernicus, see Paul Kunitzsch, *Peter Apian und Azophi: Arabische Sternbilder in Ingolstadt im frühen 16. Jahrhundert*, Bayerische Akademie der Wissenschaften, Philosophisch-Historische Klasse, Sitzungsberichte, Jahrgang 1986, Heft 3, Verlag der Bayerischen Akademie der Wissenschaften, München, 1986.

40. See for example similar contacts in the mathematical field as illustrated by Cifoletti, "Creation of the History of Algebra," and other works of Rashed on the subject.

41. Uffizi, Gabinetto dei Disegni e Stampe, U1454.

42. I have already devoted an article to this astrolabe. See G. Saliba, "A Sixteenth-Century Drawing of an Astrolabe Made by Khafīf Ghulām ʿAlī b. ʿĪsā (c. 850 A.D.)," *Nuncius, Annali di Storia della Scienza* 6 (1991): 109–119.

43. The relationship between those two astrolabists was already known to al-Nadīm, *Fihrist*, p. 451.

44. For examples of widespread influence of Islamic astrolabes on their European counterparts, see King, *In Synchrony II*, p. 41ff.

45. This astrolabe was once at the Time Museum, in Rockford Illinois, and has since been moved to a private collection. A picture of it was published in the catalogue of the Time Museum. See A. J. Turner, *Catalogue of the Collection, The Time Museum*, vol. I, *Time Measuring Instruments*, Part I, *Astrolabes Astrolabe Related Instruments*, Rockford, 1985, p. 65. See also King, *In Synchrony*, II, p. 1010, 6.2.h.

46. David King, *In Synchrony with the Heaven*, vol. II, *Instruments of Mass Calculations*, Brill, Leiden, 2005

47. David King, *In Synchrony II*, p. 398f.

48. For samples of such designs see Yousif Muḥammad Ghulām, *The Art of Arabic Calligraphy*, published by the author, 1982, pp. 72, 100, 120–121 and passim.

49. I wish to express my gratitude for this information on the tulip craze and its Ottoman origins to my colleague and friend Professor Jeanne Nuechterlein.

50. In fact there are such monographs devoted to the subject. In particular see Angelo de Gubernatis, *Matériaux pour servir a l'histoire des études orientales en Italie*, Paris, 1876, and Fück, Johann, *Die Arabischen Studien in Europa bis in den Anfang des 20. Jahrhundert*, Leipzig, 1955, and more recently, John Robert Jones, *Learning Arabic in Renaissance Europe (1505–1624)*, London University Dissertation, No. DX195516, 1988. Dannenfeldt's article "The Renaissance Humanists" remains very useful as well.

51. See *Dictionary of Scientific Biography.*, s.v. "Leo Africanus," and *EI*², s.c. "Leo Africanus."

52. *EI*², s.v. "Leo Africanus."

53. For more information on this very interesting person, see the short discussion of his association with Copernican astronomy in Swerdlow and Neugebauer, *Mathematical Astronomy*, p. 16f., and the interesting articles by Peter Barker and Bernard Goldstein, "Patronage and Production of *De Revolutionibus*," *Journal for the History of Astronomy* 34 (2003), pp. 345–368, esp. 348, and by Bernard Goldstein, "Kepler and Hebrew Astronomical Tables," *Journal for the History of Astronomy* 32 (2001), 130–136. See also the very informative biographical note about him in Michaud's *Biographie Universelle*, 1847, vol. 44, which also puts him in contact with another very interesting orientalist by the name of Ambrosio Teseo (1469–1539) the slightly older contemporary of Copernicus who also knew Arabic and frequented northern Italy towards the turn of the sixteenth century, and S. Riezler, "Widmanstetter, Johann Albrecht," in *Allgemeine Deutsche Biographie* (Leipzig, 1875–1912), vol. xlii, pp. 357–361. I wish to express my gratitude to Noel Swerdlow, who alerted me to Bernard Goldstein's articles regarding Widmanstadt in the *Journal for the History of Astronomy* and to Bernard Goldstein who supplied the exact references and quotations as well as the reference to Riezler.

54. Michaud, Louis Gabriel, *Biographie Universelle*, Paris, 1847, vol. 44, pp. 56, 570f.

55. For information about this patriarch, see Yūḥannā ʿAzzô, "Risālat al-baṭriyark Ighnāṭyūs Niʿmeh," *al-Mashriq* 31 (1933): 613–623, 730–737, 831–838.

56. Laurentiana, Ms. Or 177, fol. 79r.

57. For the exploits of this patriarch in Italy, see Robert Jones, *Learning Arabic in Renaissance Europe*, pp. 41–44.

58. The works of Robert Jones have been helpful in documenting the details about this press. See Robert Jones, *Learning Arabic in Renaissance Europe* and "The Medici Oriental Press (Rome 1584–1614) and the Impact of its Arabic Publications on Northern Europe," in *The 'Arabick' Interest of the Natural Philosophers in Seventeenth-Century England*, ed. G. Russell, Leiden, 1994, pp. 88–108. For more on this press and on Ignatius Niʿmatallāh's role, see G. J. Toomer, *Eastern Wisdome and Learning*, Oxford, 1996.

59. For this requirement, see Ursula Weisser, "Avicenna: Influence on Medical Studies in the West," in *Encyclopedia Iranica*, vol. III, pp. 107–110, esp. 109, col. 2.

60. On Copernicus's advise on the calendar, see Swerdlow and Neugebauer, p. 31.

61. On the extensive role of the Patriarch on that committee, see G. Coyne, M. Hoskin, and O. Pedersen, *Gregorian Reform of the Calendar: Proceedings of the Vatican Conference to Commemorate its 400th anniversary 1582–1982*, Vatican, 1983, pp. 137, 148, 215, 216, 217, 218, 221, 232, 235.

62. See Andreas Vesalius, *On the Fabric of the Human Body*, Book I, San Francisco, 1998, p. xlvii.

Chapter 7

1. See Sachau, *Chronology*, p. x.

2. An example of the appeal of the conflict paradigm can be seen in the work of the distinguished physicist Pervez Hoodbhoy, *Islam and Science*. As for Ghazālī's deleterious influence, Sachau's opinion is still quoted in almost all sources dealing with Islamic intellectual history.

3. See De Vaux, "Les sphères célestes," and Nau, *Livre de l'ascension de l'esprit*,

4. From the modern period we see people like Huff, and even Sabra and King, almost always referring to the works of Ibn al-Shāṭir as the climax of astronomical thought, implying of course that they were the last flicker in a dying civilization, and that post Ibn al-Shāṭir period may not be worth the attention. See Toby Huff, *The Rise of Early Modern Science: Islam, China and the West*, Cambridge University Press, 1995, p. 47 n.1 and passim. Sabra, *Appropriation*, esp. pp. 238–242, where he has a section devoted to the issues of decline, and where he quotes David King on similar ideas.

5. A recent and well balanced assessment of this factor in the decline of Islamic science, discussed with other factors as well, has been elegantly summarized by Aḥmad Yūsuf al-Ḥassan, "Factors behind the Decline of Islamic Science after the

Sixteenth Century," in *Islam and the Challenge of Modernity*, ed. Sharifah Shifa al-Attas, Kuala Lumpur, 1996, pp. 351–389, esp. 374–376.

6. As a result the names of Hulagu and his grandfather Gengis Khan are usually followed by the expression "May God curse him," as in Abū al-Fidā's, *al-Mukhtaṣar fī Akhbār al-Bashar*, Cairo, 1907, vol. 3, p. 122, vol. 4, p. 2. See also *EI*[2] for a good survey of the sources that describe the fall of Baghdad and its devastation.

7. See Hill, *Book of Knowledge*, and al-Hassan, *al-Jām*ʿ.

8. For the works of Banū Mūsā, see the English translation by Donald Hill, *The Book of Ingenious Devices (Kitāb al-ḥiyal) by the Banū (sons of) Mūsā bin Shākir*, Dordrecht, 1979, and the edition of the Arabic text by Aḥmad Yūsuf al-Hassan, *Kitāb al-ḥiyal by the Banū (sons of) Mūsā bin Shākir*, Aleppo, 1981.

9. See Carra de Vaux, "Le Livre des appareils pneumatiques et des machines hydroliques par Philon de Byzance," *Notices et Extraits des Manuscrits de la bibliothèque Nationale* 38 (1903): 27–237.

10. Hero of Alexandria, *The Pneumatics of Hero of Alexandria*, London, 1971.

11. See Saliba, "The Function of Mechanical Devices."

12. See al-Ḥassan, *al-Jāmi*ʿ, p. 5.

13. See Ibn Abī Uṣaybiʿa, ʿUyūn, vol. I, p. 207, where he reports about al-Kindī's affliction at the hands of Banū Mūsā as having been caused by al-Mutawakkil's fascination for the moving devices of Banū Mūsā (*istihtār al-mutawakkil bi-l-ālāt al-mutaḥarrika*).

14. Ibn al-Nafīs, Abū al-Ḥasan ʿAlāʾ al-Dīn b. Abī al-Ḥazm al-Qarshī al-Dimashqī (d. 1288), *Kitāb Sharḥ Tashrīḥ al-Qānūn*, ed. Silmān Qaṭṭāya, Cairo, 1988, p. 293–294.

15. See *Dictionary of Scientific Biography*, s.v. Kamāl al-Dīn al-Fārisī.

16. John Hayes, ed., *The Genius of Arab Civilization: Source of Renaissance*, New York, 1975, p. 215.

17. Saliba, *A History*, p. 144.

18. Saliba, "The Ultimate challenge,"

19. For a detailed account of this observatory and the stories surrounding its founding and its functioning, see Aydīn Sayīlī, *The Observatory in Islam*, Ankara, 1960, pp. 189–223, and Saliba, "Horoscopes and Planetary Theory"

20. For translations of this treatise, see Jourdain, *Mémoire* and Tekeli, "Al-Urdîʾnin."

21. On the Marāgha, Samarqand, and Jai Singh II observatories, see Sayīlī, *The Observatory in Islam*, pp. 358–361; G. R. Kaye, *Hindu Astronomy*, Calcutta, 1924, p. 5; Kaye, *Astronomical Observatories of Jai Singh*, Calcutta, 1918. For the direct indebtedness of

Jai Singh to the Marāgha observatory and the results obtained at that observatory, see Bose et al., *Concise History*, p. 101f.

22. For the field of scientific instruments in general, see the works of David King, *Islamic Astronomical Instruments*, London, 1987, and in particular section VII of that study on the universal astrolabe. More recently see his remarkable study of the Mecca-centered world map, in David King, *World-Maps for Finding the Direction and Distance to Mecca: Innovation and Tradition in Islamic Science*, Leiden, 1999; King, *In Synchrony with the Heavens: Studies in Astronomical Timekeeping and Instrumentation in Medieval Islamic Civilization*, Leiden, 2004; François Charette, *Mathematical Instrumentation in Fourteenth-Century Egypt and Syria: The Illustrated Treatise of Najm al-Dīn al-Miṣrī*, Leiden, 2003, and now King's *In Synchrony with the Heavens*, vol. II, *Instruments of Mass Calculation*, Brill, 2005.

23. For a reference to this Italian lexicographer, see the Arabic copy of the *Almagest*, now kept at Tunis, Bibliothèque Nationale, no. 7116, which has a signed statement on the flyleaf, in the hand of Ibn Maʿrūf in which he quotes Calepino. A picture of that note is now published in G. Saliba, "The World of Islam and Renaissance Science and Technology," in Catherine Hess, ed. *The Arts of Fire: Islamic Influences on Glass and Ceramics of the Italian Renaissance*, Los Angeles, 2004, pp. 55–73, esp. 71.

24. See, for example, Ms. 2994, preserved in Nurosmania Library, Istanbul.

25. The most recent application of this analysis can be seen in the work of Toby Huff, *The Rise of Early Modern Science*, who also quotes Needham and Weber on similar ideas.

26. I have already discussed these connections between the European academies and the discovery or the New World, as well as the connection of Galileo to all that activity in a relatively obscure journal in the context of a debate with the modern historian of science Toby Huff. See G. Saliba, "Flying Goats and Other Obsessions: A Response to Toby Huff's 'Reply,'" *Bulletin of the Royal Institute for Inter-Faith Studies* 4, no. 2 (2002): 129–141, especially p. 135f. This debate is now available on World Wide Web.

27. Now see the study of David Freedberg, *The Eye of the Lynx: Galileo, His Friends, and the Beginnings of Modern Natural History*, Chicago, 2002.

28. For a fuller appreciation of this fact, and its implications for the rise of modern science, see Joseph Needham, *Within the Four Seas: The Dialogue of East and West*, Toronto, 1969.

Bibliography

Aaboe, Asger. "On the Babylonian Origin of Some Hipparchian Parameters." *Centaurus* 4 (1955–56): 122–125.

ʿAbd al-Ghanī, Musṭafā Labīb, *Dirāsāt fī tārīkh al-ʿulūm ʿinda al-ʿArab,* volume 1. Dār al-Thaqāfa, Cairo, 2000.

Abū al-Fidā, Ismāʿīl (d. 1331). *al-Mukhtaṣar fī Akhbār al-Bashar.* Cairo, 1907.

Abū al-Fidāʾ. *Géographie d'Aboulféda, Taqwīm al-Buldān,* ed. M. Reinaud. Paris, 1840.

Abū Maʿshar al-Balkhī (d. 886). *al-Madkhal ilā ʿilm aḥkām al-nujūm.* Jārullah (Carullah) Ms. 1058, published in facsimile. Frankfurt, 1985.

Abuna, Albert. *Adab al-Lugha al-Ārāmīya.* Beirut, 1970.

Akhawayn, Muḥyī al-Dīn Muḥammad b. Qāsim, al- (fl. c. 1498). *al-Ishkālāt fī ʿIlm al-Hayʾa* (Problems in Science of Astronomy). Austrian National Library, Vienna, Arabic, 1422.

Almagest. al-Ḥajjāj's Arabic translation. British Library manuscript, Add. 7474.

Almagest. al-Ḥajjāj's Arabic translation. Leiden University, Or. 680.

Almagest. Isḥāq-Thābit translation. Bibliothèque Nationale de Tunis, no. 7116.

Anonymous (Andalusian). *Kitāb al-Hayʾa.* Osmania University Library, Hyderabad, Arabic Ms. 520RH.

Aristotle. *On the Heavens I & II,* ed. S. Leggatt. Aris & Phillips, 1995.

Aristotle. *On the Heavens.* Loeb, 1939. Reprinted in 1960.

Aouad, M. *Averroès (Ibn Rushd): Commentaire Moyen à la Rhétorique d'Aristote, Édition critique du texte arabe et traduction française.* Paris, 2002.

ʿAskarī, al-Ḥasan b. ʿAbdallāh, Abū Hilāl, al- (993). *Kitāb al-awāʾil.* Beirut, 1997.

Averroes (Ibn Rushd). *Tafsīr mā baʿd al-ṭabīʿa,* ed. M. Bouyges. Beirut, 1948.

Averroes (Ibn Rushd). *Commentaire Moyen à la Rhétorique d'Aristote, Édition critique du texte arabe et traduction française.* Paris, 2002.

ʿAzzô, Yūḥannā. "Risālat al-baṭriyark Ighnāṭyūs Niʿmeh." *al-Mashriq* 31 (1933): 613–623, 730–737, 831–838.

Baghdādī, Abū al-Barakāt al-. *Kitāb al-Muʿtabar.* Hyderabad, 1938.

Baghdādī, Muwaffaq al-Dīn Abū Muḥammad b. Yūsuf, ʿAbd al-Laṭīf al- (1231). *al-Ifāda wa-l-Iʿtibār,* ed. A. Ghassān Sabānū. Dār Ibn Zaydūn (Beirut) and Dār Qutayba (Damascus), 1984.

Banū Mūsā. *Kitāb al-ḥiyal by the Banū (sons of) Mūsā bin Shākir,* ed. A. Y. al-Hassan. Aleppo, 1981.

Banū Mūsā. *The Book of Ingenious Devices (Kitāb al-ḥiyal) by the Banū (sons of) Mūsā bin Shākir.* Dordrecht, 1979.

Bar Hebraeus. See Nau.

Barker, Peter, and Bernard Goldstein. "Patronage and Production of *De Revolutionibus.*" *Journal for the History of Astronomy* 34 (2003): 345–368.

Bergstrasser, Gotheil. *Ḥunain b. Isḥāq, Über die Syrischen und Arabischen Galenübersetzungen.* Leipzig, 1925.

Bīrūnī, Abū al-Raiḥān (1048). *al-Āthār al-Bāqiya ʿan al-Qurūn al-Khāliya* (*Chronology of Ancient Nations*), ed. E. Sachau. London, 1879.

Bīrūnī, Abū al-Raiḥān, Muḥammad b. Aḥmad al- (1048). *Kitāb al-tafhīm li-awāʾil ṣināʿat al-tanjīm* (The Book of Instruction in the Elements of the Art of Astrology). London, 1934.

Bose, D. M., S. N. Sen, and B. V. Subbarayappa, eds. *A Concise History of Science in India.* New Delhi, 1971.

Boyer, Carl. *A History of Mathematics.* New York, 1968.

Brubaker, Leslie, ed. *Byzantium in the Ninth Century: Dead or Alive?* Ashgate, 1998.

Brunschvig, R., and G. E. Grunebaum. *Classicisme et Déclin Culturel dans l'histoire de l'Islam.* Paris, 1957.

Burnett, Charles, Jan Hogendijk, Kim Plofker, and Michio Yano, eds. *Studies in the History of the Exact Sciences in Honor of David Pingree.* Boston, 2004.

Būzjānī, Abū al-Wafāʾ, al- (997). *Mā yaḥtāj ilaih al-Ṣunnāʿ min ʿilm al-handasa.* Baghdad, 1979.

Būzjānī, Abū al-Wafāʾ, al-. *Mā yaḥtāj ilaih al-kuttāb wa-l-ʿummāl wa-ghairihim min ʿilm al-ḥisāb.* In A. S. Saidan, *Abū al-Wafāʾ al-Būzjānī: ʿilm al-ḥisāb al-ʿarabī.* Amman, 1971.

Cambridge History of Arabic Literature: Religion, Learning and Science in the Abbasid Period, ed. M. Young et al. Cambridge University Press, 1990.

Charette, François. *Mathematical Instrumentation in Fourteenth-Century Egypt and Syria: The Illustrated Treatise of Najm al-Dīn al-Miṣrī.* Brill, 2003.

Cheikho, Louis. *"Risālat al-Khujandī fī mayl wa-ʿarḍ al-balad."* *Mashriq* 11 (1908): 60–69.

Cifoletti, Giovanna. "The Creation of the History of Algebra in the Sixteenth-Century." In *L'Europe mathématique—Mythes, histoires, identités,* ed. C. Goldstein et al. Editions de la Maison des sciences des l'hommes, 1996.

Copernicus, Nicolaus. *De Revolutionibus: Faksimiles des Manuskriptes.* Hildesheim, 1974.

Coyne, G. V., M. A. Hoskin, and O. Pedersen. *Gregorian Reform of the Calendar: Proceedings of the Vatican Conference to Commemorate its 400th anniversary 1582–1982.* Vatican, 1983.

d'Alverny, Marie-Thérèse. *Avicenne en Occident.* Paris, 1993.

Daniel, Norman. *Islam and the West: The Making of an Image.* Oxford University Press, 1960, 1993.

Dannenfeldt, Karl. "The Renaissance Humanists and the Knowledge of Arabic." *Studies in the Renaissance* 2 (1955): 96–117.

Debarnot, Marie-Thérèse. "The Zīj of Ḥabash al-Ḥāsib: A Survey of Ms. Istanbul Yeni Cami 784/2." In King and Saliba, *From Deferent to Equant.* New York Academy of Sciences, 1987.

Della Vida, Giorgio Levi. *Ricerche sulla formazione del più antico fondo deu manoscritti orientali della biblioteca Vaticana.* Studi e Testi, Biblioteca Apostolica Vaticana, Citta del Vaticano, 1939.

de Vaux, Baron Carra. "Les spheres célestes selon Nasîr-Eddîn Attûsî." In Paul Tannery, *Recherches sur l'histoire de l'astronomie ancienne.* Gauthier-Villars, 1893.

de Vaux, Baron Carra. "Le Livre des appareils pneumatiques et des machines hydroliques par Philon de Byzance. *Notices et Extraits des Manuscrits de la bibliothèque Nationale* 38 (1903): 27–237.

Dictionary of Scientific Biography. New York, 1970–1990.

Dobrzycki, Jerzy, and Richard L. Kremer. "Peurback and the Marāgha astronomy? The Ephemerides of Johannes Angelus and Their Implications." *Journal for the History of Astronomy* 27 (1996): 187–237.

Dorotheus Sidonius. See Pingree, 1976.

Dzielska, Maria. *Hypatia of Alexandria.* Harvard University Press, 1995.

Eche, Youssef. *Les Bibliotheques Arabes.* Damas, 1967.

EHAS. See *Encyclopedia of the History of Arabic Sciences.*

EI². See *Encyclopedia of Islam.*

Encyclopaedia Britannica.

Encyclopedia Iranica. http://www.iranica.com/articlenavigation/index.html.

Encyclopedia Italiana, Storia della Scienza, volume III. Rome, 2002.

Encyclopedia of Islam, second edition. Brill, 1986–2004. Cited as *EI²*.

Encyclopedia of the History of Arabic Sciences, ed. R. Rashed in collaboration with Régis Morelon. Routledge, 1996. Cited as *EHAS.*

Farghānī, Ibn Kathīr, al- (fl. 861). *Jawāmi' 'ilm al-nujūm.* Amsterdam, 1669.

Freedberg, David. *The Eye of the Lynx: Galileo, His Friends, and the Beginnings of Modern Natural History.* University of Chicago Press, 2002.

Fihrist. See al-Nadīm.

Fück, Johann. *Die Arabischen Studien in Europa bis in den Anfang des 20. Jahrhundert.* Leipzig, 1955.

Ghars al-Dīn, Aḥmad b. Khalīl al-Ḥalabī (d. 1563). *Tanbīh al-nuqqād 'alā mā fī al-hay'a al-mashhūra min al-fasād* (Alerting the Critics to the Corruption of the Well-Known Astronomy). Istanbul, Yeni Jāmi', Ms. 1181.

Ghazālī, Abū Ḥāmid (d. 1111), al-. *The Incoherence of the Philosophers,* ed. M. Marmura. Provo, 1997.

Ghulām, Yousif Muḥammad. *The Art of Arabic Calligraphy.* Published by the author, 1982.

Goldstein, Bernard. *Al-Biṭrūjī: On the Principles of Astronomy.* Yale University Press, 1971.

Goldstein, Bernard. "Copernicus and the Origin of His Heliocentric Universe." *Journal for the History of Astronomy* 33 (2002): 219–235.

Goldstein, Bernard. "Ancient and Medieval Values for the Mean Synodic Month." *Journal for the History of Astronomy* 34 (2003): 65–74.

Goldstein, C., J. Gray, and J. Ritter, eds. *L'Europe mathématique—Mythes, histoires, identités. Mathematical Europe—Myth, History, Identity.* Editions de la Maison des sciences des l'hommes, 1996.

Goodman L. E. "The Translation of the Greek materials into Arabic." In *Cambridge History of Arabic Literature: Religion Learning and Science in the Abbasid Period.* Cambridge University Press, 1990.

Grafton, Anthony. "Michael Maestlin's Account of Copernican Planetary Theory." *Proceedings of the American Philosophical Society* 117, no. 6 (1973): 523–550.

Grant, Edward, ed. *A Source Book in Medieval Science.* Harvard University Press, 1974.

Green, V. H. H. *Renaissance and Reformation: A Survey of European History Between 1450 and 1660.* London. 1954. Reprinted in 1975.

Gregory, Timothy. *A History of Byzantium.* Blackwell, 2005.

Grignaschi, Mario. "Les "Rasā'il Arisṭāṭālīs ilā-l-Iskandar" de Sālim Abū-l-ʿAlā' et l'Activité Culturelle a l'Époque Omayyade." *Bulletin d'Études Orientales* 19 (1965–66): 7–83.

Grunebaum, Gustav Von. *Islam: Essays in the Nature and Growth of a Cultural Tradition.* Greenwood, 1981.

Gubernatis, Angelo de. *Matériaux pour servir a l'histoire des études orientales en Italie.* E. Leroux, 1876.

Gutas, Dimitri. "Paul the Persian on the Classifications of the Parts of Aristotle's Philosophy: A Milestone between Alexandria and Baghdad." *Der Islam* 60 (1983): 31–267.

Gutas, Dimitri. *Greek Thought, Arabic Culture.* Routledge, 1998.

Ḥabash al-Ḥāsib, Aḥmad b. ʿAbdallāh (c. 850). See Kennedy, Kunitzsch, and Lorch, 1999.

Ḥajjāj b. Maṭar (c. 830). *Almagest* (Arabic translation). Leiden University, Ms. Or 680.

HAMA. See Neugebauer, 1975.

Hartner, Willy. "Copernicus, the Man, the Work, and Its History." *Proceedings of the American Philosophical Society* 117, no. 6 (1973): 413–422.

Hartner, Willy. "Naṣīr al-Dīn al-Ṭūsī's Lunar Theory." *Physis* 11 (1969): 289–304.

Hāshimī, ʿAlī b. Sulaimān al- (ninth century). *Kitāb fī ʿilal al-zījāt.* Bodleian Ms. Selden A. 11.

Haskins, Charles Homer. *The Renaissance of the 12th Century.* Cambridge, Massachusetts, 1927.

Ḥassan, Aḥmad Y. al-, ed. *al-Jāmiʿ bain al-ʿilm wa-l-ʿamal al-nāfiʿ fī ṣināʿat al-ḥiyal (Combining Theory and Useful Practice in the Craft of Mechanical Arts).* Aleppo, 1979. See also Hill 1974.

Ḥassan, Aḥmad Y. al-, and Donald Hill. *Islamic Technology: An illustrated History.* UNESCO and Cambridge University Press, 1986.

Ḥassan, Aḥmad Yūsuf al-. "Factors behind the Decline of Islamic Science after the Sixteenth Century." In *Islam and the Challenge of Modernity,* ed. S. Shifa al-Attas. Kuala Lumpur, 1996.

Hayes, John, ed. *The Genius of Arab Civilization: Source of Renaissance.* New York University Press, 1975.

Heinen, Anton. *Islamic Cosmology.* Orient Institute, Beirut, 1982.

Hero of Alexandria. *The Pneumatics of Hero of Alexandria.* Macdonald, 1971.

Hess, Catherine, ed., with contributions by Linda Komaroff and George Saliba. *The Arts of Fire: Islamic Influences on Glass and Ceramics of the Italian Renaissance.* J. Paul Getty Museum, 2004.

Hill, Donald. *The Book of Ingenious Devices (Kitāb al-Ḥiyal) by Banū (sons of) Mūsā bin Shākir.* Reidel, 1979.

Hill, Donald. *The Book of Knowledge of Ingenious Mechanical Devices.* Reidel, 1974 (translation of Jazarī).

Homā'ī, Jalāl al-Dīn. *al-Tafhīm li-awā'il ṣinā'at al-tanjīm.* Babak, Teheran.

Hoodbhoy, Pervez. *Islam and Science: Religious Orthodoxy and the Battle for Rationality.* Zed, 1991.

Hoodbhoy, Pervez. *Muslims and Science: Religious Orthodoxy and the struggle for Rationality.* Vanguard, Lahore, 1991.

Huff, Toby. *The Rise of Early Modern Science: Islam, China and the West.* Cambridge University Press, 1995.

Ibn Abī Uṣaybi'a, Aḥmad b. Qāsim (1270). *'Uyūn al-Anbā' fī Ṭabaqāt al-Aṭibbā',* ed. R. Müller. Königsberg, 1884.

Ibn al-Athīr, 'Izz al-Dīn Abū al-Ḥasan (d. 1223). *al-Kāmil fī al-Tārīkh.* Dār al-Kitāb al-'Arabi, Beirut, 1995.

Ibn al-Baiṭār. *al-Jāmi' li-Mufradāt al-Adwiya wa-l-Aghdhiya.* Bulaq, 1874.

Ibn al-Haitham, al-Ḥasan Abū 'Alī (d. 1049). "On the Elevation of the Pole." Leiden 1643, British Museum Ms. Add. 3034, dated 1646.

Ibn al-Haitham, al-Ḥasan Abū 'Alī (d. 1049). *al-Shukūk 'alā Baṭlamyūs (Dubitationes in Ptolemaeum),* ed. A. Sabra and N. Shehaby. Dār al-Kutub, Cairo, 1971.

Ibn al-Nafīs, Abū al-Ḥasan 'Alā' al-Dīn b. Abī al-Ḥazm al-Qarshī al-Dimashqī (d. 1288). *Kitāb Sharḥ Tashrīḥ al-Qānūn,* ed. S. Qaṭṭāya. Cairo, al-Hay'a al-Maṣrīya, 1988.

Ibn al-Shāṭir, ʿAlāʾ al-Dīn (1375). *Kitāb nihāyat al-sūl fī taṣḥīḥ al-uṣūl* (The Ultimate Quest in the Rectification of [Astronomical] Principles) (Bodleian. Ms. Marsh 139) ed. George Saliba (forthcoming).

Ibn al-Ukhuwwa, Muḥammad b. Muḥammad b. (1329). *Maʿālim al-qurbah fī aḥkām al-ḥisbā,* ed. R. Levey. Cambridge, 1938.

Ibn Khaldūn, ʿAbd al-Raḥmān (1406). *The Muqaddimah.* Princeton, 1958.

Ibn Mamātī, Asʿad b. Muhadhdhab (1209). *Qawānīn al-Dawāwīn,* ed. A. Atiya. maṭbaʿat Miṣr, Cairo, 1943.

Ibn Qutayba. *ʿUyūn al-akhbār.* Dār al-Kitāb, Beirut, 1997.

Ibn Qutayba. *Kitāb al-anwāʾ.* Dār al-maʿārif al-Osmāniya, Hyderabad, 1956.

Itlīdī, Muḥammad Diyāb al- (seventeenth century). *Iʿlām al-nās bi-mā waqaʿa li-l-barāmika maʿa banī al-ʿabbās.* Beirut, 1990.

Jāḥiẓ, ʿAmr b. Baḥr, al- (869). *Kitāb al-Bukhalāʾ.* Dār al-Fikr, Beirut, n.d.

Jahshiyārī, Muḥammmmad b. ʿAbdūs, al- (d. 942). *Kitāb al-Wuzarāʾ wa-l-Kuttāb.* Dār al-Fikr, Beirut, 1988.

Jazarī, Ismāʿīl Abū al-ʿIzz, al- (c. 1206). *al-Jāmiʿ bain al-ʿilm wa-l-ʿamal al-nāfiʿ fī ṣināʿat al-ḥiyal (Combining Theory and Useful Practice in the Craft of Mechanical Arts),* ed. A. Y. al-Ḥassan. Aleppo, 1979. See also Hill, 1974; al-Ḥassan, 1979.

Job of Edessa (c. 817). *Book of Treasures,* ed. A. Mingana. Heffer, 1935.

Jones, John Robert. *Learning Arabic in Renaissance Europe (1505–1624).* Dissertation DX195516, London University, 1988.

Jones, John Robert. "The Medici Oriental Press (Rome 1584–1614) and the Impact of its Arabic Publications on Northern Europe." In *The 'Arabick' Interest of the Natural Philosophers in Seventeenth-Century England,* ed. G. Russell. Brill, 1994.

Jones, Alexander. "Later Greek and Byzantine Astronomy." In *Astronomy before the Telescope,"* ed. C. Walker. St. Martin's Press, 1996.

Jourdain, A. *Mémoire sur l'observatoire de Méragah et sur Quelques Instruments Employés pour Observer.* Paris, 1870.

Jumaylī, Rashīd al-. *Ḥarakat al-tarjama fī al-mashriq al-islāmī fī al-qarnain al-thālith wa-l-rābiʿ li-l-hijra,* al-kitāb. Tripoli, 1982.

Kaye, G. R. *Hindu Astronomy.* Calcutta, 1924.

Kennedy, E. S. "A Survey of Islamic Astronomical Tables." *Transactions of the American Philosophical Society,* New Series, 46, no. 2 (1956): 123–177.

Kennedy, E. S. "Late Medieval Planetary Theory." *Isis* 57 (1966): 365–378.

Kennedy, E. S. "The Arabic Heritage in the Exact Sciences." *al-Abḥāth* 23 (1970): 327–344. Reprinted in *Studies in the Islamic Exact Sciences by E. S. Kennedy, Colleagues and Former Students,* ed. D. King and M. Kennedy. American University, Beirut, 1983.

Kennedy, E. S., and Imad Ghanim. *The Life and Work of Ibn al-Shāṭir.* Aleppo, 1976.

Kennedy, E. S., Fuad I. Haddad, and David Pingree. *The Book of the Reasons Behind Astronomical Tables.* Scholars' Facsimile and Reprints, New York, 1981.

Kennedy, E. S. *Studies in the Islamic Exact Sciences by E. S. Kennedy, Colleagues and Former Students,* ed. D. King and M. Kennedy. American University, Beirut, 1983.

Kennedy, E. S., P. Kunitzsch, and R. P. Lorch, eds. *The Melon-Shaped Astrolabe in Arabic Astronomy.* Steiner, 1999.

Khafrī, Shams al-Dīn (1550). See Saliba, 1994, 1997, 2000.

Kharaqī, ʿAbd al-Jabbār, al- (1139). *Muntahā al-idrāk fī taqāsīm al-aflāk.* Bibliothèque Nationale de France, Arabe 2499.

Khwārizmī, Muhammad b. Mūsā, al-. *Kitāb al-jabr wa-l-muqābala.* See Rosen, 1831.

Khwārizmī al-Kātib, Muḥammad b. Aḥmad b. Yūsuf, al- (997). *Mafātīḥ al-ʿulūm,* ed. G. van Vloten. Leiden, 1895.

King, David. "Ibn al-Shāṭir." In *Dictionary of Scientific Biography,* 1975.

King, David, and George Saliba. *From Deferent to Equant. Annals of the New York Academy of Sciences* 500 (1987).

King, David. *Islamic Astronomical Instruments.* Variorum, 1987.

King, David. *Astronomy in the Service of Islam.* Aldershot, 1993.

King, David. *World-Maps for Finding the Direction and Distance to Mecca: Innovation and Tradition in Islamic Science.* Brill, 1999.

King, David. *In Synchrony with the Heavens: Studies in Astronomical Timkeeping and Instrumentation in Medieval Islamic Civlization*, volume I: *The Call of the Muezzin.* Leiden, 2004.

King, David. *In Synchrony with the Heavens: Studies in Astronomical Timkeeping and Instrumentation in Medieval Islamic Civlization*, volume II: *Instruments of Mass Calculations.* Brill, 2005.

Kunitzsch, Paul. *Peter Apian und Azophi: Arabische Sternbilder in Ingolstadt im frühen 16. Jahrhundert.* Bayerische Akademie der Wissenschaften, Philosophisch-Historische Klasse, Sitzungsberichte, Jahrgang 1986, Heft 3, Verlag der Bayerischen Akademie der Wissenschaften, 1986.

Kuntz, Marion. *Guillaume Postel: Prophet of the Restitution of All Things, His Life and Thought.* Kluwer, 1981.

Legacy of Islam, ed. T. Arnold and A. Guillaume. Oxford University Press, 1931.

Lemerle, Paul. *Le Premier Humanisme Byzantin: Notes et remarques sur enseignement et culture à Byzance des origins au Xe siècle.* Presses Universitaires de France, 1971.

Lewis, Bernard, ed. *The World of Islam.* Thames and Hudson, 1976.

Mango, Cyril, ed. *The Oxford History of Byzantium.* Oxford University Press, 2002.

Masʿūdī, Abū al-Ḥasan ʿAlī b. Ḥusain b, ʿAlī, al- (956). *Murūj al-Dhahab (Les Prairies d'Or),* ed. C. Barbier de Meynard. Paris, 1874.

Mavroudi, Maria. *A Byzantine Book on Dream Interpretation: The Oneirocriticon of Achmet and Its Arabic Sources.* Brill, 2002.

Mercier, André, ed. *Islam und Abendland.* Herbert Lang, 1976.

Meyerhof, Max. "Von Alexandrien nach Bagdad: Ein Beitrag zur Geschichte des philosophischen und medizinischen Unterrichts den Araben." *Sitzungsberichte der Berliner Akademie der Wissenschaften,* Philologisch-historische Klasse, 1930: 389–429.

Meyerhoff, Max. "Science and Medicine." In the *Legacy of Islam,* ed. T. Arnold and A. Guillaume. Oxford University Press, 1931.

Michaud, Louis Gabriel. *Biographie Universelle.* Paris, 1847–.

Mingana. See Job of Edessa.

Morelon, Régis. "Eastern Arabic Astronomy between the Eighth and the Eleventh Centuries." In *EHAS.*

Morelon, Régis. *Thābit Ibn Qurra: Oeuvres d'Astronomie.* Belles Lettres, Paris, 1987.

Morrison, Robert. "Quṭb al-Dīn al-Shīrāzī's Use of Hypotheses." *Journal for the History of Arabic Science* 13 (2005): 21–140.

Mumtaḥan, The Verified Astronomical Tables for the Caliph al-Ma'mūn, by Yaḥyā b. Abī Manṣūr, photographic print of Escorial Ms. Árabe 927, by Fuat Sezgin, Institute for the History of Arabic-Islamic Science, Frankfurt, 1986.

Nadīm, Abū al-Faraj Muḥammad b. Abī Yaʿqūb Isḥāq, al- (987). *al-Fihrist,* ed. Y. ʿAlī Ṭawīl. Beirut, 1996.

Nallino, Carlo A. *ʿIlm al-Falk: Tārīkhuhu ʿind al-ʿArab fī al-qurūn al-wusṭā.* Rome, 1911.

Nau, F. [Bar Hebraeus]. *Livre de l'ascension de l'esprit sur la forme du ciel et de la terre.* Émile Bouillon, 1899.

Nau, F. "Le traité de l'astrolabe plan de Sévère Sébokt, publié pour la première fois d'aprés un Ms. de Berlin." *Journal Asiatique* 13 (1899): 56–101, 238–303.

Nau, F. "La plus ancienne mention orientale des chiffres indiens." *Journal Asiatique* 16 (1910): 225–227.

Nau, F. "Notes d'astronomie syrienne." *Journal Asiatique,* 2e ser. t. xvi (1910): 225–226.

Nau, F. "La traité sur les "constellations" ecrit, en 661, par Sévère Sébokht Évèque de Qennesrin." *Revue de l'Orient Chretien,* 3e serie, 7, no. xxvii (1929–30): 327–338.

Needham, Joseph. *Within the Four Seas: The Dialogue of East land West.* University of Toronto Press, 1969.

Neugebauer, Otto. "Studies in Byzantine Astronomical Terminology." *Transactions of the American Philosophical Society,* New Series, 50 (1960): 1–45

Neugebauer, Otto. "Thābit Ben Qurra 'On the Solar Year' and 'On the Motion of the Eighth Sphere,' translation and commentary." *Proceedings of the American Philosophical Society* 106 (1962): 64–299.

Neugebauer, Otto. "On the Planetary Theory of Copernicus." *Vistas in Astronomy* 10 (1968): 89–103.

Neugebauer, Otto. *A History of Ancient Mathematical Astronomy.* Springer-Verlag, 1975. Cited as *HAMA.*

O'Leary, De Lacy. *How Greek Science Passed to the Arabs.* Routledge, 1949. Reprinted in 1964.

Oxford Dictionary of Byzantium. Oxford University Press, 1991.

Paret, R. *Der Islam und das griechische Bildungsgut.* Mohr, 1950.

Paschos, E. A., and P. Sotiroudis. *The Schemata of the Stars: Byzantine Astronomy from AD 1300.* World Scientific, 1998.

Pingree, David. "Gregory Chioniades and Paleologan Astronomy." *Dumbarton Oaks Papers* 18 (1964): 133–160.

Pingree, David. "The Fragments of the Works of Yaʿqūb ibn Ṭāriq." *Journal of Near Eastern Studies* 26 (1968): 97–125.

Pingree, David. "The Fragments of the Works of al-Fazārī." *Journal of Near Eastern Studies* 29 (1970): 103–123.

Pingree, David, and E. S. Kennedy. *Astrological History of Māshāʾallāh.* Harvard University Press, 1971.

Pingree, David. "The Greek Influence on Early Islamic Mathematical Astronomy." *Journal of the American Oriental Society* 93 (1973): 32–43.

Pingree, David. *Dorotheus Sidonius Carmen Astrologicum*. Teubner, 1976.

Pingree, David. *The Astronomical Works of Gregory Chioniades*. J. C. Gieben, 1985.

Postel, Guillaume (1581–1981). *Guillaume Postel (1581–1981)*. Actes du Colloque International d'Avranches, 5–9 Septembre, 1981, Paris, 1985.

Ptolemy, Claudius. *The Almagest*. See *Almagest;* Toomer, 1984.

Qifṭī, ʿAlī b. Yūsuf (1248), al-, *Taʾrīkh al-ḥukamāʾ*. Dietrich'sche Verlagsbuchlandlung, Leipzig, 1903.

Ragep, F. Jamil. *Naṣīr al-Dīn al-Ṭūsī's Memoir on Astronomy*. Springer-Verlag, 1993.

Ragep, F. Jamil. "Ṭūsī and Copernicus: The Earth's Motion in Context." *Science in Context* 14 (2001): 145–163.

Ragep, F. Jamil. "Freeing Astronomy from Philosophy: An Aspect of Islamic Influence on Science." *Osiris* 16 (2001): 49–71.

Ragep, F. Jamil. "ʿAlī Qushjī and Regiomontanus: Eccentric Transformations and Copernican Revolutions." *Journal for the History of Astronomy* 36 (2005): 359–371.

Rashed, Roshdi. "l'Idée de l'Algèbre Selon Al-Khwārizmī." *Fundamenta Scientiae* 4 (1983) 87–100.

Rashed, Roshdi. *l'Art de l'Algèbre de Diaphonte*. al-Hayʿa al-Miṣrīya, Cairo, 1975. French translation Paris 1984.

Rashed, Roshdi. "Problems of the Transmission of Greek Scientific Thought into Arabic: Examples from Mathematics and Optics." *History of Science* 27 (1989): 199–209. Reprinted in Rashed, *Optique et mathématique*. Variorum, 1992.

Rashed, Roshdi. *Optique et mathématique*. Variorum, 1992.

Rhazes (Rāzī, Muḥammad b. Zakarīya, Abū Bakr, al- (925)). *De variolis et morbillis*. London, 1760.

Rāzī, Muḥammad b. Zakarīya, Abū Bakr, al- (925). *al-Shukūk ʿalā Jālīnūs* (*Doubts contra Galen*), ed. M. Mohaghegh. Tehran, 1993.

Ritter, Helmut. "l'Orthodoxie a-t-elle une part dans la Décadence?" In *Classicisme et Déclin Culturel dans l'histoire de l'Islam,* ed R. Brunschvig and G. Grunebaum. Paris, 1957.

Roberts, Victor. "The Solar and Lunar Theory of Ibn al-Shāṭir: A Pre-Copernican Copernican Model." *Isis* 48 (1957): 428–432. Reprinted in *Studies in the Islamic Exact Sciences,* ed. D. King and M.-H. Kennedy. American University, Beirut, 1983.

Rosen, Frederic. *The Algebra of Mohammed ben Musa*. London, 1831. Reprint: Olms, 1986.

Rosenthal, F. *The Classical Heritage in Islam.* Routledge, 1965.

Russell, Gül, ed. *The 'Arabick' Interest of the Natural Philosophers in Seventeenth-Century England.* Brill, 1994.

Ryssel, V. "Die Astronomischen Briefe Georgs des Araberbischofs." *Zeitschrift für Assyriologie und verwandte Gebiete* 8 (1893): 1–55.

Sabra, A. I. "The Scientific Enterprise." In *The World of Islam,* ed. B. Lewis. Thames and Hudson, 1976.

Sabra, A. I. "The Appropriation and Subsequent Naturalization of Greek Science in Medieval Islam: A Preliminary Statement." *History of Science* 25 (1987): 223–243.

Sabra, A. I. "Situating Arabic Science: Locality versus Essence." *Isis* 87 (1996): 654–679.

Sachau, Edward. Bīrūnī's *Chronology of Ancient Nations.* William H. Allen, 1879.

Ṣafadī, Ṣalāh al-Dīn Khalīl b. Aybak, al- (d. 1362). *Al-Ghayth al-musajjam fī sharḥ lāmiyat al-ʿajam.* Beirut, 1997.

Ṣafadī, Ṣalāḥ al-Dīn Khalīl b. Aybak, al- (d. 1363). *Kitāb al-wāfī bi-l-wafayāt.* Wiesbaden, 1981.

Saidan, A. S. *Abū al-Wafāʾ al-Būzjānī: ʿilm al-ḥisāb al-ʿarabī.* Amman, 1971.

Saidan, A. S. *The Arithmetic of al-Uqlīdisī.* Reidel, 1978.

Saliba, George. "The Original Source of Quṭb al-Dīn al-Shīrāzī's Planetary Model." *Journal for the History of Arabic Science* 3 (1979): 3–18.

Saliba, George. "The Function of Mechanical Devices in Medieval Islamic Society." *Annals of the New York Academy of Sciences* 441 (1985): 141–151.

Saliba, George. "The Determination of the Solar Eccentricity and Apogee According to Muʾayyad al-Dīn al-ʿUrḍī (d. 1266 A.D.)." *Zeitschrift für Geschichte der Arabisch-Islamischen Wissenschaften* 2 (1985): 47–67. Reprinted in Saliba, *A History.*

Saliba, George. "The Role of the *Almagest* Commentaries in Medieval Arabic Astronomy: A Preliminary Survey of Ṭūsī's Redaction of Ptolemy's *Almagest.*" *Archives Internationales d'Histoire des Sciences* 37 (1987): 3–20. Reprinted in Saliba, *A History.*

Saliba, George. "Theory and Observation in Islamic Astronomy: The work of Ibn al-Shāṭir of Damascus." *Journal for the History of Astronomy* 18 (1987): 35–43. Reprinted in Saliba, *A History.*

Saliba, George. "Arabic Astronomy in Byzantium." *Journal for the History of Astronomy* 20 (1990): 211–215.

Saliba, George. *The Astronomical Work of Muʾayyad al-Dīn al-ʿUrḍī (d. 1266): A Thirteenth Century Reform of Ptolemaic Astronomy,* ʿUrḍī's *Kitāb al-Hayʾa.* Beirut, 1990, 1995. Third corrected edition, 2001.

Saliba, George, and E. S. Kennedy. "The Spherical Case of the Ṭūsī Couple." *Arabic Sciences and Philosophy* 1 (1991): 285–291. Reprinted with minor mistakes in *Naṣīr al-Dīn al-Ṭūsī: Philosophe et savant du xiiie siècle,* ed. N. Pourjavadi and Z. Vesel. Institut Français de Recherche en Iran and Presses Universitaires d'Iran, Teheran, 2000.

Saliba, George. "A Sixteenth-Century Drawing of an Astrolabe Made by Khafīf Ghulām ʿAlī b. ʿĪsā (c. 850 A. D.)." *Nuncius, Annali di Storia della Scienza* 6 (1991): 109–119.

Saliba, George. "Copernican Astronomy in the Arab East: Theories of the Earth's Motion in the Nineteenth Century." In *Transfer of Modern Science and Technology to the Muslim World,* ed. E. Ihsanoglu. Istanbul, 1992.

Saliba, George. "The Role of the Astrologer in Medieval Islamic Society." *Bulletin d'Études Orientales* 44 (1992): 45–68. Reprinted in *Magic and Divination in Early Islam,* ed. E. Savage-Smith. Ashgate-Variorum, 2004.

Saliba, George. "Al-Qushjī's Reform of the Ptolemaic Model for Mercury." *Arabic Sciences and Philosophy* 3 (1993): 161–203.

Saliba, George. "A Sixteenth-Century Arabic Critique of Ptolemaic Astronomy: The Work of Shams al-Dīn al-Khafrī." *Journal for the History of Astronomy* 25 (1994): 15–38.

Saliba, George. "Early Arabic Critique of Ptolemaic Cosmology: A Ninth-Century Text on the Motion of the Celestial Spheres." *Journal for the History of Astronomy* 25 (1994): 115–141.

Saliba, George. *A History of Arabic Astronomy: Planetary Theories During the Golden Age of Islam.* New York University Press, 1994.

Saliba, George. "Paulus Alexandrinus in Syriac and Arabic." *Byzantion* 65 (1995): 440–454.

Saliba, George. "Arabic Planetary Theories after the Eleventh Century AD." In *Encyclopedia of the History of Arabic Science.* Routledge, 1996.

Saliba, George. "A Redeployment of Mathematics in a Sixteenth-Century Arabic Critique of Ptolemaic Astronomy." In *Perspectives arabes et médiévales sur la tradition scientifique philosophique grecque. Actes du Colloque de la S. I. H. S. P. A. I.* (Société internationale d'histoire des sciences et de la philosophie arabe et Islamique), ed. A. Hasnawi, A. Elamrani-Jamal, and M. Aouad. Peeters, 1997.

Saliba, George. "Persian Scientists in the Islamic World: Astronomy from Maragha to Samarqand." In *The Persian Presence in the Islamic World,* ed. R. Hovannisian and G. Sabagh. Cambridge University Press, 1998.

Saliba, George. "Rethinking the Roots of Modern Science: Arabic Manuscripts in European Libraries." Occasional Paper, Georgetown Center for Contemporary Arab Studies, 1999.

Saliba, George. "Critiques of Ptolemaic Astronomy in Islamic Spain." *al-Qanṭara* 22 (1999): 3–25.

Saliba, George. "Competition and the Transmission of the Foreign Sciences: Ḥunayn at the Abbāsid Court." *Bulletin of the Royal Institute for Inter-Faith Studies* 2 (2000): 85–101.

Saliba, George. "The Ultimate Challenge to Greek Astronomy: *Ḥall mā lā Yanḥall* of Shams al-Dīn al-Khafrī (d. 1550)." In *Sic Itur Ad Astra: Studien zur Geschichte der Mathematik und Naturwissenschafte,* ed. M. Folkert and R. Lorch. Harrassowitz Verlag, 2000.

Saliba, George. "Science before Islam." In *The Different Aspects of Islamic Culture,* volume 4: *Science and Technology in Islam,* ed. A. Y. al-Hassan et al., part 1, *The Exact and Natural Sciences.* UNESCO, 2001.

Saliba, George. "Islamic Astronomy in Context: Attacks on Astrology and the Rise of the *Hay'a* Tradition." *Bulletin of the Royal Institute for Inter-Faith Studies* 4 (2002): 25–46.

Saliba, George. "Greek Astronomy and the Medieval Arabic Tradition." *American Scientist,* July-August 2002: 360–367.

Saliba, George. "Alternative all'astronomia tolemaica." In *Storia della Scienza,* volume III, ed. S. Petruccioli. Rome, 2002.

Saliba, George. "Flying Goats and Other Obsessions: A Response to Toby Huff's 'Reply.'" *Bulletin of the Royal Institute for Inter-Faith Studies* 4, no. 2 (2002): 129–141.

Saliba, George. "Aristotelian Cosmology and Arabic Astronomy." In *De Zénon d'Élée à Poincaré,* ed. R. Morelon and A. Hasnawi. Peeter, 2004.

Saliba, George. "Reform of Ptolemaic Astronomy at the Court of Ulugh Beg." In *Studies in the History of the Exact Sciences in Honor of David Pingree,* ed. C. Burnett et al. Boston, 2004.

Saliba, George. "The World of Islam and Renaissance Science and Technology." in *The Arts of Fire: Islamic Influences on Glass and Ceramics of the Italian Renaissance,* ed. C. Hess. Los Angeles, 2004.

Saliba, George. "Re-visiting The Astronomical Contacts between the World of Islam and Renaissance Europe: The Byzantine Connection." Forthcoming.

Saliba, George. "Whose Science Was Arabic Science in Renaissance France?" http://www.columbia.edu/~gas1/project/visions/case1/sci.1.html.

Sarris, Peter. "The Eastern Roman Empire from Constantine to Heraclius (306–641)." In *The Oxford History of Byzantium,* ed. C. Mango. Oxford University Press, 2002.

Sarton, George. *Introduction to the History of Science.* Williams and Wilkins, 1927.

Savage-Smith, Emilie, ed. *Magic and Divination in Early Islam*. Ashgate-Variorum, 2004.

Sayīlī, Aydīn. *Ghiyâth al-Dîn al-Kâshî's letter on Ulugh Bey and the Scientific Activity in Samarqand*. Ankara, 1985.

Sayīlī, Aydīn. *The Observatory in Islam*. Ankara, 1960.

Sezgin, Fuat. *Geschichte des Arabischen Schrifttums*. Leiden, 1967–.

Shahid, Irfan. "Islam and Byzantium in the IXth century: The Baghdad, Constantinople Dialogue." In *Cultural Contacts in Building a Universal Civilization: Islamic Contributions,* ed. E. Ihsanoglu. Istanbul, 2005.

Shaibānī. See Ibn al-Athīr.

Shīrāzī, Quṭb al-Dīn al-. *al-Tuḥfa al-Shāhīya*. Paris, BnF Arabe, 2516.

Ṣūfī, Abd al-Raḥmān, al- (d. 986). *Ṣuwar al-Kawākib*. Hyderabad, 1953.

Swerdlow, Noel. "Aristotelian Planetary Theory in the Renaissance: Giovanni Batista Amico's Homocentric Spheres." *Journal for the History of Astronomy* 3 (1972): 36–48.

Swerdlow, Noel. "The Derivation and First Draft of Copernicus's Planetary Theory: A Translation of the Commentariolus with Commentary." *Proceedings of the American Philosophical Society* 117, no. 6 (1973): 423–512.

Swerdlow, Noel. "Copernicus's Four Models of Mercury." In *Studia Copernicana XIII,* ed. O. Gingerich and J. Dobrzycki. Warsaw, 1975.

Swerdlow, Noel, and Otto Neugebauer. *Mathematical Astronomy in Copernicus's De Revolutionibus*. Springer-Verlag, 1984.

Swerdlow, Noel. "Jābir Ibn Aflaḥ' s Interesting Method for Finding the Eccentricities and Direction of the Apsidal Line of a Superior Planet." In King and Saliba, *From Deferent to Equant*. New York Academy of Sciences, 1987.

Swerdlow, Noel. "Astronomy in the Renaissance." In *Astronomy before the Telescope,* ed. C. Walker. St. Martin's Press, 1996.

Ṭabari, Ibn Jarīr (932). *Tārīkh al-rusul wa-l-mulūk*. Beirut, 1987.

Tannery, Paul. *Recherches sur l'histoire de l'astronomie ancienne*. Gauthier-Villars, 1893.

Tannūkhī, al-Muhassin b. ʿAlī, al- (994). *Nishwār al-muḥāḍara wa akhbār al-mudhākara,* ed. A. Shaljī. Beirut, 1971–1973.

Taşköprülü-zade (d. 1561). *al-Shaqāʾiq al-nuʿmānīya fī ʿulamāʾ al-dawla al-ʿuthmānīya*. Istanbul, 1985.

Ṭawīl, Yusuf ʿAlī, ed. *Fihrist al-Nadīm*. Beirut, 1996.

Tekeli, Sevim. "Al-Urdî'nin 'Risalet-ün Fi Keyfiyet-il-Ersad' Adli makalesi." *Araştirma* 7 (1970): 57–98.

Thābit b. Qurra. See Neugebauer, 1962; Morelon, 1987.

Tihon, A. "L'astronomie byzantine (du Ve au XVe siècle)." *Byzantion* 51 (1981): 603–624.

Tihon, A. *Études d'astronomie byzantine.* London, 1994.

Toll, Christopher. "Arabische Wissenschaft und Hellenistisches Erbe." In *Islam und Abendland,* ed. A. Mercier. Frankfurt, 1976.

Toomer, Gerald. *Ptolemy's Almagest.* Springer-Verlag, 1984.

Toomer, G. J. *Eastern Wisdome and Learning.* Oxford University Press, 1996.

Treadgold, Warren. "The Struggle for Survival (641–780)." In *The Oxford History of Byzantium,* ed. C. Mango. Oxford University Press, 2002.

Turner, A. J. *Catalogue of the Collection, The Time Museum,* volume I: *Time Measuring Instruments, Part I, Astrolabes Astrolabe Related Instruments.* Rockford, 1985.

Ṭūsī, Naṣīr al-Dīn (d. 1274). See de Vaux, 1893; Ragep, 1993.

Ṭūsī, Naṣīr al-Dīn (d. 1274). *Taḥrīr al-majisṭī.* Bibliotheque Nationale de France, arabe 2485, and India Office, Loth, 741.

Ṭūsī, Naṣīr al-Dīn, al-. *The Rawḍatu't-Taslīm: commonly called Taṣawwurāt.* Persian text, ed. W. Ivanow. Leiden, 1950.

Ṭūsī, Naṣīr al-Dīn, al-. *Tajrīd al-'i'tiqād.* Cairo, 1996.

Ṭūsī, Naṣīr al-Dīn al-. *Contemplation and Action: The Spiritual Autobiography of a Muslim Scholar.* I. B. Tauris, 1998.

Ṭūsī, Naṣīr al-Dīn, al-. *Awṣāf al-ashrāf.* Beirut, 2001.

Ullman, Manfred. *Die Medizin im Islam.* Brill, 1970.

Uqlīdisī (c. 952). See Saidan, 1978.

'Urḍī (d. 1266). See Saliba, 1993.

Vesalius, Andreas. *On the Fabric of the Human Body,* Book I. San Francisco, 1998.

Abū al-Wafā'. See Būzjānī.

Al-Wāfī bi-l-wafayāt. See Ṣafadī.

Walker, Christopher, ed. *Astronomy before the Telescope.*" St. Martin's Press, 1996.

Weill, Georges, and François Secret. *Vie et caractère de Guillaume Postel.* Milan, 1987.

Weisser, Ursula. "Avicenna: Influence on Medical Studies in the West." In *Encyclopedia Iranica* III, pp. 107–110.

Wiedeman, E., with T. W. Juynbol. "Avicennas Schrift über ein von ihm ersonnenes Beobachtunginstrument." *Acta Orientalis* xi, no. 5 (1926): 81–167.

Yāqūt, al-Ḥamwī, Shihāb al-Dīn Abū ʿAbdallāh (1228). *Muʿjam al-buldān*. Beirut, 1979.

Zīj al-Maʾmūnī al-Mumtaḥan, al-. Escorial. Arabic Ms. 927.

Index